Mass Spectroscopy

# ANALYTICAL CHEMISTRY BY OPEN LEARNING

ACOL (Analytical Chemistry by Open Learning) is a well established series of open learning books and computer based training packages. This open learning material covers all of the important techniques and fundamental principles of analytical chemistry. The following are available from Wiley:

**Books**

Samples and Standard
Sample Pretreatment
Measurement, Statistics and Computation
Chromatographic Separations
Gas Chromatography
High Performance Liquid Chromatography
Thin Layer Chromatography
Ultraviolet and Visible Spectroscopy
Fluorescence and Phosphorescence
Atomic Absorption and Plasma Spectroscopy
Mass Spectrometry
Radiochemical Methods
Thermal Methods
Microprocessor Applications
Chemometrics: Experimental Design
Environmental Analysis
Quality in the Analytical Chemistry Laboratory
Modern Infrared Spectroscopy
Biological Applications of Infrared Spectroscopy

**Software**

High Performance Liquid Chromatography
Gas Chromatography

Series Editor: David J. Ando (University of Greenwich)

# Mass Spectrometry

Analytical Chemistry by Open Learning
Second Edition

Author:
JAMES BARKER
*University of Greenwich*

Editor:
DAVID J. ANDO
*University of Greenwich*

Authors of First Edition:
REGINALD DAVIS
*University of Kingston*

MARTIN J. FREARSON
*University of Hertfordshire*



Published on behalf of ACOL (University of Greenwich)
by
JOHN WILEY & SONS
Chichester • New York • Brisbane • Singapore • Toronto

Published by John Wiley & Sons Ltd,
Baffins Lane, Chichester,
West Sussex PO19 1UD, England

National 01243 779777
International (+44) 1243 779777
e-mail (for orders and customer service enquiries): cs-books@wiley.co.uk
Visit our Home Page on http://www.wiley.co.uk
or http://www.wiley.com



*Other Wiley Editorial Offices*

John Wiley & Sons, Inc., 605 Third Avenue,
New York, NY 10158-0012, USA

WILEY-VCH Verlag GmbH, Pappalallee 3,
D-69469 Weinheim, Germany

Jacaranda Wiley Ltd, 33 Park Road, Milton
Queensland 4064, Australia

John Wiley & Sons (Asia) Pte Ltd, Clementi Loop #02-01,
Jin Xing Distripark, Singapore 129809

John Wiley & Sons (Canada) Ltd, 22 Worcester Road,
Rexdale, Ontario M9W 1L1, Canada

***Library of Congress Cataloging-in-Publication Data***

Barker, James, Ph.D.
Mass spectrometry—2nd ed. / author, James Barkers : editor.
David J. Ando.
p. cm.—(Analytical Chemistry by Open Learning)
Rev. ed. of: Mass spectrometry / authors, Reg Davis, Martin Frearson. c1987.
Includes bibliography reference and index.
ISBN 0-471-96764-5 (cloth : alk. paper).—ISBN 0-471-96762-9 (paper : alk. paper)
1. Mass spectrometry—Programmed instruction. 2. Chemistry, Analytical—Programmed instruction. I. Ando. D. J. (David J.) II. Davis, Reg. Mass spectrometry. III. Title.
IV. Series: Analytical Chemistry by Open Learning (Series)
QD96.M3D38 1998
543'.0873—dc21

98-3127
CIP

***British Library Cataloguing in Publication Data***

A catalogue record for this book is available from the British Library

ISBN 0 471 96764 5
ISBN 0 471 96762 9

To Jill

# THE UNIVERSITY OF GREENWICH ACOL PROJECT

This series of easy-to-read books has been written by some of the foremost lecturers in Analytical Chemistry in the United Kingdom. These books are designed for training, continuing education and updating of all technical staff concerned with Analytical Chemistry.

These books are for those interested in Analytical Chemistry and instrumental techniques who wish to study in a more flexible way than traditional institute attendance, or to augment such attendance.

ACOL also supply a range of training packages which contain computer software together with the relevant ACOL book(s). The software teaches competence in the laboratory by providing experience of decision making in such an environment, often based on the simulation of instrumental output, while the books cover the requisite underpinning knowledge.

The Royal Society of Chemistry uses ACOL material to run regular series of course based on distance learning and regular workshops.

Further information on all ACOL materials and courses may be obtained from:

The ACOL-BIOTOL Office, University of Greenwich, Unit 42, Butterly Avenue, Dartford Trade Park, Dartford, DA1 1JG. Tel: 0181-331-7533, Fax: 0181-331-9672.

# How to Use an Open Learning Book

Open Learning books are designed as a convenient and flexible way of studying for people who, for a variety of reasons, cannot use conventional education courses. You will learn from this book the principles of one subject in Analytical Chemistry, but only by putting this knowledge into practice, under professional supervision, will you gain a full understanding of the analytical techniques described.

To achieve the full benefit from an open learning text you need to carefully plan your place and time of study.

- Find the most suitable place to study where you can work without disturbance.

- If you have a tutor supervising your study discuss with this person the date by which you should have completed the text.

- Some people study perfectly well in irregular bursts; however, most students find that setting aside a certain number of hours each day is the most satisfactory method. It is for you to decide which pattern of study suits you best.

- If you decide to study for several hours at once, take short breaks of five or ten minutes every half hour or so. You will find that this method maintains a higher overall level of concentration.

Before you begin a detailed reading of this book, familiarise yourself with the general layout of the material. Have a look at the course contents list at the front of the book and flip through the pages to get a general impression of the way the subject is dealt with. You will find that there is space on the pages to make comments alongside the text as you study—your own notes for highlighting points that you feel are particularly important. Indicate in the margin the points you would like to discuss further with a tutor or fellow student. When you come to revise, these personal study notes will be very useful.

Π When you find a paragraph in the text marked with a symbol such as is shown here, this is where you get involved. At this point you are directed to do certain things, e.g. draw graphs, answer questions, perform calculations, etc. Do make an attempt at these activities. If necessary, cover the succeeding response with a piece of paper until you are ready to read on. This is an opportunity for you to learn by participating in the subject, and although the text continues by discussing your response, there is no better way to learn than by working things out for yourself.

We have introduced self-assessment questions (SAQs) at appropriate places in the text. These SAQs provide you with a way of finding out if you understand what you have just been studying. There is a space on the page for your answer and for any comments you want to add after reading the author's response. You will find the author's response to each SAQ at the end of the book. Compare what you have written with the response provided and read the discussion and advice.

At intervals in the text you will find a Summary and a list of Objectives. The Summary will emphasise the important points covered by the material you have just read, while the Objectives will give you a checklist of tasks you should then be able to achieve.

You can revise the book, perhaps for a formal examination, by re-reading the Summary and the Objectives, and by working through some of the SAQs. This should quickly alert you to areas of the text that need further study.

At the end of this book you will find, for reference, lists of commonly used scientific symbols and values, units of measurement, and also a periodic table.

# Contents

# Study Guide

The first Mass Spectrometry Book in the ACOL series was written by Reg Davis and Martin Frearson. The revision for the Second Edition of this text has been carried out by James Barker, currently AMS Manager at CBAMS, York, UK*, who has added to and reorganised the earlier material to bring it completely up to date.

This text is designed to provide you with a working knowledge of Mass Spectroscopy (MS). It starts off by introducing you to a simple mass spectrum and then describes how mass spectral information can be presented and interpreted. The next chapter goes on to look at some different types of ion sources and how such sources produce different fragmentation patterns. You will look at the principles behind the operation of various common types of mass analyser, such as the double-focusing, quadrupole, time-of-flight, ion-trap and Fourier-transform instruments. Methods of ion detection and computer-aided data processing will also be discussed. The characterisation of compounds through their isotopic abundances and presence of metastable ions, factors affecting the stability of ion fragments, methods of bond cleavage, and some of the fragmentation patterns of common functional groups will also be outlined. Useful tandem techniques, such as gas chromatography (GC)/MS, liquid chromatography (LC)/MS and MS/MS will be covered, as well as a discussion of the rapidly growing area of atomic (or elemental) mass spectrometry.

Mass spectrometry is still, in some respects, a fairly empirical subject. We are still some way from understanding and fully predicting the fragmentation patterns of many complex organic (and biological) molecules. However, from a knowledge of the basic concepts of

*The Centre for Biomedical Accelerated Mass Spectrometry (CBAMS), Sand Hutton, York, YO41, 1LZ,UK

organic chemistry, and with the help of computer-database searches and a little intuition, you should be able to characterise and formulate structures for many relatively simple organic molecules. Together with complimentary techniques such as infrared (IR), ultraviolet (UV)/visible, and nuclear magnetic resonance (NMR) spectroscopies, mass spectrometry holds the key to qualitative and quantitative chemical analysis.

Some of the most important relevant concepts in organic chemistry are outlined in this text, but it is assumed that your understanding of chemistry will be equivalent to that of a student holding an HNC in Chemistry, while your knowledge of physics and mathematics will be at least to the standard of GCSE or 'O' level.

The material is constructed to be used at two levels, namely by those studying for the Licentiate of the Royal Society of Chemistry (LRSC) qualification (or its equivalent), and by those who wish to obtain a more detailed knowledge of mass spectrometry, beyond that required for LRSC. Those using the text as preparation for their LRSC examinations are not expected to study the whole Unit. In this case, they are advised to leave out the following sections:

2.3–2.7, 3.7–3.9, 5.4–7.4, 7.6.2–7.6.7, 7.9–7.11, and Chapter 9 (except the parts concerning Inductively Coupled Plasma Mass Spectrometry (ICPMS)).

You may find that the coverage given in other textbooks clarifies certain aspects of this important analytical technique. For this reason, a bibliography has been provided. It is hoped that you will look at at least one of these other texts sometime during your study of this Unit. This should help to provide you with a suitable balance in your approach to the subject.

# Supporting Practical Work

Many analytical laboratories are equipped with a mass spectrometer. The technique is often used in conjunction with a variety of other analytical methods, such as nuclear magnetic resonance, infrared, and ultraviolet spectroscopies, or chromatography..

The aims of the practical work are as follows:

(a) To provide the basic experience of using a spectrometer for recording mass spectra.

(b) To give experience in the use of mass spectrometry in both qualitative and quantitative analysis, with particular emphasis being placed on good analytical practice.

In order to achieve these aims, you could repeat the mass analysis of some of the examples presented throughout the text, and then compare your spectra with the ones published, noting the differences in the spectra and trying to account for these from the knowledge that you have acquired.

# Bibliography

Most general analytical and spectroscopy textbooks have a chapter on mass spectrometry. These include the following:

F.W. Fifield and D. Kealey, *Principles and Practice of Analytical Chemistry*, 4th Edn, Blackie, London, 1995.

D.A. Skoog, F.J. Holler and D.M. West, *Analytical Chemistry, An Introduction*, 6th Edn, Saunders College Publishing, Philadelphia, 1994.

G. Svehla (Ed.), *Vogel's Qualititative Inorganic Analysis*, 7th Edn, Longman Scientific and Technical, London, 1996.

There are also many books specialising in mass spectrometry. These include the following:

J.H. Beynon and A.G. Brenton, *Introduction to Mass Spectrometry*, University of Wales Press, Cardiff, 1982.

E. De Hoffman, J. Charette and V. Stroobant, *Mass Spectrometry*, John Wiley & Sons, Chichester, 1996.

In addition, there are several journals which specialise in mass spectrometry. These include:

*The Journal of Mass Spectrometry*, John Wiley & Sons.

*The International Journal of Mass Spectrometry and Ion Processes*, Elsevier.

*Mass Spectrometry Reviews*, John Wiley & Sons.

*The Journal of the American Society for Mass Spectrometry*, John Wiley & Sons.

*Rapid Communications in Mass Spectrometry*, John Wiley & Sons.

*The Mass Spectrometry Bulletin* (which includes a two-yearly review), The Royal Society of Chemistry.

In addition, the published proceedings of various mass spectrometry conferences can often provide a useful source of specialist information.

# Acknowledgements

Figures 2.1d, 2.4c, 4.2g and 6.1a are redrawn from R.A.W. Johnstone and M.E. Rose, *Mass Spectrometry for Chemists and Biochemists*, 2nd Edn, Cambridge University Press, Cambridge, 1996. Permission has been requested.

Figures 2.2b, 2.3b, and 3.3g are redrawn from D.H. Williams and I. Howe, *Principles of Organic Mass Spectrometry*, McGraw-Hill, Maidenhead, 1972. Permission has been requested.

Figures 2.2c, 3.9b, 7.11a–7.11d, 8.2f–8.2m and 8.4a are redrawn , and Appendices 1 and 2 are reproduced from E. de Hoffman, J. Charette and V. Stroobant, *Mass Spectrometry: Principles and Applications*, John Wiley & Sons, Chichester, 1996. Permission has been requested.

Figure 2.2d is redrawn from S.B. Martin, J.H. Karam, P.H. Forsham and J.B. Knight, *Biomed. Mass Spectrom.* **1,** 320 (1974). Permission has been requested.

Figure 2.4b is redrawn from *Fast Atom Bombardment Spectra*, Kratos Analytical, Manchester. Permission has been requested.

Figure 3.4a is redrawn from B.J. Millard, *Quantitative Mass Spectrometry*, Heyden, London, 1978/79. Permission has been requested.

Figure 3.4b is redrawn from J.R. Chapman, *Computers in Mass Spectrometry*, Academic Press, London, 1978. Permission has been requested.

Figures 5.3a, 5.3b and 7.6e are redrawn from R. Davis and C.H.J. Wells, *Spectral Problems in Organic Chemistry*, International Textbook Company (Blackie Group), London, 1984. Permission has been requested.

Figure 7.6b, 7.10b and 7.10c are redrawn from H. Budzikiewicz, C. Djerassi and D.H. Williams, *Mass Spectrometry of Organic Compounds*, Holden-Day Inc., San Francisco, 1967. Permission has been requested.

Figure 7.7c is redrawn from D.L. Pavia, G.M. Lampman and G.C. Kris, Jr, *Introduction to Spectroscopy*, Saunders and Co., London, 1979. Permission has been requested.

Figures 8.1a, 8.2d and 8.2e are redrawn from instrument manuals published by VG Instruments Group Ltd, Altrincham, 1996. Permission has been requested.

Figures 8.2b and 8.2c are redrawn from M.L. Vestal, *Eur. Spectrosc. News*, No. 63, 22 (1985/86). Permission has been requested.

Figures 9.1a–9.1c, and 9.1e are redrawn from F.A. White and G.M. Wood, *Mass Spectrometry: Applications in Science and Engineering*, Wiley-Interscience, New York, 1986. Permission has been requested.

Figure 9.1d is redrawn from L.G. Earwaker, *Vacuum*, **45,** 783 (1994). Permission has been requested.

Figure 9.2a is redrawn from W. Kutschera, *AIP Conf. Proc.*, 329 (Resonance Ionization Spectroscopy 1994), 22 (1995). Permission has been requested.

# 1. Introduction

This part of the Unit will introduce you to some of the basic concepts of mass spectrometry. After a brief historical summary, the chapter will go on to describe simple ionisation and fragmentation of organic molecules. The student will be introduced to a simple mass spectrum and, given some basic ideas and terminology, then shown how to interpret it.

## 1.1 MASS SPECTROMETRY: ITS CONCEPTS AND BEGINNINGS

Mass spectrometry (MS) is an analytical technique in which atoms or molecules from a sample are ionised (usually positively), separated according to their *mass-to-charge ratio* ($m/z$), and then recorded. The instrument used to carry out this measurement is called a mass spectrometer.

Π Can you describe the principles behind the operation of any other analytical instrument that you have used?

Mass spectrometry is quite different from other common methods of analysis such as infrared, ultraviolet/visible and nuclear magnetic resonance spectroscopies. All of these techniques are based on the principle of absorption of electromagnetic radiation to transform the molecule from a ground-state energy level to an excited-state one. The molecule will eventually return to its ground-state level, and thus all of the methods are non-destructive. The sample remains intact after analysis.

However, since mass spectrometry forms and utilises **ions** rather than molecules, the conversion back to molecules (atoms) in the sample is therefore impossible. Mass spectrometry is therefore a destructive method of analysis.

Mass spectrometry had its birth in 1886 when Goldstein, a German physicist, discovered positive ions in a low-pressure electrical discharge tube. In 1898, Wien showed that a beam of these ions could be deflected by using electric and magnetic fields. Between 1912 and 1919, Thompson and Aston managed to resolve ion trajectories corresponding to mass differences of less than 10%, thus enabling them to confirm the existence of two distinct neon isotopes. In 1922, Aston received the Nobel Prize for his work and by 1924 he had characterised the isotopic abundances of around fifty elements. Reliable mass spectrometers, for use in the petroleum industry, became available in the early 1940s.

Today, mass spectrometers are used in a wide variety of disciplines. For example, by calculating mass measurements very precisely, the binding energies of atoms may be determined, and, via isotope-ratio measurements, trace elements in biological systems may be assayed.

## 1.2 IONISATION AND FRAGMENTATION

### 1.2.1 Ionisation

In a mass spectrometer used for **organic analysis**, we are usually dealing with the analysis of positive ions derived from molecules rather than atoms. This is carried out by using a variety of methods, which we shall look at in Chapter 2. Simple ionisation of a molecule is shown below in the following equation:

$$\mathrm{M:} \longrightarrow \mathrm{M}^{+\bullet} + \mathrm{e} \qquad (1.1)$$

The ion $M^{+\bullet}$ is known as the *molecular ion*. As mentioned previously, the ion's *mass* ($m$) and *charge* ($z$) are measured by the mass spectrometer and it is this ratio that is important. Often ions with only a single positive charge are formed, and thus the mass to charge ratio is +1. This is therefore equal to the mass of the ion itself. The mass of the ion is, of course, related to the relative molecular mass of the molecule.

Π Can you recall the meanings of the following terms: mass number ($A$), atomic number ($Z$), relative atomic mass ($A_r$), atomic weight, relative molecular mass ($M_r$), and unified atomic mass unit (amu)?

Mass number is the number of protons and neutrons in the nucleus.

Atomic number is the number of protons in the nucleus.

The unified atomic mass unit (amu, sometimes referred to as u) is defined such that the mass of the $^{12}C$ atom is exactly 12 × (amu). The value of the amu is $1.661 \times 10^{-27}$g.

The relative atomic mass (sometimes called the atomic weight) of an atom is often quoted in amu and is equal to the following ratio:

$$[(\text{mass of an atom})/(\text{mass of a } ^{12}_{6}\text{C atom})] \times 12$$

Thus, all atoms are assigned masses relative to $^{12}C$. Fortunately, many elements have average relative atomic masses (sometimes called average atomic weights) which are very close to integral numbers (in amu). This is a result of the abundances of the isotopes of these elements. This means that when we are roughly calculating molecular masses, we can assign integral values to most of the common elements present in organic molecules.

The relative molecular mass is thus the sum of the average relative atomic masses of the constituent atoms in a compound.

As an example, the average relative atomic mass of oxygen is 15.994 (very close to 16), as a result of the abundance of the $^{16}O$ isotope (99.756%). Similarly, carbon has an average atomic weight of 12.011, as a result of the abundance of the $^{12}C$ isotope (98.90%). This abundance has valuable ramifications for mass spectrum identification (see Section 1.2.5).

**SAQ 1.2a**

Calculate *m/z* values (to the nearest whole numbers) for the following molecular ions:

(a) $C_6H_6^{+\bullet}$

(b) $C_2H_5NH_2^{+\bullet}$

(c) $CH_3OH^{+\bullet}$

(d) $C_6H_5COOH^{+\bullet}$

(e) $C_3H_6^{+\bullet}$

(f) $C_2H_5CN^{+\bullet}$

(g) $C_3H_7SN^{+\bullet}$

(h) $CH_3CHONH_2^{+\bullet}$

(i) $(CH_3O)_3P^{+\bullet}$

(j) $C_6H_5NH_2^{2+}$

**SAQ 1.2b**

Express the mass of an electron in atomic mass units and use this result to calculate the mass difference (in amu) between the following pairs, given that the mass of an electron is $9.110 \times 10^{-31}$g and 1 amu is $1.661 \times 10^{-27}$g. Take the relative molecular masses to the nearest whole numbers.

(a) $C_2H_6$ and $C_2H_6^{+\bullet}$

(b) $C_{60}$ and $C_{60}^{+\bullet}$

Since an electron has a very small mass indeed, i.e. $5.49 \times 10^{-4}$ of the mass of a proton, the mass loss in the atom is negligibly small upon ionisation, when compared to the (mass) resolution of most mass spectrometers (say, 1 amu in $10^4$ amu). Therefore, we can assume for the purposes of identification that the mass of a molecular ion is equivalent to the mass of the parent molecule from which it is derived.

You may have noticed in equation (1.1) that there were other labels used with the molecule and molecular ion, namely the symbols ':' and '$+\bullet$'. The are used to help us remember the electron configuration of the molecule. Virtually all organic molecules have an even number of electrons, so the designation in this case is M:. These electrons occupy bonding or non-bonding orbitals. Thus, if upon ionisation one electron is removed, then the resultant molecular ion will have an odd electron configuration, i.e. one unpaired electron is left. The molecular ion is therefore designated $M^{+\bullet}$. In other words, it is a radical cation. Note that *all* molecular ions have an odd-electron configuration.

### 1.2.2 Fragmentation

If we are forming a singly charged molecular ion from a neutral molecule, then we must reach the first ionisation potential of the molecule. This value depends on the nature of the highest occupied molecular orbital, since it is from the latter that the electron will be removed.

You may recall that orbitals in organic molecules are commonly of five kinds, namely $\sigma$-bonding ($\sigma$), $\sigma$-antibonding ($\sigma^*$), $\pi$-bonding ($\pi$), $\pi$-antibonding ($\pi^*$) and non-bonding (n) orbitals.

Π The first ionisation potentials of most organic molecules lie in the range 8–15 eV. How does this compare with the first ionisation potential of a metal atom such as iron?

Metal atoms often have slightly lower ionisation potentials; the value for iron is 7.9 eV.

The process of ionisation (e.g. electron-impact ionisation—see Chapter 2) often uses an energy which is far greater than that needed to simply ionise the molecules. This excess energy can be transferred to the molecular ion as it is created, in the form of rotational, translational, vibrational or electronic energy. The latter two categories are the most important for mass spectrometry since it is the *fragmentation* of these excited states that invariably produces a new ion and a neutral particle.

We shall now consider the molecular ion, $M^{+\bullet}$, fragmenting into new products, namely a *fragmentation ion* and a neutral particle. It can only follow two *fragmentation pathways*. Note that the positive charge and odd-electron status need to be conserved in both mechanisms.

The corresponding equations are as follows:

$$M^{+\bullet} \longrightarrow A^{+} + N^{\bullet} \tag{1.2}$$

$$M^{+\bullet} \longrightarrow B^{+\bullet} + N \tag{1.3}$$

Equation (1.2) shows fragmentation to an even-electron-fragment cation and an odd-electron neutral species. The neutral species may be an atom such as H•, or a radical such as OH•.

Equation (1.3) shows fragmentation to an odd-electron ion and an even-electron neutral species. In this case, the neutral species may be a stable molecule such as $C_2H_4$ or CO, or an even-electron species such as an enol, $CH_2CHOH$.

**SAQ 1.2c**

Let us take a simple molecule, e.g. $C_2H_6$. Draw three fragmentation pathways for the molecular ion, with each involving only C—H bond cleavage but which lead to the production of a neutral atom, a neutral radical and a neutral molecule, respectively.

The correct response to the above SAQ should show the formation of two complimentary ion pairs. The first of these is either $C_2H_5^+$ or $H^+$, while the second pair is either $C_2H_4^{+\bullet}$ or $H_2^{+\bullet}$. It is quite common to observe such complementary pair formation in mass spectrometry, although it must be remembered that both of the fragmentation pathways are different.

Compare these pathways with those shown in equations (1.2) and (1.3).

Note that in both of the pathways which lead to the formation of $C_2H_5^+$ and $H^+$, only one bond is broken in each case and no new bonds are formed. In the pathway leading to the formation of $C_2H_4$ and $H_2^{+\bullet}$, two bonds are broken and two bonds are formed:

$$\begin{array}{l} H_2C{-}H^{+\bullet} \\ \;\;\; | \\ H_2C{-}H \end{array} \longrightarrow H_2C{=}CH_2 + H{-}H^{+\bullet} \qquad (1.4)$$

In the formation of $C_2H_4^{+\bullet}$ and $H_2$ (see equation (1.5)), the answer is not so obvious. Even though $C_2H_4^{+\bullet}$ will not have sufficient electrons to form four C—H single bonds and a C=C double bond, it is usually represented in the following way:

$$\begin{array}{l} H_2C{-}H^{+\bullet} \\ \;\;\; | \\ H_2C{-}H \end{array} \longrightarrow H_2C{=}CH_2^{+\bullet} + H{-}H \qquad (1.5)$$

The fragmentation does not have to stop here, provided that the fragment ions that are formed possess insufficient internal energy.

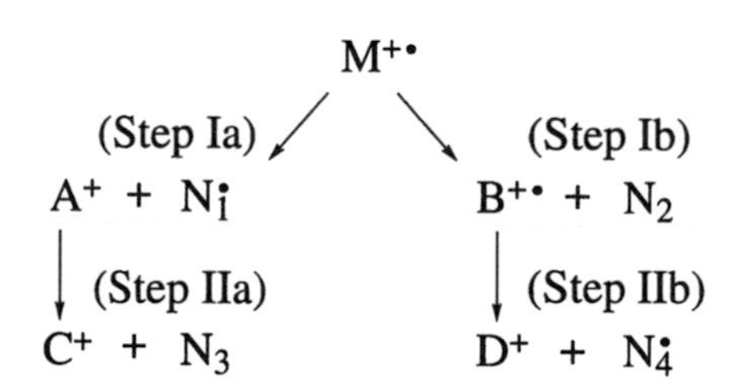


Again, these secondary fragmentation pathways may involve loss of either even- or odd-electron neutrals species, either in competition or as unique pathways. All of these fragmentation pathways form a range of ions which in turn gives rise to the associated *mass spectrum* of the compound in question. Therefore, a knowledge of typical fragmentation pathways allows us to construct a characteristic *fragmentation pattern*.

The fragmentation pathway and the corresponding products are influenced by four main factors:

(a) the strengths of the bonds which are to be broken;

(b) the stability of the products of fragmentation, both ions and neutral species;

(c) the internal energies of the fragmenting ions;

(d) the time interval between ion formation and ion detection.

In general terms, fragmentations involving the cleavage of only one bond are fast processes, whereas those involving both bond breaking and bond making are relatively slow.

### 1.2.3 Rules for Assignment of Fragments

It may not always be clear whether to write a fragment ion with an odd-electron ($^{+\bullet}$) or even-electron ($^{+}$) configuration. The following provides some helpful tips for ascription:

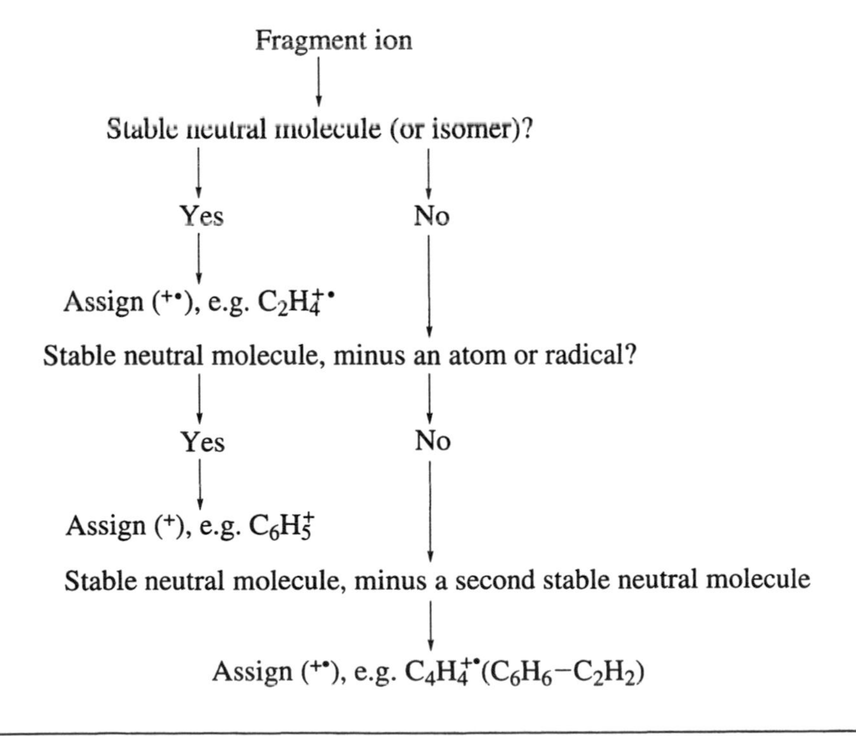

### 1.2.4 The Nitrogen Rule

This is a universal rule covering **both** molecular ions and fragment ions with a single charge. It states the following:

*all odd-electron ions have even* m/z *values unless they contain an odd number of nitrogen atoms.* (For the purposes of this rule, zero is regarded as an even number.)

For example, $H_2NOCCONH_2^{+\bullet}$ has an *m/z* value of 88 (even), while $C_6H_5NH_2^{+\bullet}$ has an *m/z* value of 93 (odd).

This rule results from the fact that nitrogen has an even atomic mass and an odd valency.

**SAQ 1.2d**

Decide whether ions of the following compounds are odd- or even-electron in nature:

(a) HI

(b) $C_7H_{15}$

(c) $C_6H_5NH_2$

(d) $CH_3CO$

(e) $C_6H_5COOC_2H_5$

(f) $C_6H_5CH_2CN$

(g) $C_6H_4(NH_2)_2$

(h) $C_6H_4$

(i) $C_7H_7$

(j) $(CH_3O)_3P$

### 1.2.5 Low-relative-abundance Peaks

Peaks of low relative abundance may occur around large peaks; these may be due to *isotope peaks* of low abundance. For example, a small proportion of methanol molecules will have $^{13}C$ (1.08% natural abundance) rather than $^{12}C$ as their constituent isotope and some hydrogen atoms may have $^{2}H$ (0.015%) rather than $^{1}H$. The subject of isotopes in mass spectrometry is very important and will be discussed in more detail in Chapter 5.

There may also be small peaks at non-integer mass values. These are in fact *multiply charged ions* (often doubly charged, since it is easier to remove just two electrons rather than a larger number). Doubly charged ions of odd mass are relatively easy to identify since they have half-integral $m/z$ values. The identification of doubly charged ions of even mass is more difficult, since these will occur at the same position in the spectrum as singly charged ions of half the mass.

## 1.3 THE SIMPLE MASS SPECTRUM

Now we have reached the stage where it is time to analyse a simple mass spectrum. A common method of displaying a mass spectrum is that of a scan trace which can be obtained directly from the recorders of most mass spectrometers. Figure 1.3a (see below) shows two scan traces of methanol, with one of these recorded at ten times the sensitivity of the other. This allows the very abundant ions and also ions of low abundance to be recorded on a single plot.

Π Examine the spectrum of methanol shown in Figure 1.3a and list the *m/z* values of the four most abundant ions. Attempt to assign formulae to these ions.

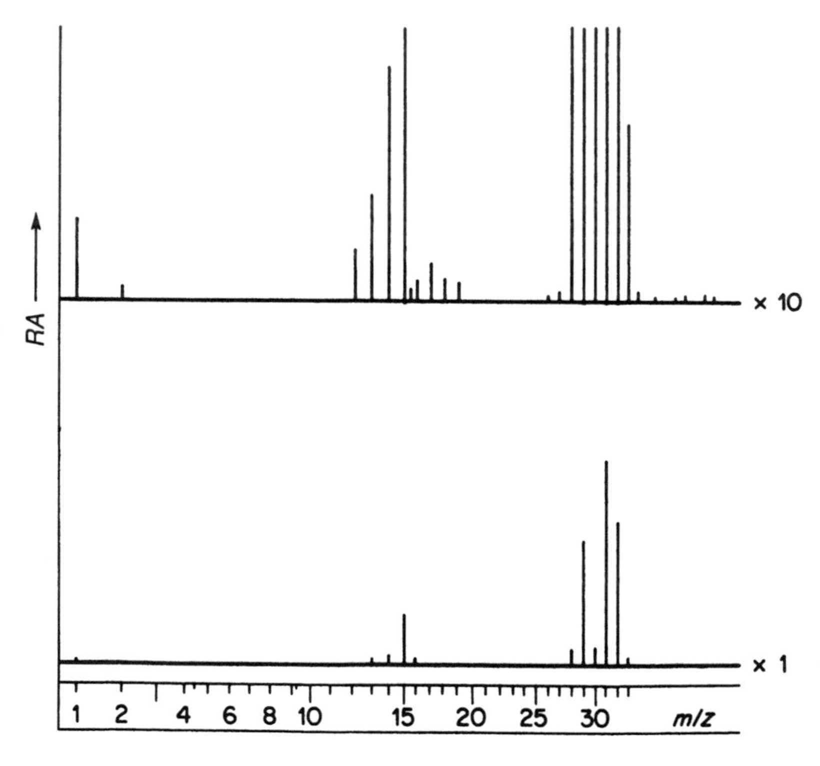


**Figure 1.3a** Mass spectra of methanol ($CH_3OH$)

The major ions are as follows:

*m/z* 32 $CH_3OH^{+\bullet}$

*m/z* 31 $CH_2OH^+$

*m/z* 29 $CHO^+$

*m/z* 15 $CH_3^+$

The most intense ion in the methanol spectrum is at *m*/*z* 31, i.e. $CH_2OH^+$, and this is termed the *base peak* of the spectrum. One way of representing mass spectra is in the form of a bar (line) diagram, with the most intense peak being assigned 100% relative abundance; the latter then forms the base against which the relative abundances of the other ions are measured.

The ion, *m*/*z* 32, is of course the molecular ion; apart from a very small line at *m*/*z* 33, it is the highest mass ion in the spectrum. The other ions are fragment ions and arise in the following way.

The ion, *m*/*z* 31, is formed by the loss of a hydrogen atom from the molecular ion. It is one of the C—H bonds that is broken in this process, rather than the O—H bond; hence its formulation is $CH_2OH^+$. Just by looking at the mass spectrum of methanol does not allow us to distinguish between C—H and O—H bond cleavages, so don't worry if you wrote down $CH_3O^+$ in your response.

The ion, *m*/*z* 29, arises as a consequence of the $CH_2OH^+$ ion losing a molecule of hydrogen and can be formulated as $CHO^+$.

The ion, *m*/*z* 15, is the $CH_3^+$ ion formed by those molecular ions which have the correct energy for fragmentation of the C—O bond, rather than C—H bond cleavage.

**SAQ 1.3a**

> Draw up a fragmentation pattern for methanol based on the above assignments and underline the most intense ion in the spectrum.

**SAQ 1.3b**

Which ion(s) could be responsible for the *m/z* values of 33 and 15.5 in the mass spectrum of methanol?

**SAQ 1.3c**

The ion $CH_3OH^{2+}$ has an *m/z* value of 16. Which singly charged ions with this value might you expect to observe in the methanol spectrum?

**SAQ 1.3d**

Why is the *m/z* axis non-linear in the scan trace?

**SAQ 1.3e**

The mass spectrum of ethanol, $C_2H_5OH$, is shown in Figure 1.3b. Use this to carry out the following:

(a) write down the *m/z* value of the molecular ion;

(b) write down the *m/z* value of the base peak;

(c) the base peak is formed from the molecular ion via a single step—write down a pathway for this process;

(d) write down two fragmentation pathways for the production of ions of *m/z* 29 which may have different formulae;

(e) write down fragmentation pathways for the following stepwise decompositions:

$$m/z\ 46 \longrightarrow m/z\ 45 \longrightarrow m/z\ 43$$

(f) assign a formula to *m/z* 18 and give a fragmentation pathway for its direct formation from the molecular ion;

(g) combine the fragmentation pathways mentioned above to give the fragmentation pattern for ethanol. →

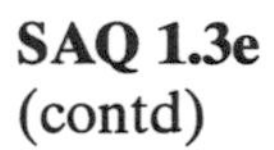

**SAQ 1.3e**
(contd)

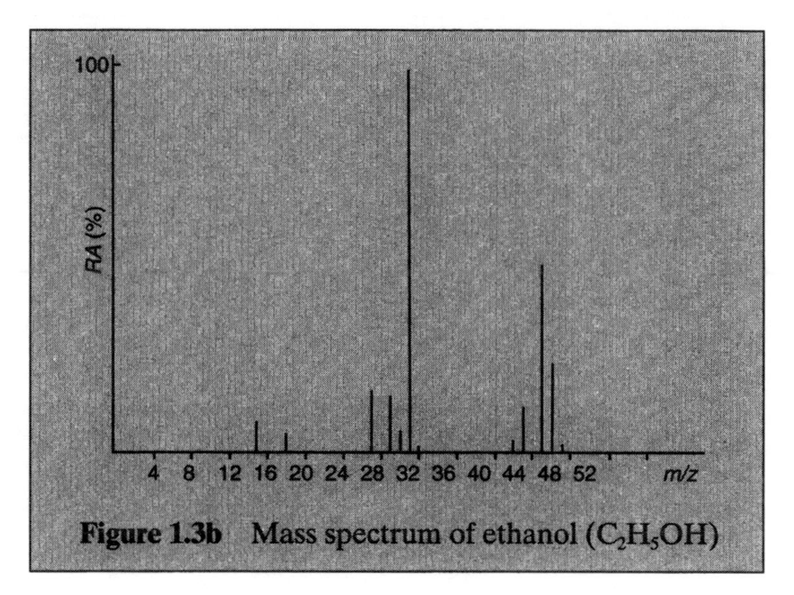


**Figure 1.3b** Mass spectrum of ethanol ($C_2H_5OH$)

**Summary**

The basic ideas of mass spectrometry have now been introduced to you. These include the ionisation and fragmentation of organic molecules, the assignment of odd- or even-electron fragment ions and neutral species and the application of the Nitrogen Rule. You have also been shown, and asked to interpret, a simple mass spectrum, and you should now be familiar with mass spectra both in the form of a scan trace or as a fragmentation pattern.

**Objectives**

When you have completed this part of the Unit you should be able to:

- explain the general principles behind mass spectrometry;
- define the terms molecular ion, fragment ion and base peak;
- explain the circumstances under which ions fragment;
- recognise the difference between, and the significance of, odd- and even-electron ions;
- calculate $m/z$ values for various ions;
- assign correct electron configurations to fragment ions;

- identify low-abundance peaks due to isotopes and multiply charged ions;
- explain what is meant by the terms fragmentation pathway and fragmentation pattern;
- understand various representations of mass spectra;
- understand and use the Nitrogen Rule;
- draw possible fragmentation pathways when given the mass spectrum of a simple molecule, and construct a fragmentation pattern.

# 2. Ion Sources and Methods of Ionisation

We have already seen in the previous chapter how simple molecules can ionise and fragment, but how does this ionisation take place? This present chapter will introduce you to the most popular methods of ion formation and discuss the advantages and disadvantages of each technique. Many modern spectrometers allow the user to simply interchange between each of the ion sources at will. We will confine ourselves to the analysis of organic substances in this chapter and look later at other special ion sources, e.g. those used principally with chromatography (Chapter 8) and in elemental analysis (Chapter 9).

## 2.1 IONISATION BY ELECTRON IMPACT

This important technique of ionisation was developed in the 1920s. It is sometimes referred to as EI or *electron ionisation* and can be used for the analysis of many gases, volatile compounds and metallic vapours. In this ion source (which we shall discuss in more detail in Chapter 3), the sample material, in the form of a gas or vapour, is passed through an inlet into an ionisation chamber, where it is bombarded by an orthogonal beam of electrons, whose energy may be varied to produce maximum ionisation efficiency. Figure 2.1a shows a schematic representation of an electron-impact ion source. Sometimes the sample is desorbed from a heated rhenium filament near the electronic beam. This method is called *desorption electron ionisation* (DEI).

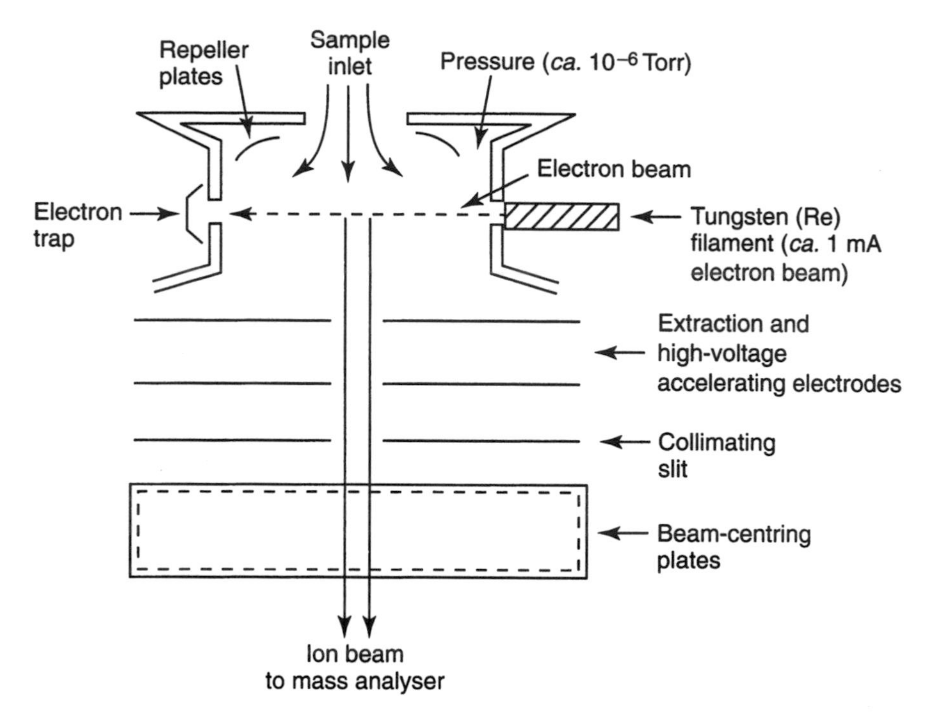


**Figure 2.1a** A schematic representation of an electron-impact ion source

For most singly charged ions this energy is in the region of 50 to 100 eV. Although the process is termed electron impact, ionisation is actually carried out, not by electrons directly hitting the molecule, but by an interaction of the fields of both the electron and the molecule, when the electrons pass close to or through the sample molecules. The process can be written as follows:

$$e_b + M \longrightarrow M^{+\bullet} + e_m + e'_b \tag{2.1}$$

where:

$e_b$ is the bombarding electron before collision;
M is the sample molecule;
$M^{+\bullet}$ is the molecular ion;
$e_m$ is the emitted electron;
$e'_b$ is the bombarding electron after collision.

The low mass and high kinetic energy of the electrons causes little increase in the translational energy of the impacted molecules; instead, they are left in highly excited vibrational and rotational states. When these ions relax, a good deal of fragmentation occurs and complex mass spectra often results. EI, for this reason, is often termed a *hard ionisation* source. EI is an ideal technique to use where unambiguous identification of the analyte is needed.

Π What is the kinetic ($KE$) that a singly charged ion will acquire if it is accelerated through a potential of $10^3$ V in an electron-impact source?

$$KE = eV \tag{2.2}$$

and therefore

$$KE = 1.6 \times 10^{-19} \times 10^3 = 1.6 \times 10^{-16}\,\text{J}$$

where $e$ is the charge on the ion and $V$ is the accelerating potential.

Π Does this kinetic energy depend upon the ion's mass?

No, only upon its charge and accelerating potential.

Π Does the velocity of the ion depend upon its mass?

The translational component of the $KE$ of an ion is a function of the ion mass, $m$, and its velocity, $v$, where

$$KE = \tfrac{1}{2}mv^2 \quad \text{or} \quad v = \sqrt{(2KE/m)} \tag{2.3}$$

Therefore, if the kinetic energy of all similarly charged ions is the same, then those ions with the largest mass must have the smallest velocities.

This ionisation process usually results in the formation of positive ions with a narrow energy spread; however, electron attachment to the molecular ion is also possible at low energies (*ca.* 1 eV), thus generating negative ions. These negative molecular ions are relatively unstable and fragment very easily. You might expect therefore that a highly informative picture could be built up of the structure of the molecular ion. Unfortunately, this is often not the case and the negative-ion spectrum is dominated by a few fragment ions.

**SAQ 2.1** Can you think of any negative ions or species that are often encountered in routine analysis and which you might expect to dominate a negative-ion mass spectrum of the molecules of which they are a part?

Π What is the energy in ($J\,mol^{-1}$) that electrons acquire as a result of being accelerated through a potential of 80 V?

Again, from equation (2.2):

$$KE = eV$$

and therefore $KE = 1.60 \times 10^{-19} \times 80$, for each electron.

Thus, per mole ($6.02 \times 10^{23}$ electrons):

$$KE = 1.6 \times 10^{-19} \times 80 \times 6.02 \times 10^{23} = 7.7 \times 10^{6}\ J\,mol^{-1}.$$

Π How does the above value compare to typical bond energies?

Bond energies often lie in the region $10^2$–$10^3$ J mol$^{-1}$. Therefore, an electron has at least three order of magnitude more energy than that required to break a chemical bond.

We can gain further valuable information from ion sources, as described in the following.

Π For example, what is the name given to the electron-beam energy value at which we start to see the molecular ion being produced from a molecule?

It is, of course, the *ionisation potential* of that molecule, i.e.

$$M \longrightarrow M^{+\bullet} + e \qquad (2.4)$$

By altering the electron-beam energy until we see the appearance of a molecular ion, we can get a rough idea of the ionisation potential for that molecule.

A similar value can also be obtained for the fragment ion. This is know as the *appearance potential* and corresponds to the overall energy for the above process and the two processes shown below.

$$M^{+\bullet} \longrightarrow F_1^{+} + N_1^{\bullet} \qquad (2.5)$$

or

$$M^{+\bullet} \longrightarrow F_2^{+\bullet} + N_2 \qquad (2.6)$$

Information gained from these potential measurements can be used to determine heats of formation of ions and bond dissociation energies.

Π Why is it important to choose electron-beam energy values well above that of the ionisation potential?

There are two reasons for this. First, at energies close to the ionisation potential, the ionisation efficiency is changing rapidly. Thus, it is preferable to have a relatively constant output for diagnostic purposes. Secondly, fragmentation becomes more pronounced at higher beam energies. The spectra of $C_6H_5COOH$, recorded at 9, 12, 15, 20, 30, and 70 eV, is shown in Figure 2.1b.

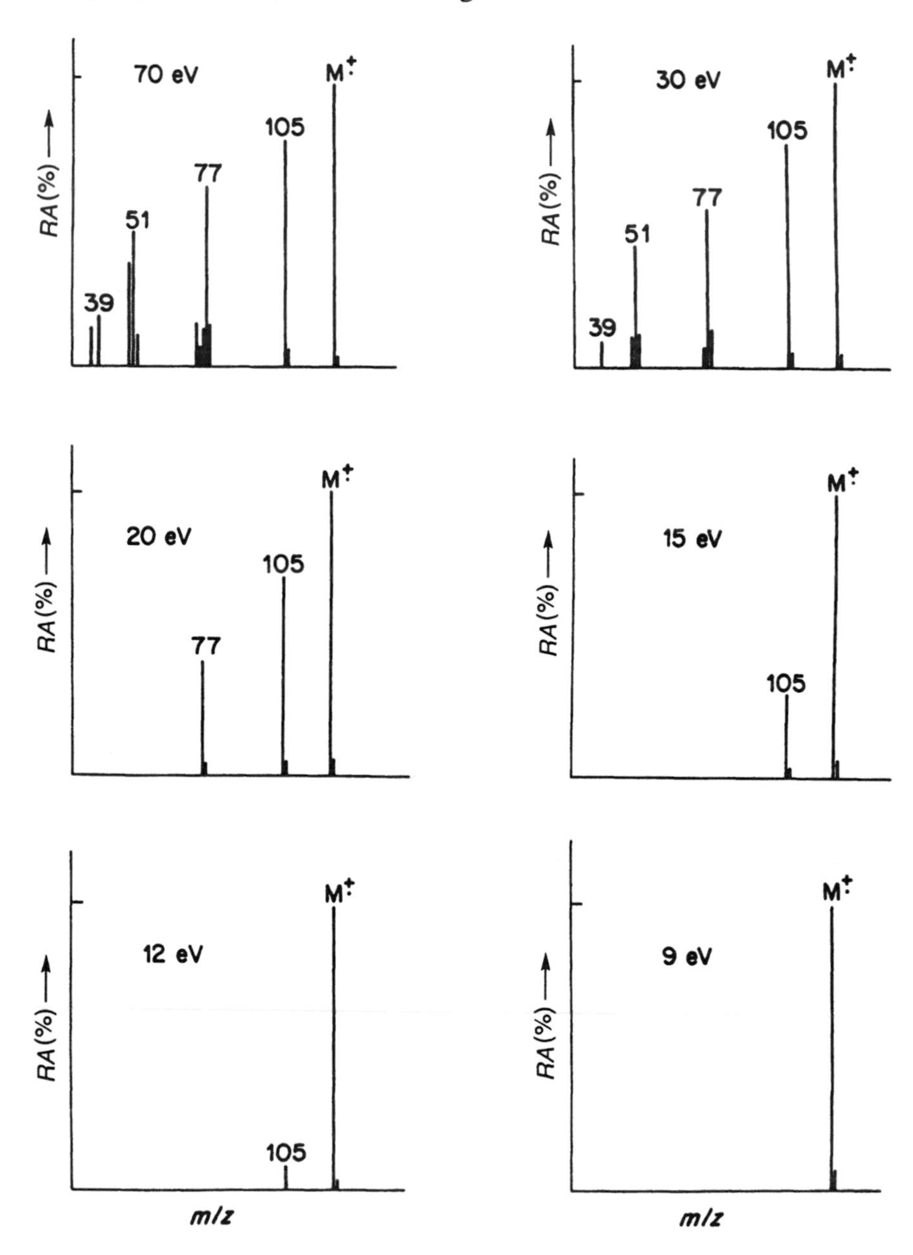


**Figure 2.1b** Mass spectra of benzoic acid ($C_6H_5COOH$) obtained by using different beam energies

As you can see, we have gained much more structural information at the higher beam energies. One point to note is that it is not always a good idea to use high energies with complex molecules . You may find that the enormous amount of structural information obtained conceals the more important features of the fragmentation pattern.

Electron impact is a popular and widely used technique because it often produces both molecular and fragment ions. It thus allows the determination of both the relative molecular mass and the molecular structure of a molecule.

However, there are some drawbacks:

(a) For some molecules, the molecular ion is not observed and thus it is difficult to determine the relative molecular mass of the compound. This occurs because all of the molecular ions fragment **before** they leave the ion source. An example of this is $C_9H_{20}$, with its mass spectrum being shown in Figure 2.1c.

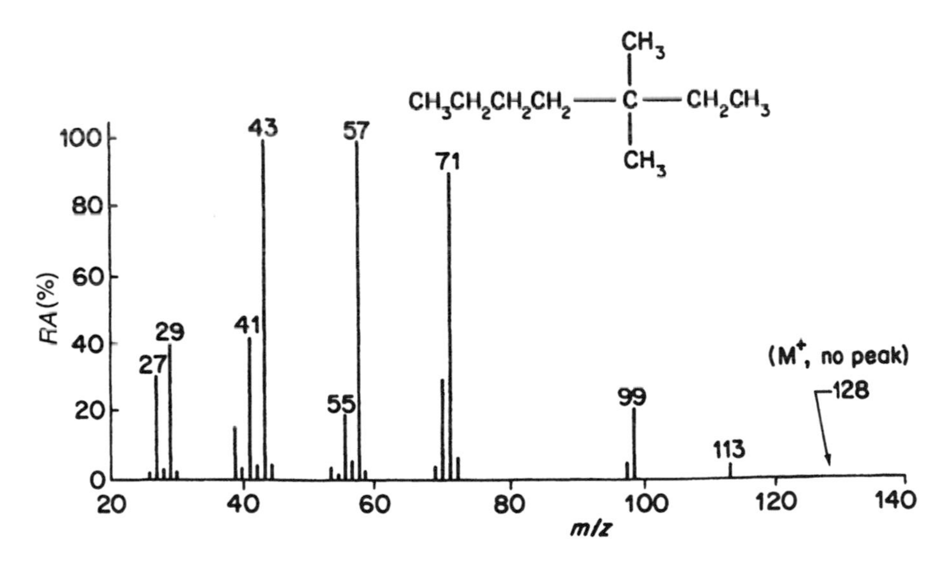


**Figure 2.1c** Mass spectrum of nonane ($C_9H_{20}$)

(b) It is often difficult to distinguish between isomers.

(c) Some compounds may undergo thermal decomposition prior to ionisation, or be very prone to fragmentation after ionisation, because of the temperatures required for vaporisation. As an example of this, note the differences seen in the mass spectra of the peptide derivative shown in Figure 2.1d, which have been recorded by using two different source temperatures.

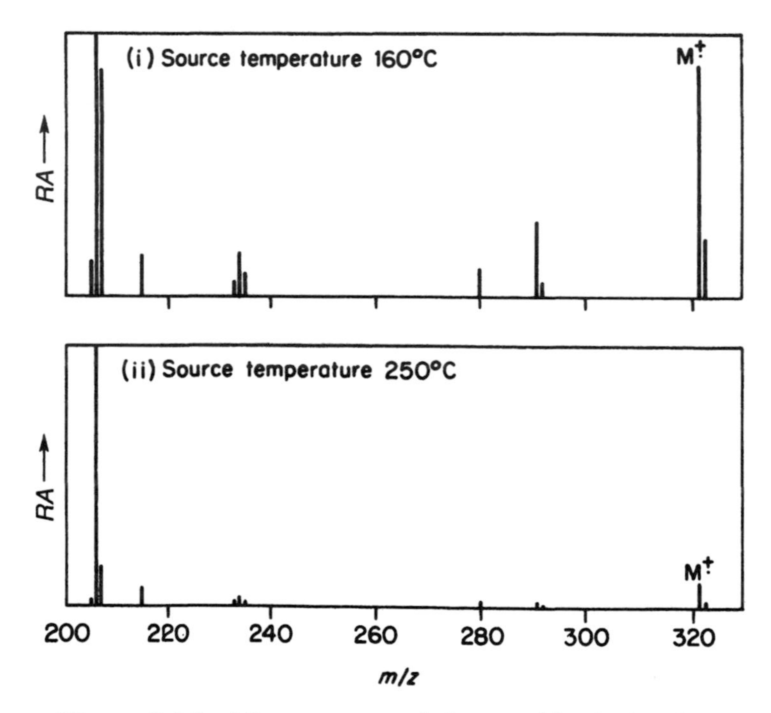


**Figure 2.1d** Mass spectra of the peptide derivative, $C_6H_5CH_2OCONHCH(CH(CH_3)_2\ CONHCH_2COOCH_3$, recorded at (i) 160° and (ii) 250°C

(d) Some compounds are simply too involatile to give a spectrum, e.g. polystyrene is involatile at temperatures up to 250°C.

(e) Occasionally, ion–molecule reactions produce peaks at higher mass numbers than that of the molecular ion. The quasi-molecular ion (as it is known), $(MH)^+$, is sometimes observed at positions which are 1 amu higher than the molecular ion. This often occurs when the sample pressure is high.

If any of these problems are encountered, then one of the following methods of ionisation may be used to a better effect.

## 2.2 CHEMICAL IONISATION

*Chemical ionisation*, or CI as it is sometimes known, has become very popular for structural elucidation since it was first developed in the mid-1960s. Here, the concept of ionisation relies on the interaction of ions with neutral molecules and the further production of new ions. In general, the amount of fragments ions produced is much less than in electron-impact ionisation, since little internal energy is imparted on the ionised molecule. For this reason, CI is termed a *soft ionisation* technique. Figure 2.2a shows a schematic representation of a chemical-ionisation ion source.

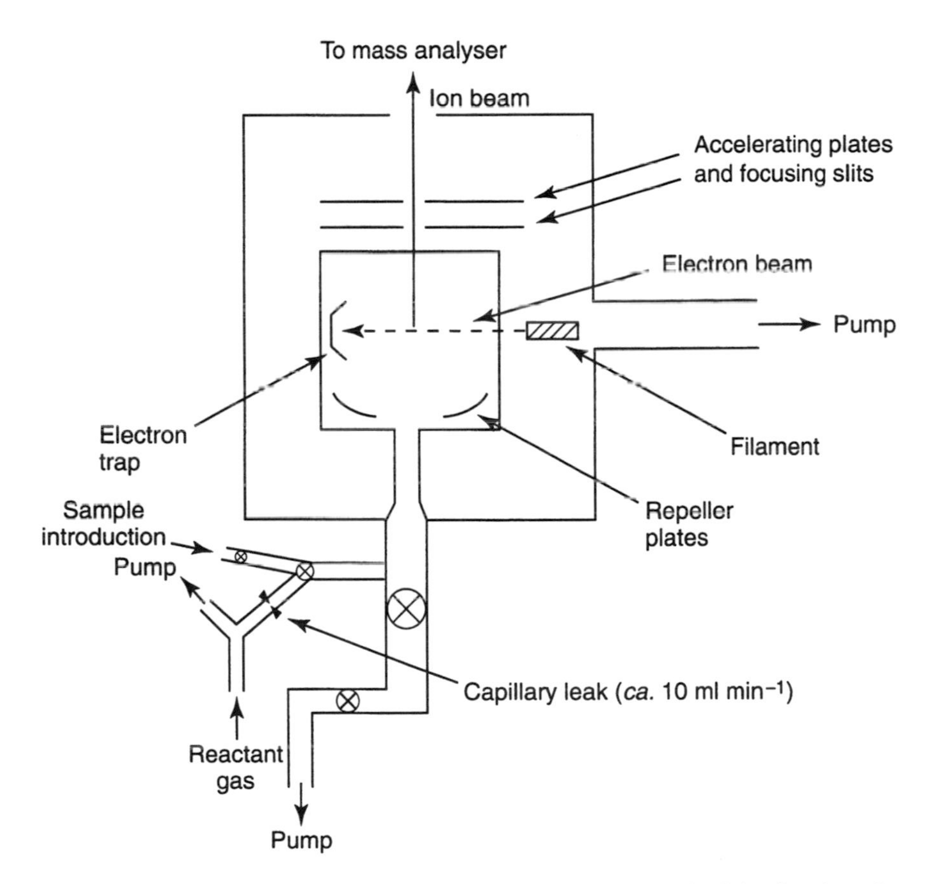


**Figure 2.2a** Schematic representation of a chemical-ionisation ion source

If we ionise methane gas at a pressure between 0.1 and 1.0 torr by electron impact, then we initially create the molecular ion:

$$CH_4 + e \longrightarrow CH_4^{+\bullet} + 2e$$

However, the methane is at a high pressure, so there is a distinct possibility of the molecular ion colliding with another methane molecule. When this happens the most likely reaction is as follows:

$$CH_4^{+\bullet} + CH_4 \longrightarrow CH_3^{\bullet} + CH_5^{+}$$

**SAQ 2.2a**

Which of the above species is acting as an acid and which as a base?

Therefore, if a small amount of the sample to be analysed is introduced into the ion source in the vapour phase, it is species such as $CH_5^+$ that act as the means of ionisation of the sample. For instance, if 'M' is the analyte then it can react as follows:

$$CH_5^+ + M \longrightarrow CH_4 + (MH)^+$$

The quasi-molecular ion, $(MH)^+$, will have an $m/z$ value which is 1 amu greater than that of the molecular ion; it also has a much lower internal energy than the molecular ion produced by electron-impact ion sources.

Π How would the mass spectra of a compound analysed by EI and CI sources differ?

Well, since the quasi-molecular ion has less energy, it will not fragment as much and thus may be observed in higher abundance in the CI spectrum rather than the EI spectrum. Figure 2.2b shows mass spectra of proline obtained by using both EI and CI ion sources.

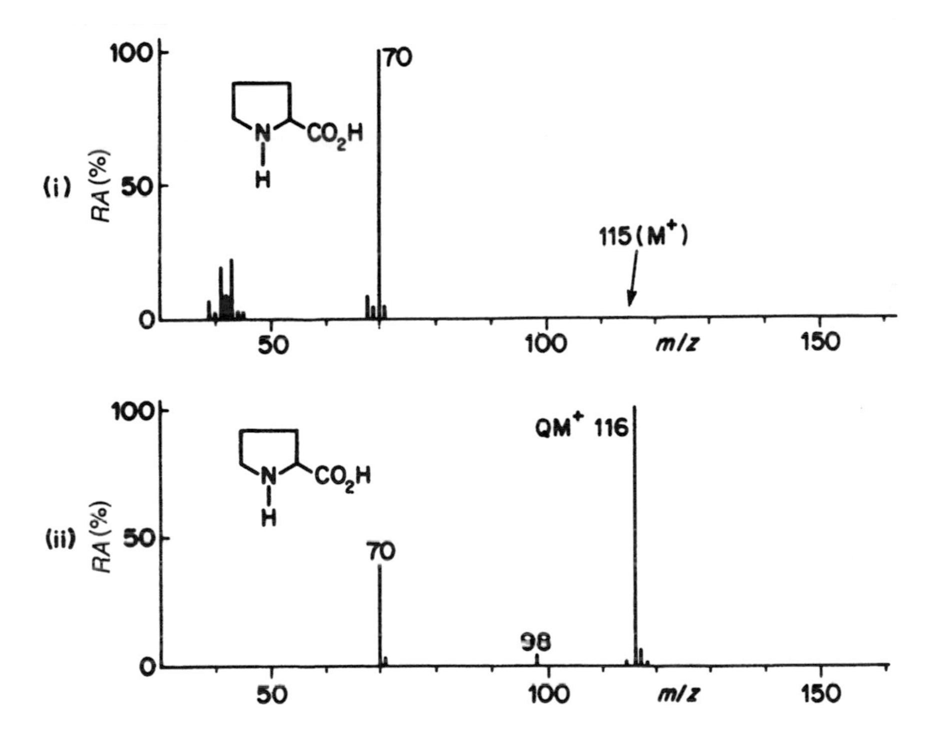


**Figure 2.2b** Mass spectra of proline obtained by using different ion sources: (i) electron impact; (ii) chemical ionisation

Modern mass spectrometers have combined CI and EI ion sources that can be readily changed by the turning of a handle (see Figure 2.2c).

However, you may ask, what if the sample, M, is ionised to $M^{+\bullet}$ by electron impact as well as the *reactant gas*, i.e. methane? Well, to minimise this problem, we make sure that there is at least 1000 times more methane than sample, so there is a much higher probability of the sample molecule colliding with a $CH_5^+$ ion than with an electron.

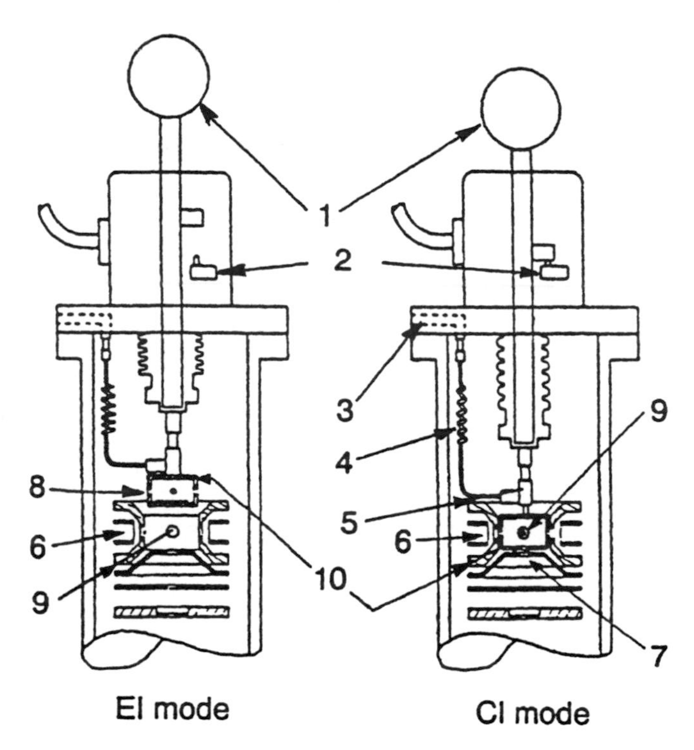


**Figure 2.2c** Combined EI and CI source: (1) EI/CI switch; (2) microswitch; (3) entrance for the reagent gas; (4) flexible capillary carrying the reagent gas; (5) diaphragm; (6) filament giving off electrons; (7) path of the ions towards the analyser inlet; (8) hole for the ionising electrons in CI mode; (9) sample inlet; (10) box with holes (also known as the 'ion volume'). Lowering the box (10) switches from the EI to CI mode; in EI mode this box serves as a 'pusher'

Interpretation of the CI spectrum has so far relied upon the fact that the relative molecular mass of the compound is one unit less than that of the quasi-molecular ion. There are exceptions to this, namely the reactant gas may undergo other ion–molecule reactions leading to other reactant ions and these may react differently with the sample. In the case of methane, the $CH_4^{+\bullet}$ ion may undergo fragmentation before it collides with a $CH_4$ molecule:

$$CH_4^{+\bullet} \longrightarrow CH_3^+ + H^\bullet$$

The fragment ion, $CH_3^+$, may itself undergo an ion–molecule reaction to form the $C_2H_5^+$ ion:

$$CH_3^+ + CH_4 \longrightarrow C_2H_5^+ + H_2$$

The $C_2H_5^+$ ion can also produce a quasi-molecular ion:

$$C_2H_5^+ + M \longrightarrow (MH)^+ + C_2H_4$$

as can ethylation of the sample molecules:

$$M + C_2H_5^+ \longrightarrow (M + C_2H_5)^+$$

Therefore, a peak situated 29 mass units above the compound's relative molecular mass can also be observed. The mass spectrum presented in Figure 2.2d shows both $(M+1)^+$ and $(M+29)^+$ ions.

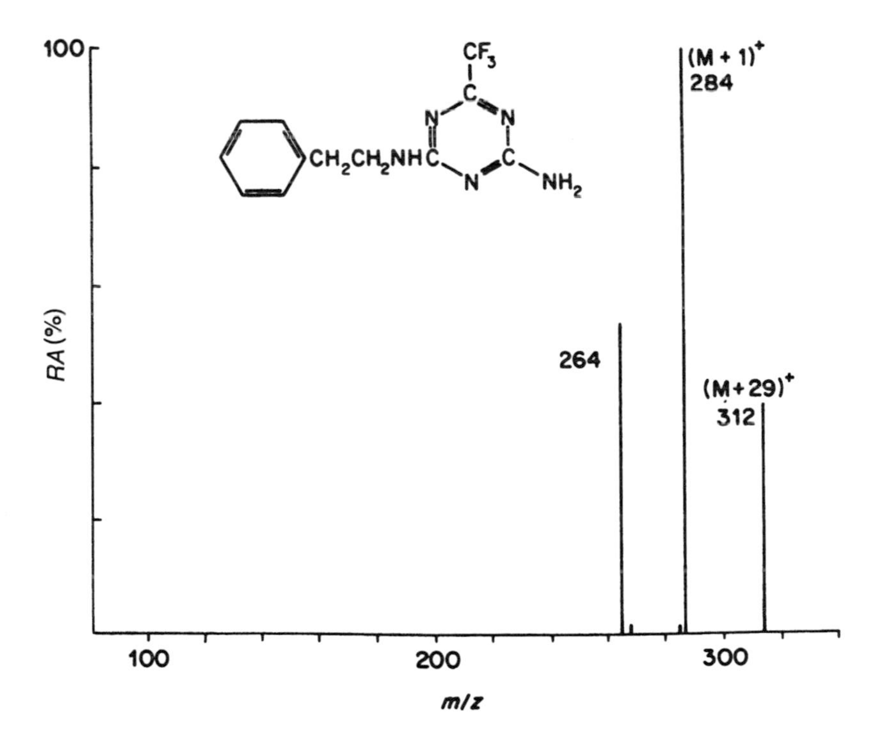


**Figure 2.2d** CI mass spectrum showing $(MH)^+$ and $(M+C_2H_5)^+$ ions

In some cases, where the proton affinity of M is lower than that of $CH_4$, the following reactions may take place, resulting in a quasi-molecular ion of 1 mass unit less than the molecular ion. This reaction is predominant in the higher alkanes.

$$CH_5^+ + M \longrightarrow (M-H)^+ + CH_4 + H_2$$

and

$$C_2H_5^+ + M \longrightarrow (M-H)^+ + C_2H_6$$

Other reactant gases that are used in CI are ammonia and isobutane ($(CH_3)_3CH$).

Π What are the major reactant ions generated by the use of these gases? Think about the stability of the ions that are generated.

The ions produced are $NH_4^+$ and $(CH_3)_3^+$.

**SAQ 2.2b**

Suggest how you might expect $NH_4^+$ and $(CH_3)_3C^+$ to react in ion–molecule reactions.

By using alternative reactant gases, different amounts of fragmentation can be observed from the same quasi-molecular ion. This is because the amount of energy transferred to a sample molecule on protonation depends on both the proton affinity of the sample and the acidity (the willingness to release a proton) of the reactant ion.

$CH_5^+$ is more acidic than $NH_4^+$, and thus the internal energy of the quasi-molecular ion formed by the $CH_5^+$ reactant ions is also greater. It follows that the amount of fragmentation will likewise be larger. Therefore, the use of different reactant gases can prove useful in structural determinations.

**SAQ 2.2c**

If you look at Figure 2.2b, you will see a major fragment ion at *m/z* 70. This represents the loss of 45 mass units from $M^{+\bullet}$ in the EI mass spectrum and 46 mass units from $(MH)^+$ in the CI spectrum. Suggest a formula for the neutral species produced and two possible structures for the quasi-molecular ion.

There is another small fragment ion in the CI spectrum at *m/z* 98. This corresponds to the loss of $H_2O$ from the quasi-molecular ion.

**SAQ 2.2d**

Suggest mechanisms for loss of water from the quasi-molecular ions you have drawn in SAQ 2.2c.

It seem that protonation does not occur at the nitrogen atom in this case. Note also that quasi-molecular ions are always even-electron ions. These are quite stable compared to their odd-electron counterparts. Fragmentation will usually involve loss of even-electron neutral species and formation of even-electron fragments. This is indeed the case for proline.

CI can be used to distinguish between isomers. In some cases for instance, the quasi-molecular ions of 5-β-3-keto steroids readily lose $H_2O$, whereas the 5-α-isomers hardly show this type of fragmentation.

**SAQ 2.2e**

Explain why the fragmentation of methy phenyl ketone, $C_6H_5COCH_3$, with isobutane reactant gas gives no fragmentation products, but with methane gives fragments, including *m/z* 43, 77, and 105.

**SAQ 2.2f**

List the advantages and disadvantages of using EI and CI as ion sources in mass spectrometry.

All CI plasmas contain electrons with low energies (1 eV) that are often derived from primary ionisation reactions. For instance, in a $CH_4/N_2O$ mixture (75/25) the following reactions occur:

$$N_2O + e \longrightarrow N_2O^{-\bullet}$$

$$N_2O^{-\bullet} + CH_4 \longrightarrow N_2 + CH_3^{\bullet} + OH^-$$

The basic hydroxide ion that is produced can be reacted with acidic compounds.

Recently, CI using negative ions has enabled researchers to detect tetrachlorodioxins with a higher sensitivity than can be achieved by using positive ions alone.

Pyrolysis of non-volatile samples can sometimes be prevented by the application of a sample solution on to rhenium wire, followed by evaporation and introduction into the CI source. The sample is rapidly desorbed by passing an electric current through the filament. The ion signal may only last for a few seconds and the observed spectrum may derive from a number of possible mechanisms. This method is known as *desorption chemical ionisation* (DCI) and is especially suitable for the analysis of small peptides, nucleic acids and certain organic salts in either the positive- or negative-ion mode.

## 2.3 FIELD IONISATION AND FIELD DESORPTION

### 2.3.1 Field Ionisation

*Field ionisation* (FI), developed in the 1960s, and *field desorption* (FD), developed in the 1970s, are two very closely related methods of ionisation, which both rely on essentially the same principle. A schematic diagram of the field-ionisation ion source is shown in Figure 2.3a. A potential difference of 10 kV is applied across the electrodes (field strength, $10^8$ V cm$^{-1}$) so that when a sample molecule in the vapour phase (pressure, $\leqslant 10^{-4}$ torr) impinges on the anode or comes very close to it, the sample becomes ionised and is accelerated towards the cathode. The cathode has a hole in it, and thus some of

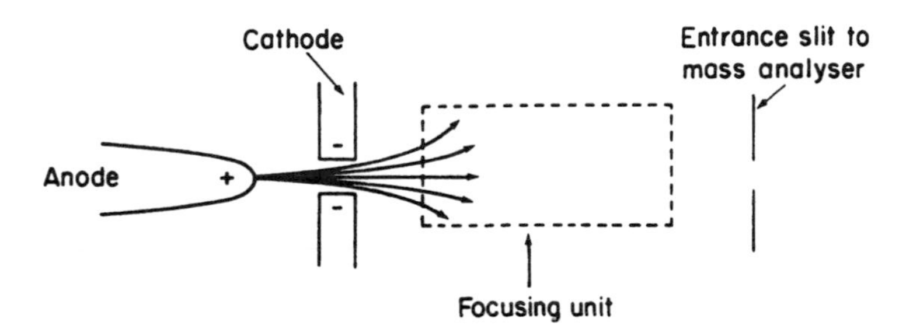


**Figure 2.3a** Schematic representation of a field-ionisation ion source

the ions can pass straight through. The focusing device beyond the cathode serves two purposes. First, it slows the ions down so that they can be easily analysed, and secondly it focuses the ion beam into the mass-analysis system.

The anode of an FI source is often referred to as the *emitter*, because ions are emitted from its vicinity. It is usually in the form of a sharp tip or a wire. Sometimes the surface area of the emitter is increased by the formation of 'whiskers' on its surface (see Section 2.3.2).

FI usually produces molecular ions, $M^{+\bullet}$, but in some cases quasi-molecular ions, $(MH)^+$, are observed. This is a result of ion–molecule reactions occurring near to the anode, where large concentrations of sample molecules are present. The uncertainty that occurs in having both types of ion produced makes the technique somewhat less attractive. In addition, the sensitivity is somewhat less than that achieved when using EI. However, no higher adducts (such as $(M+C_2H_5^+)$) are produced and the molecular ions have much lower internal energies than those produced by EI.

Π What does this indicate regarding the amount of fragment ions that are produced?

Well, it means that there is less fragmentation than in EI mass spectra, but the scans can still be structurally informative. Note that the structurally significant ions in the mass spectrum of xanthosine at *m/z* 152 and 133 (see Figure 2.3b) are easier to pick out in the FI spectrum, when compared to the EI spectrum.

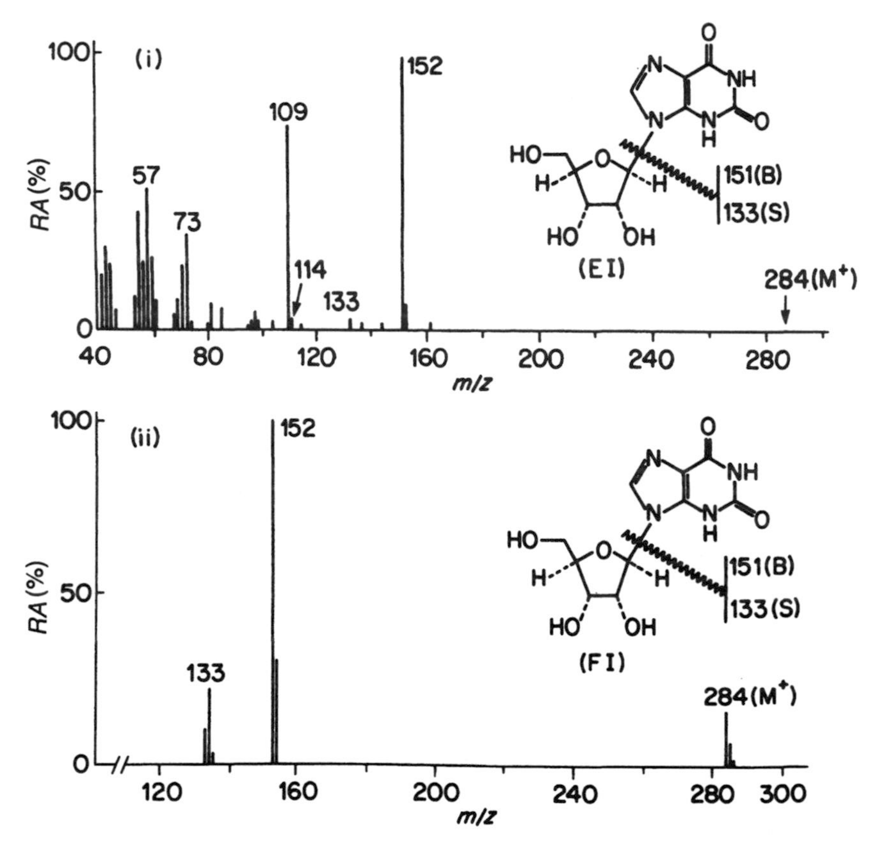


**Figure 2.3b** Mass spectra of xanthosine obtained by using different ion sources: (i) electron impact; (ii) field ionisation

**SAQ 2.3a**

List the advantages and disadvantages of FI over EI or CI ion sources.

### 2.3.2 Field Desorption

The big advantages of using field-desorption (FD) ionisation is that it can deal with involatile or thermally unstable compounds. The gas-phase sources we have met earlier are usually restricted to compounds that have boiling points of less than about 500°C. This often means that there is an upper mass limit of about $10^3$. We don't have this problem with FD. Field desorption operates on the same principles as field ionisation in that it uses an emitter. This emitter (a Re or W filament) is held at a high temperature with respect to the cathode, and the sample (present as a solid) is placed on its surface. Ions are thus desorbed directly from the solid towards the cathode.

The anode is first prepared by covering it with a carbon coating (obtained from the pyrolysis of benzonitrile in a high electric field) in the form of micro-needles or 'whiskers' (*ca.* 0.001 cm in length). This enlarges the surface area for contact with the sample and increases the ionisation efficiency. The sample is applied as a solution and is then evaporated to produce a solid deposit which is introduced into the ion source by a probe and a vacuum interlock. Field desorption is one of the 'softest' ionisation techniques discussed in this text. It generally produces only molecular and/or quasi-molecular ions with few fragments, and thus it is of little use in determining the structures of uncharacterised compounds.

One further drawback with FD is the transitory nature of the ion beam that is produced. Unlike EI, where a 1 mg of sample may last for a few hours before exhaustion, an FD ion beam may decay in a matter of minutes, thus making it difficult for fine-tuning the ion source.

FD has now largely been replaced by other desorption techniques. However, it remains an excellent method of ionising high-molecular-mass, non-polar compounds.

**SAQ 2.3b**

List the advantages and disadvantages of FD over the other techniques mentioned.

## 2.4 FAST-ATOM-BOMBARDMENT

*Fast-atom-bombardment* (FAB) desorption was developed in the 1980s and is now commonplace in most mass spectrometry laboratories. In this technique, a beam of fast-moving neutral atoms is directed on to a metal plate which has been coated with the sample. Much of the high kinetic energy of the atoms is transferred to the sample molecules upon impact. The energy can be dissipated in various ways, some of which can lead to ionisation of the sample.

The bombarding ions (as a beam of *ca.* 5 keV) are usually rare gases, either Xe or Ar. These atoms are first ionised and then passed through an electric field to increase their kinetic energy. After acceleration, the fast-moving ions pass into a chamber containing further gas atoms in which collisions of ions and atoms lead to charge exchange:

$$Ar^{+\bullet}(\text{fast}) + Ar(\text{thermal}) \longrightarrow Ar(\text{fast}) + Ar^{+\bullet}(\text{thermal})$$

The fast atoms that are formed retain most of the original kinetic energy of the fast ions and continue in the same direction.

Π How can any remaining fast ions or thermal ions be removed before sample bombardment?

A negatively charged deflector plate can be used to direct the ions away.

Positive or negative ions may be formed in the bombardment process, with extraction of the ions from the sample plate being dependent on whether the extraction slits are held at a positive or negative voltage. The rapid sample heating from the noble-gas beam reduces sample fragmentation. A schematic representation of an FAB source is shown in Figure 2.4a.

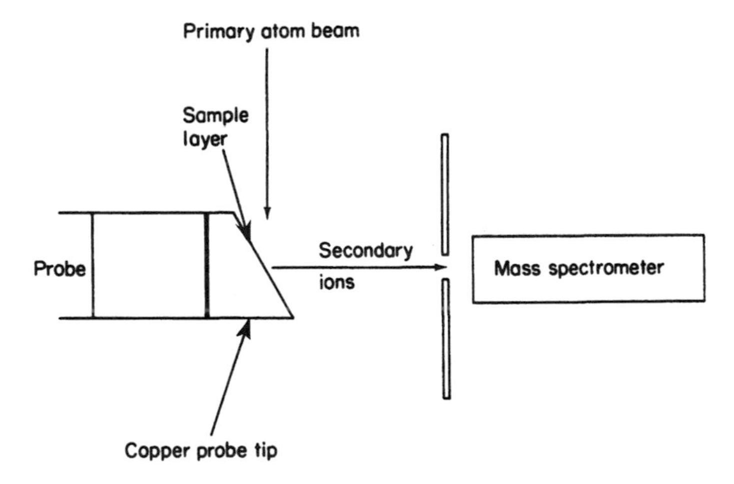


**Figure 2.4a** Schematic representation of an FAB source

The sample is usually applied to the plate in the form of a solution of an inert involatile liquid (matrix material), such as glycerol, *m*-nitrobenzylic alcohol (good for non-polars), diethanolamine (negative ions), and thioglycerol (dithiothreitol/dithioerythritol (5/1 wt/wt). The liquid matrix helps to reduce the lattice energy, which must be overcome to desorb an ion from a condensed phase, and also provides a means of 'mending' the damage done by bombardment. In order to achieve efficient ionisation, it is important to form a monolayer of sample at the surface of the matrix material and to make sure that the sample is well dissolved. The sample is applied to a probe (as in FD) and then introduced to the ion source via a vacuum interlock.

**SAQ 2.4a**

What problems could we have when introducing a matrix material into the ion source?

The spectra that are produced show the presence of molecular, quasi-molecular, and fragment ions. Figure 2.4b shows a spectrum of β-chaconine ($M_r$ 705), which has been obtained by using an FAB source.

Π What are the principle advantages of an FAB source?

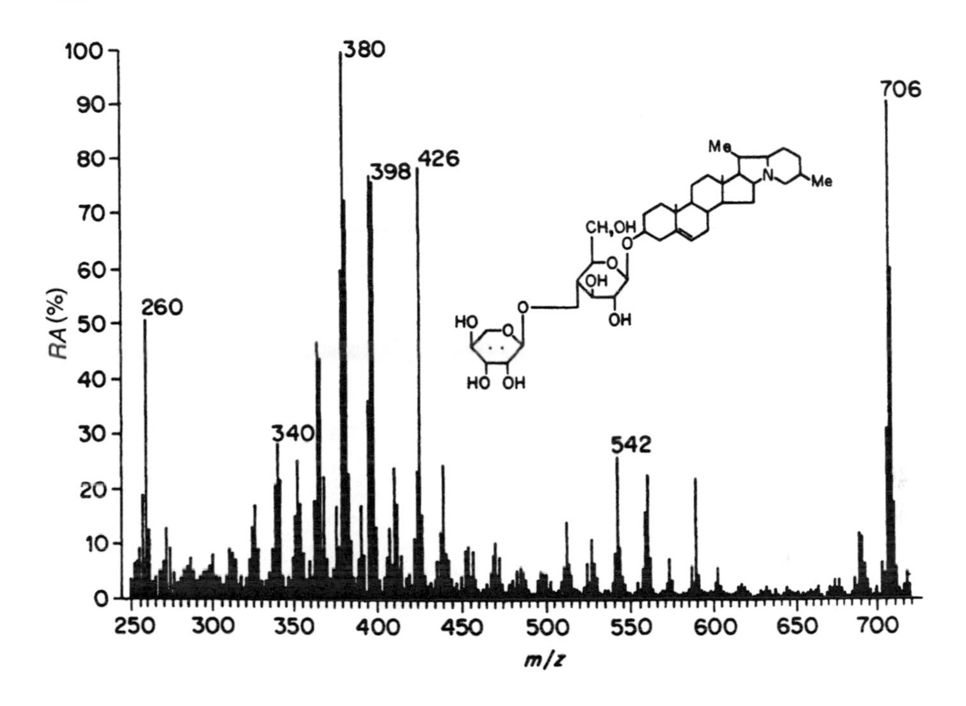


**Figure 2.4b** Mass spectrum of β-chacoline obtained by using an FAB source

(a) Ions can be generated from samples at room temperature; volatilisation is not required.

(b) Molecular, quasi-molecular, and (fragment) positive or negative ions may be produced.

(c) The lifetime of the sample is long (>20 min) and beam currents are quite stable.

(d) The sample sensitivity is high.

(e) The source is especially applicable to biomolecules of high molecular weight ($>10^4$) and thermally unstable compounds.

(f) Instrumentation is relatively simple, with source voltages <10 kV being required.

**SAQ 2.4b**

Figure 2.4c shows mass spectra of D-glucose ($M_r$ 180) obtained by using EI, FD, and FI ion sources. Compare the different features of each spectrum, and suggest origins for the ions at *m/z* 163, 145, and 127 in the FI spectrum.

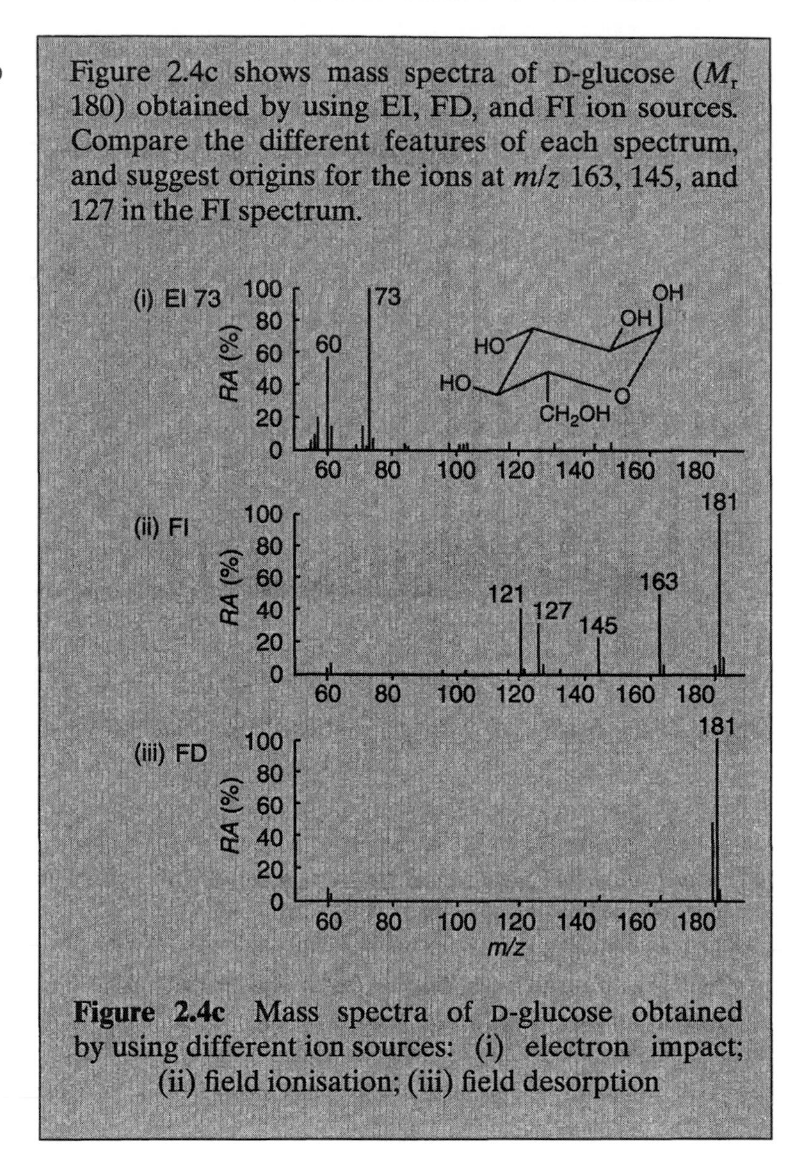


**Figure 2.4c** Mass spectra of D-glucose obtained by using different ion sources: (i) electron impact; (ii) field ionisation; (iii) field desorption

A complementary technique that uses ions rather than atoms is secondary-ion mass spectrometry (SIMS). In liquid-SIMS the sample dissolved in a liquid matrix (as in FAB) and then bombarded with $Cs^+$ ions at 30 keV. This method is claimed to give better sensitivity than FAB.

## 2.5 PLASMA DESORPTION

*Plasma desorption* (PD) has now been largely superseded by laser desorption methods. In this technique, the sample is deposited on to a small aluminised nylon foil. A $^{252}Cf$ source is then brought into the vicinity of the sample, whereupon the fission fragments from the nuclear decay, having energies up to several MeVs, impart shock waves in the sample, thus inducing the desorption of neutral species and ions.

## 2.6 LASER DESORPTION

*Laser desorption* (LD) can produce gaseous ions relatively efficiently. If laser pulses (a few ns each) of $10^6$–$10^{10}$ W cm$^{-2}$ are focused on to a sample surface (often a solid) with an area of *ca.* $10^{-3}$–$10^{-4}$ cm$^2$, then a few pmols of compound are desorbed in the form of neutral molecules and ions. Reactions among these molecules themselves may occur in the dense vapour-phase region near the sample surface. By altering the wavelength, selective ionisation can be achieved. Time-of-flight (TOF) analysers (see Section 3.7) or simultaneous detection analysers are needed to detect the short-lived signals.

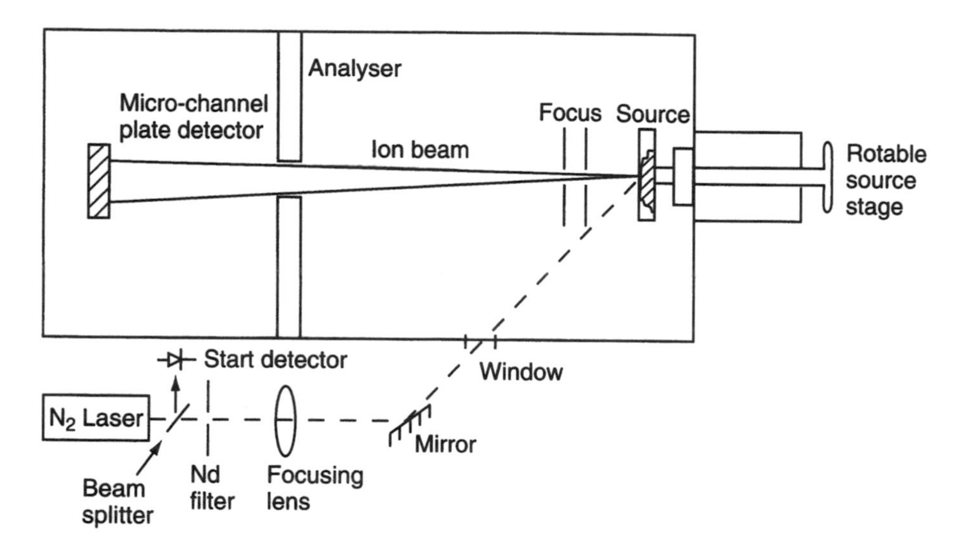


**Figure 2.6a** Schematic representation of a MALDI–TOF mass spectrometer

Matrix-assisted laser desorption ionisation (MALDI) has recently become a very popular technique in the analysis of bimolecules (see Figure 2.6a). Here, an $N_2$ UV laser generates a 4 ns pulse at 337 nm through a neutral-density filter in conjunction with a variable iris (needed to vary the energy); the ion source is maintained at 25 kV. (A biased micro-channel plate, known as a reflectron, may be placed prior to the detector to cut down on peak broadening.)

The solid sample is dissolved in a suitable matrix solvent and the solution evaporated. The laser pulse electronically excites the matrix molecules, thus resulting in desorption of the ions formed by proton transfer between the excited matrix–analyte mixture. This process is illustrated in Figure 2.6b.

Two of the most commonly used MALDI matrices in peptide analysis are 2,5-dihydroxybenzoic acid (DHB) and α-cynao-4-hydroxycinnamic acid (4-HCCA). Both of these matrices produce relatively few ions above $m/z$ 400 and do not complicate the mass spectrum by forming adducts with peptide ions.

Sinapinic acid (3,5-dimethoxy-4-hydroxycinnamic acid) is also used as

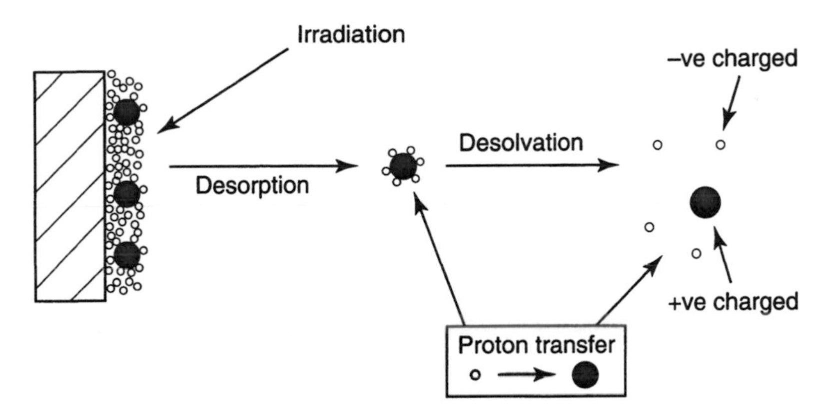


**Figure 2.6b** Illustration of the principles behind MALDI: (•) product; (○) matrix

a matrix; a MALDI–TOF spectrum of bovine serum albumin in this solvent is shown in Figure 2.6c.

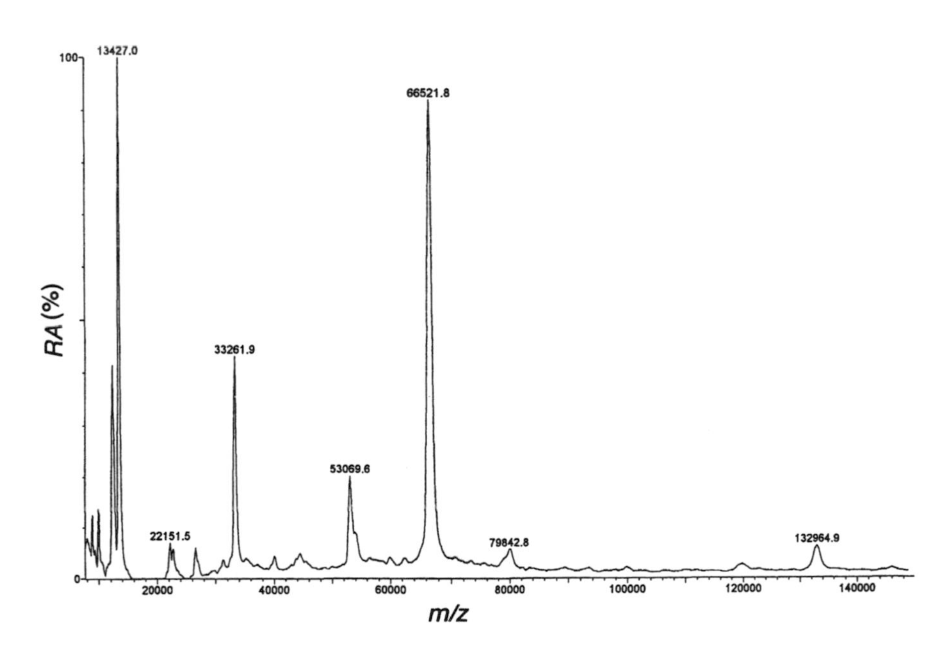


**Figure 2.6c** MALDI–TOF mass spectrum of bovine serum albumin in a sinapinic acid matrix

The advantages of the MALDI method are as follows:

(a) It is not necessary to match the absorption frequency of each analyte with the wavelength.

(b) The MALDI process is independent of the size of the compounds and their absorption properties, e.g. proteins with masses greater than 300 000 Da can be analysed.

(c) There is a much higher sensitivity, e.g. a nicotinic acid matrix can allow the detection of a few pmols of certain proteins.

(d) Sample ion clusters are prevented from forming by using excess matrix (to analyte) molecules.

Π Why might sample ion clusters be a problem?

They may hinder the appearance of molecular ions.

## 2.7 OTHER ION SOURCES

*Electrohydrodynamic ionisation sources* (EHMSs) use a sharply pointed liquid meniscus as the emitter, with the ions being formed by field evaporation.

*A pulsed-field desorption* (PFD) source uses negative-voltage pulses of *ca.* 100 ns duration which are applied to the cathode. This permits the electrical field (where ionisation takes place) to be kept relatively low, and as a result the fragmentation pattern is further reduced.

Other ion sources, such as *inductively coupled plasmas* (ICPs) and *spark sources*, will be described in Chapter 9, while *electrospray/thermospray* ionisation techniques will be discussed in Chapter 8.

## 2.8 METHODS FOR SAMPLE INTRODUCTION

The purpose of the inlet system is to permit introduction of a representative sample into the ion source with a minimal loss of vacuum. The three main types of inlet used are *batch inlets (cold, all-glass heated inlet systems (AGHISs)* and *septums), direct-insertion probe inlets*, and *chromatographic inlets*.

Π Why do we need different types of inlets?

We may need to analyse samples in different physical and chemical forms, i.e. solids, liquids, gases, pure compounds, and mixtures.

For instance, the EI source requires a thermally stable sample in the vapour phase. In addition, during sample introduction the pressure inside the spectrometer should be kept as low as possible. Gases and liquids with a high vapour pressure at room temperature are usually admitted to the spectrometer via a *cold inlet*, shown in Figure 2.8a. This consists of a reservoir to hold the sample and a series of valves connected to the ion source and an auxiliary vacuum system. By a series of manipulations, the sample can be transferred to the reservoir, and from there, to the ion source. A fritted glass disc is placed in the inlet line to control the leak of gas into the ion source.

If we are dealing with an involatile liquid, the whole system, including the transfer line, is heated.

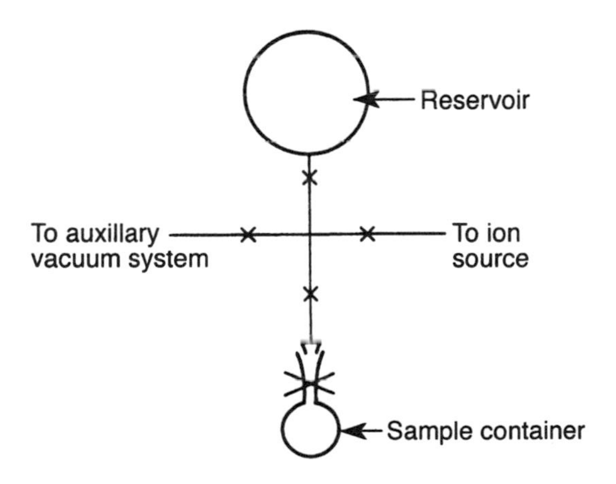


**Figure 2.7a** Schematic representation of a cold-inlet system; **X** represents a vacuum tap

**SAQ 2.8a**

When handling hot samples, why is it preferable to construct the whole inlet system out of glass rather than metal? (An 'all-glass heated inlet system' is often called an *AGHIS* system.)

Another device which is often used for liquids is the *septum inlet*. This consists of a heated evacuated reservoir which is protected from the external atmosphere by a rubber septum, similar to that used in gas chromatography (GC) analysis. The sample is injected through the septum by means of a hypodermic syringe.

Solids are best introduced into the ion source by using the *direct-insertion probe inlet*. This has previously been mentioned in connection with FD and FAB sources and is shown in Figure 2.8b. It consists of a retractable metal rod which is inserted into the ion source via a vacuum lock. The ion source and the probe tip can be heated (and cooled), taking care not to outgas the sample too quickly and thus run into vacuum problems. Heating between 50–200°C at a pressure of $10^{-6}$ torr is usually sufficient to volatilise most organic solids. One advantage of this probe is that much less sample is needed for analysis. The low pressure in the ionisation area and the proximity of the sample to the ionisation source often make it possible to obtain spectra of thermally unstable compounds before major decomposition has had time to occur. The low pressure also leads to greater concentrations of relatively non-volatile compounds in the ionisation

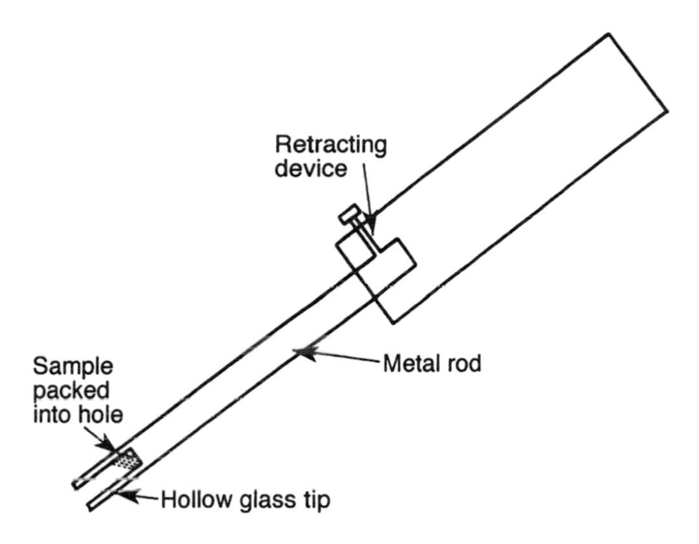


**Figure 2.7b** Schematic representation of a direct-insertion probe for use with solid samples (NB: not drawn to scale)

area. Therefore, studies of steroids and carbohydrates (among others) would be permitted.

A number of modern spectrometers have *chromatographic inlet* systems connected to either gas or liquid chromatographs, thus enabling the analysis of mixtures. This will be dealt with in more detail in Chapter 8.

**SAQ 2.8b**

Which inlet system best meets the requirements needed for analysis of the following:

(a) a mixture of insect sex pheromones which can be separated by liquid chromatography;

(b) a relatively non-volatile organometallic compound or polymeric solid;

(c) a high-vapour-pressure gas?

## Summary

This chapter has covered many of the most popular techniques of ionisation which are encountered in organic mass spectrometry, namely electron-impact, chemical ionisation, fast-atom-bombardment, field ionisation, field desorption, plasma desorption, and laser desorption. We have also discussed some of the advantages and disadvantages of each method, along with a brief interpretation of the spectra that can be obtained by using such ionisation methods.

We finished the chapter by looking at the various inlet systems (batch, direct-insertion probe and chromatographic) which are used for analysing solids, liquids, and gases, and commented on their strengths and drawbacks.

## Objectives

After studying this chapter you should be able to:

- describe the theory and operational principles behind electron-impact ionisation, chemical ionisation, field ionisation, field desorption, plasma desorption, laser desorption and fast-atom-bombardment;

- discuss the advantages and disadvantages of each method of ionisation;

- describe the simple differences seen in the spectra of compounds analysed by using these various methods of ionisation;
- describe the inlet systems (batch, direct-insertion probe and chromatographic) used for admitting solids, liquids, and gases to the mass spectrometer;
- recognise any limitations of each inlet system;
- decide which is the inlet system of choice for different types of samples.

# 3. The Mass Spectrometer

This part of the Unit will introduce you to the instrumental side of mass spectrometry. In the previous chapter, we examined some different types of ion sources and the contrasting spectra that were obtained from each of these. This present chapter will start by looking at thc mass spectrometer system as a whole and then go on to look in detail at some commonly encountered mass analysers. We will discuss the advantages and limitations of each system, with particular emphasis being placed upon mass resolution, scan rate and multiple-ion monitoring.

## 3.1 THE MASS SPECTROMETER SYSTEM: AN OVERVIEW

Π What are the basic components of a mass spectrometer?

We have already discussed the first component, i.e. the ion source. The other main parts are the *mass analyser*, the *detector* and the *recorder*. These are shown in Figure 3.1a, arranged in the order that the molecular-ion beam sees them.

The ion source, as you have already seen, simply converts molecules from the sample into ions (usually positively charged) and accelerates them into a mass analyser. The mass analyser distinguishes ions by nature of their mass-to-charge ($m/z$) ratios, and not just simply by their mass, as the name might imply. However, most ions carry only a single positive charge and therefore, to a first approximation, a simple mass-spectrum scan characterises the abundance of each ion in relation to its mass.

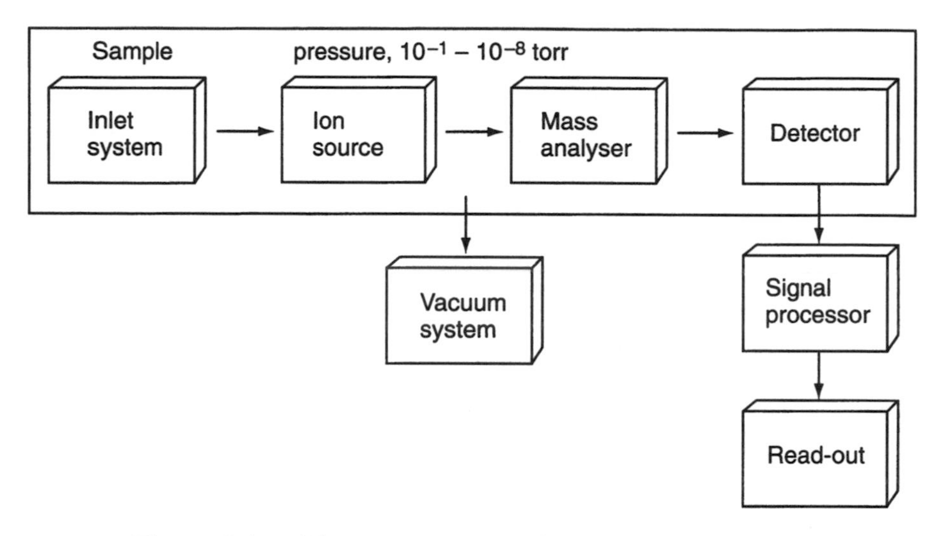


**Figure 3.1a** The components of a mass spectrometer

Once the ions have been analysed, they then pass to a detector; the information gained can then be recorded in several different ways.

Let us start our investigation of the mass spectrometer by examining a 'single-focusing magnetic-sector mass spectrometer with an electron-impact ion source'. You may think that this instrument is an incredibly complex device, but in fact it is one of the more simple and commonly encountered instruments. Figure 3.1b shows a schematic diagram of this type of mass spectrometer.

The whole instrument is kept at a low pressure, *ca.* $10^{-6}$ torr (1 torr = 133.3 N m$^{-2}$). This is for a variety of reasons.

Π Can you think of any reasons why the mass spectrometer needs to be kept at a low pressure?

First, the ion-source filament would be rapidly oxidised at normal air pressure, and would quickly burn out. Imagine it to be similar to the filament in a light bulb.

Secondly, a lower pressure would make it easier for less-volatile liquids and solids (used in desorption ion sources), with very low vapour pressures at stp, to vaporise. In addition, since many ion

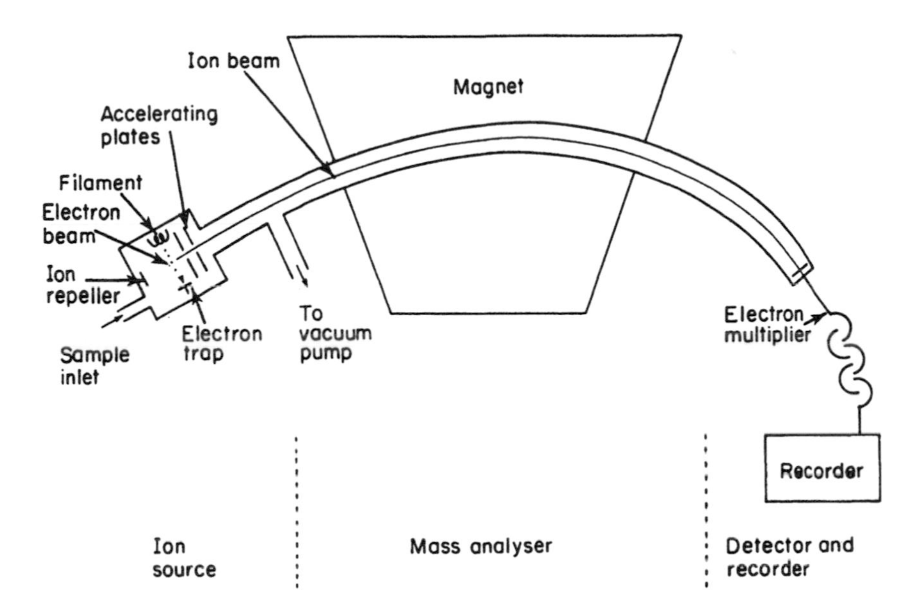


**Figure 3.1b** Schematic diagram of a single-focusing mass spectrometer with an electron-impact ion source

sources are heated (to *ca.* 200°C), a vacuum would help to remove any gases which may be dissolved in the sample material or the metallic interior of the source itself.

Thirdly, a low pressure will prevent ions from colliding with atmospheric gas molecules and so losing their charge to these molecules. The ions would therefore not be accelerated by the high-voltage plates and thus not reach the detector.

**SAQ 3.1a**

> Which gaseous ion produced from the atmosphere might you expect to be observed in a mass spectrometer operating at a pressure that was too high?

Fourthly, the ion source has a *memory effect*. For instance, if we have analysed a sample with a high vapour pressure and we now want to analyse a sample with a very low vapour pressure, we would need to wait some time for the vapour of the previous sample to be dispersed. The efficiency for ionisation is only about 1%, so we would have a lot of uncharged sample molecules floating about. If we had the system under vacuum, then the pump would remove these unwanted molecules in a matter of minutes.

Continuing with our examination of the mass-spectrometer system, you may remember that ionisation in the EI source (see Figure 2.1a) is brought about by bombardment of gaseous sample molecules with high-energy electrons directed at right angles to the sample inlet. These electrons are emitted from a heated tungstem or rhenium wire *filament*. On the opposite side of the ion source is an *electron trap*, which is held at a positive potential with respect to this filament. This device draws the electrons across the sample volume and catches any negatively charged ions and unused electrons. *Collimating magnets* (not shown in the figure) increase the pathlength of the electron beam, so that a maximum number of sample molecules are interacted with. Also present in the EI source are the *repeller* plate and the *extraction* plate. The extraction plate is held at a relative negative charge to the repeller, which therefore causes the positively charged ions to be accelerated away from the inlet area and towards the other acceleration plates. These accelerating plates are held at separate, and high potentials (4000 – 8000 V), to that maintained between the repeller and the extraction plates, and accelerate the ion beam towards focusing and collimating slits and then to the mass analyser.

**SAQ 3.1b**

Sometimes, EI mass spectrometers are used to produce negative ions. What alterations to the polarity of the ion-source components would have to be made in order for negative-ion mass spectrometry to take place?

When the ions enter the mass-analyser region, analysis (in this case) is obtained via the use of a magnetic field. It so happens that when an electrical current is passed through a magnetic field, it follows a path that is curved. Our beam of ions can be thought of as an electrical current, and therefore when they pass through the mass analyser, the ions followed a curved flight path. The magnet is often constructed so that the ion beam traverses the circumference of a circle whose sector angle is 60° or 90°. For this reason, these type of spectrometers are often referred to as *magnetic-sector* instruments.

Let us now examine the forces that affect an ion of mass $m$ and charge $z$ as it passes through a magnetic field of strength $B$. Its velocity is given by $v$, and the radius of the circle whose circumference which the ion traverses is given by $r$.

The centrifugal force ($F$) felt by the ion is given by the following formula:

$$F = mv^2/r \quad (3.1)$$

There needs to be an equally strong force directed towards the centre of the circle counterbalancing this and thus preventing the ion flying off at a tangent. This is achieved by applying a magnetic field. The magnitude of the force (F) is given by the following formula:

$$F = Bzv \qquad (3.2)$$

These forces are equal and in opposite directions in order to ensure that the ions continue to follow the same path. Thus, we can write:

$$mv^2/r = Bzv \qquad (3.3)$$

Π Can you remember two ways in which we can express the formula for the kinetic energy of ions which have passed through a potential difference of $V$? (Hint: consider equations (2.2) and (2.3) in the previous chapter.)

The gain in kinetic energy is proportional to the potential difference through which the ions have passed. Therefore:

$$\tfrac{1}{2}mv^2 = zV \qquad (3.4)$$

Π Now, can you remember which ratio mass spectrometers measure?

The answer is $m/z$.

Therefore, it would be a good idea if we could rearrange these two equations in some way so as to get a formula in which $m/z$ was the subject. This will involve a little bit of algebra.

The 'trick' in this case is to remember that the velocity of the ion as it traverses the magnetic field is governed solely by the kinetic energy it acquires during acceleration, so let's try and eliminate $v$ from both formulae. (Apart from $m$, which we will need as part of our final subject, $m/z$ it is the only term that occurs in both equations.)

Rearranging, by making $v^2$ the subject of equation (3.4), gives the following:

$$v^2 = 2zV/m \qquad (3.5)$$

Let us now make $v$ the subject of equations (3.3). This gives:

$$v^2 = Bzvr/m \tag{3.6}$$

Cancelling the $v$ on both sides of the equation gives:

$$v = Bzr/m \tag{3.7}$$

By squaring both sides of the equation we obtain:

$$v^2 = B^2z^2r^2/m^2 \tag{3.8}$$

Now, 'equate' equations (3.5) and (3.8) in $v^2$; this gives:

$$2zV/m = B^2z^2r^2/m^2 \tag{3.9}$$

Next we can rearrange, and cancel the $z$ and $m$ terms to give:

$$2Vm^2/m = B^2z^2r^2/z$$

from which

$$2Vm = B^2zr^2 \tag{3.10}$$

Remember that we need to make $m/z$ the subject, so:

$$m/z = B^2r^2/2V \tag{3.11}$$

Finished—that was somewhat laborious!

Π So far so good then, but what does this mean in terms of our ion beam?

It means that for any particular fixed value of the magnetic field strength, $B$, and accelerating voltage, $V$, only ions of one particular mass/charge ration ($m/z$) will follow the required circular path of radius $r$ to reach the detector. All other ions with different $m/z$ values will be trying to follow circles of different radii and will not therefore reach the detector (see Figure 3.1c).

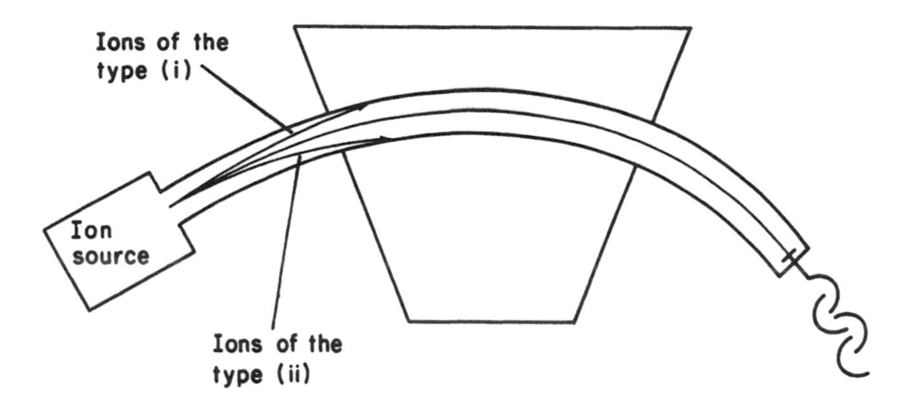


**Figure 3.1c** Flight path of ions in a mass spectrometer

**SAQ 3.1c**

Figure 3.1c shows the trajectories of ions of two different masses, but with similar charges, that do not follow the correct flight path to reach the detector. One of these follows a flight path (i) that has a radius larger than that required, while the other takes a flight path (ii) with a radius smaller than that required. Which of the two ions is the heavier?

**SAQ 3.1d**

What accelerating potential ($V$) will be required to direct a singly charged methyl ion through the exit slit of a mass spectrometer if the magnet has a field strength of 0.250 t and a radius of curvature of the ion through the magnetic field of 10.5 cm?

Π How can we make ions that have different $m/z$ ratios follow the correct trajectory and reach the detector? Note that the radius of curvature will be fixed for every magnet.

From equation (3.11) we can easily see that there are two choices, i.e. we can:

(a) vary $V$ and keep $B$ constant

or

(b) vary $B$ and keep $V$ constant.

In (a) we would be using a *voltage scanning* spectrometer and in (b) a *magnetic scanning* spectrometer. It turns our that although voltage scanning instruments are the cheapest and were the first to be used, magnetic scanning instruments are more sensitive and are therefore used on all commercially available spectrometers.

We shall leave our discussion of the detector until Chapter 4.

**SAQ 3.1e**

As a revision exercise, can you list the main components of a mass spectrometer, labelling them in the order in which the ion beam would see them?

## 3.2 MASS RESOLUTION

The ability of a mass analyser to attribute a specific $m/z$ value to a particular ion is called its *resolving power*.

For instance, what is the resolving power needed to identify ions differing by 1 amu in the region of 4000 amu, e.g. ions with values of 4000 and 4001 amu?

Resolving power is defined as the 'mass to be measured divided by the difference in masses to be identified', i.e. $m/\Delta m$. In this example, the resolving power is therefore:

$4000/(4001-4000) = 4000$, i.e. 1 amu in $4 \times 10^3$ amu.

Single-focusing mass spectrometers, such as the one shown in Figure 3.1b have a mass resolution of about 5000.

**SAQ 3.2a**

Let us suppose that we are using an instrument with a resolving power of 4000, but find that the mass region we are interested in is 400. How accurately could we measure such masses?

It may help us to think graphically as well as mathematically when it comes to resolution.

Π For example, how many of the pairs of peaks in Figure 3.2a do you think are clearly resolved?

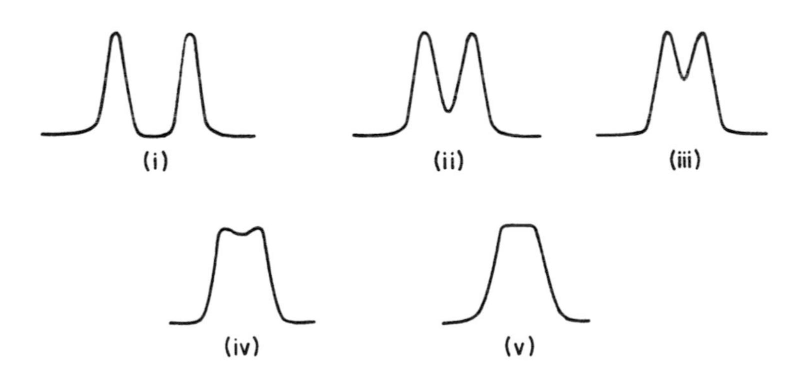


**Figure 3.2a** Schematic representations of two neighbouring peaks

Some people may only be happy with the resolution shown in (i), while others may think that (ii) is quite satisfactory. In order to obtain a standard definition of resolution, and hence resolving power, mass spectrometrists only accept a pair of peaks as being adequately resolved if the height of the overlapping portion between them is

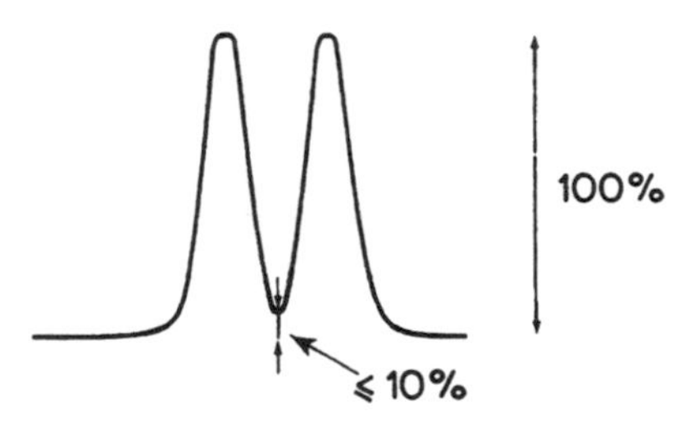


**Figure 3.2b** Illustration of the 10%-valley definition of resolution

10%, or less, of the peak height (see Figure 3.2b). This is known as the *10%-valley definition.*

Now, let us return to resolving power and discuss some examples.

Π What is the nominal and accurate relative molecular mass of the methane molecular ion, $CH_4^{+\bullet}$? Assume that we are only dealing with $^{12}C$ and $^{1}H$ isotopes and that the relative atomic mass ($A_r$) of $^{12}C$ is 12.00000 and that of $^{1}H$ is 1.00783.

For methane ($CH_4$), the nominal figure is $12 + (4 \times 1)$, i.e. 16; the more accurate figure is $12.00000 + (4 \times 1.00783) = 16.03132$ for methane, and $16.03132 - 0.00055 = 16.03077$ for the molecular ion.

Π Do you remember SAQ 1.2b, where we did a similar calculation?

**SAQ 3.2b**

> The nominal mass of the ethanol molecular ion, $CH_3OH^{+\bullet}$, is 32. Calculate its accurate mass, assuming, as before, that we are dealing with the predominant isotopes for each, i.e. $^{12}C$, $^{1}H$, and $^{16}O$ ($^{16}O$, $A_r$ = 15.994 92).

The relative atomic mass ($A_r$) values for the four most common elements found in organic chemistry are as follows:

$^{12}C$ 12.000 0

$^{1}H$ 1.007 83

$^{16}O$ 15.994 92

$^{14}N$ 14.003 07

**SAQ 3.2c**

The molecular ion of a colourless gas is examined by mass spectrometry and gives a nominal mass of 32, which is similar to the nominal mass of methanol. However, its accurately measured mass is found to be 31.9898 ± 0.0001. Use the above list of accurately measured $A_r$ values to identify the molecule.

It is therefore now easy to see that even molecules with similar nominal masses have different accurately determined masses when measured with a *high resolution spectrometer*.

**SAQ 3.2d**

If we had a mixture of methanol and oxygen in the mass spectrometer, what resolving power would be needed to distinguish between the molecular ions?

**SAQ 3.2e**

Calculate the resolving power needed to distinguish between the following pairs of ions. If the resolution of the magnetic analyser was 5000, which of these pairs could be differentiated?

(a) $C_7H_{14}^{+\bullet}$ and $C_6H_{10}O^{+\bullet}$

(b) $C_{23}H_{46}^{+\bullet}$ and $C_{22}H_{42}O^{+\bullet}$

(c) $C_{39}H_{78}^{+\bullet}$ and $C_{38}H_{74}O^{+\bullet}$

## 3.3 DOUBLE-FOCUSING MASS SPECTROMETER

Π Do you remember the three properties of an ion that determine the way it behaves in a mass spectrometer? (Refer to equations (3.3) and (3.4).)

These are the ion's mass, charge and velocity.

Π One of these properties was assumed to be constant for any single ion after acceleration and passage through a magnetic field, and could therefore be eliminated from these equations. Which property was this? (Equations (3.5) and (3.8) should help here.)

It is the velocity of the ion in question.

Do you remember from Section 2.1 that we said that if the kinetic energy of all of the ions with a similar charge was the same, then it followed that the heaviest ions moved the slowest? Sometimes, *velocity focusing* is used as a means of characterising ions. However, it turns out that the kinetic energies of all ions with the same $m/z$ values are not exactly the same. There is, in fact, a spread of kinetic energies among the ions, as shown in Figure 3.3a.

Π How does this spread arise?

(a) The ions follow the Boltzmann distribution of thermal energies from the molecules from which they were formed.

(b) Field inhomogeneity in the ion source. Some ions with the same $m/z$ values experience slightly different potentials depending on exactly where in the ion source they are formed.

Figure 3.3a also shows three curves with kinetic-energy spreads of differing magnitudes.

Π For which of the three curves are accurate measurements of maximum kinetic energy most easily and least easily achieved?

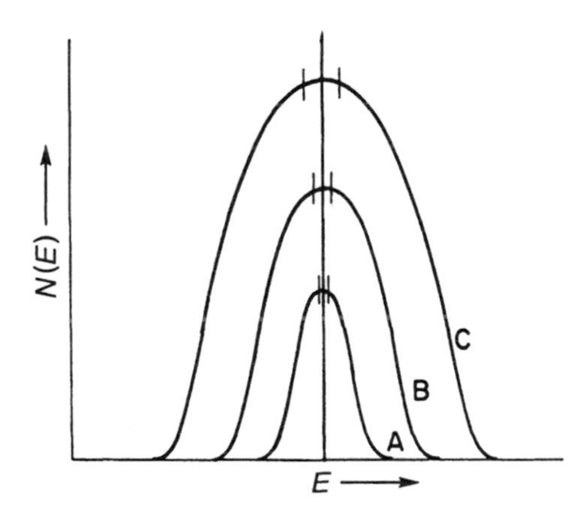


**Figure 3.3a** Plots of the number of ions of energy $E$ ($N(E)$) against $E$

Curve A has the sharpest maximum and is therefore most accurately measured, while Curve C, which has a relatively flat top, is the least accurately measured.

If we remember the principle on which the magnetic analyser functions, i.e. all ions of the same $m/z$ value have the same kinetic energy (i.e. the same velocity), it is easily shown that any variations in kinetic energy will be assumed by the magnet to be a variation in mass. Hence, Figure 3.3a could be redrawn with the abscissa as a mass/charge scale rather than as an energy scale; this is shown in Figure 3.3b.

Therefore, ions wiht the greatest spread of kinetic energies will have the largest uncertainty in the measurement of mass. It is this fact that limits the resolving power of the mass spectrometer.

We must therefore make the peaks sharp as in curve A in Figure 3.3a. This can be achieved by placing an extra focusing device between the ion source and the magnetic analyser, which reduces the kinetic-energy spread. This takes the form of a pair of curved metal plates, called the *electrostatic analyser* or *electric sector*. These have an electrical potential maintained across them. Such an arrangement is shown in Figure 3.3c.

When an ion passes between a pair of plates, it will feel a force acting upon it. Assuming that the ions are positive, they will be attracted to the negative plate, and thus the ion beam will be deflected towards the latter.

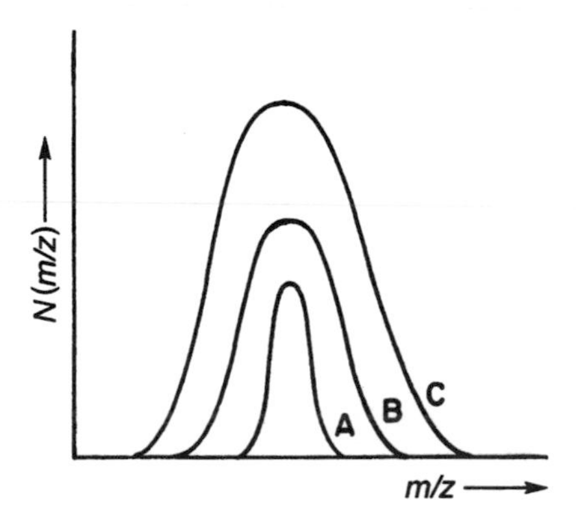


**Figure 3.3b** Plots of the number of ions of mass-to-charge ratio $m/z$ ($N(m/z)$) against $m/z$

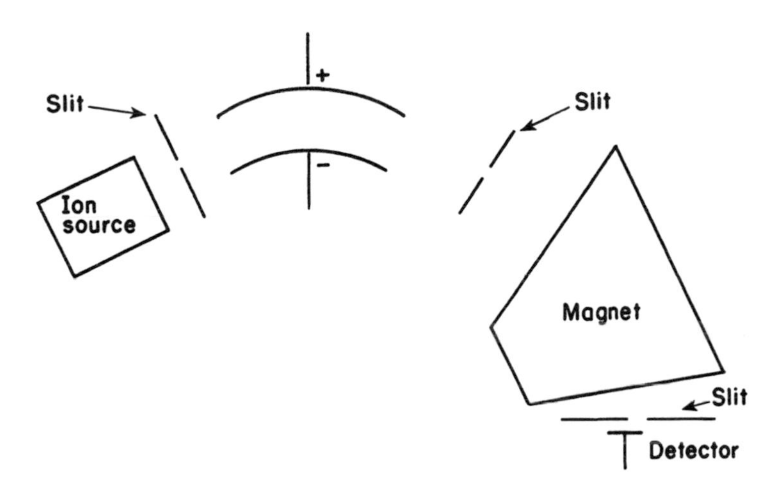


**Figure 3.3c** Schematic diagram of a mass spectrometer which uses both electrostatic and magnetic focusing

Π What is the magnitude of this force ($F$)?

It is simply the product of the charge of the ion ($z$) and the electric field strength between the plates ($E$, in units of V m$^{-1}$):

$$F = zE \tag{3.12}$$

If the electric force is balanced by the centrifugal force that the ions attain as a consequence of their velocity $v$ they will be deflected from their original flight paths into a circular trajectory which is coincident with the circumference of a circle of radius $R$, as shown in Figure 3.3d.

Therefore, we can equate both the electrical and centrifugal forces, as follows:

$$zE = mv^2/R \tag{3.13}$$

In addition, if you remember that we can also equate the kinetic energies (as shown previously) as follows:

$$\tfrac{1}{2}mv^2 = zV \tag{3.4}$$

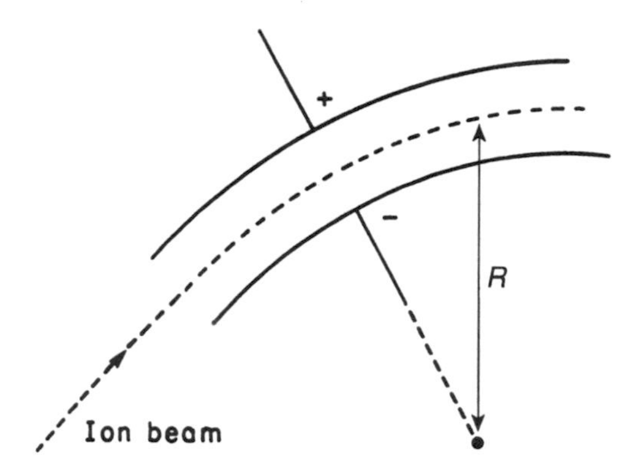


**Figure 3.3d** Passage of an ion through an electrostatic analyser

we obtain:

$$mv^2 = 2zV = zER \quad (3.14)$$

Thus, by cancelling $z$:

$$R = 2V/E \quad (3.15)$$

Π Will ions of any $m/z$ values, which possess similar kinetic energies, follow the same trajectories?

Yes, only $E$ and $V$ are contained in the equation. Ions of any $m/z$ values, which have been accelerated through a potential $V$ and passed through a field $E$, all follow a curve of radius $R$, provided that they have the same kinetic energies.

If their kinetic energies are different, then trajectories similar to those shown in Figure 3.3e will be observed.

**SAQ 3.3**

> In Figure 3.3e, which of the flight paths, a, b, or c, will the ions of highest kinetic energy follow?

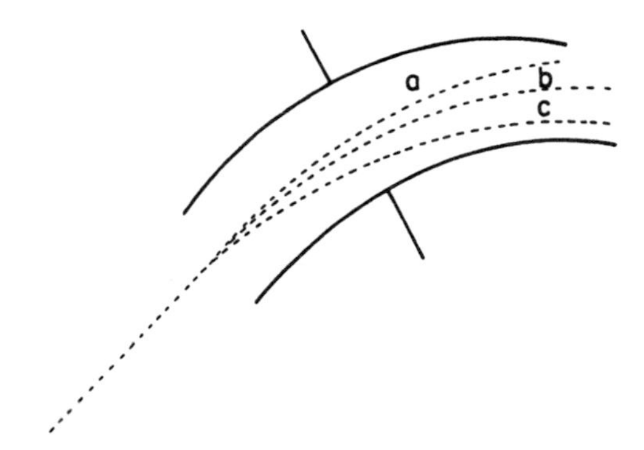


**Figure 3.3e** Flight paths of ions of different kinetic energies in an electrostatic analyser

If you refer back to Figure 3.3c, you will notice that there is a slit placed between the electrostatic analyser and the magnetic analyser. This means that only ions of the required kinetic energy, i.e. those that follow a circle of radius $R$, are allowed to pass through (see Figure 3.3f).

An instrument that operates on these principles is known as a *double-focusing* mass spectrometer (Figure 3.3f). In these devices, the electric sector acts as an energy analyser, while the magnetic sector acts as a

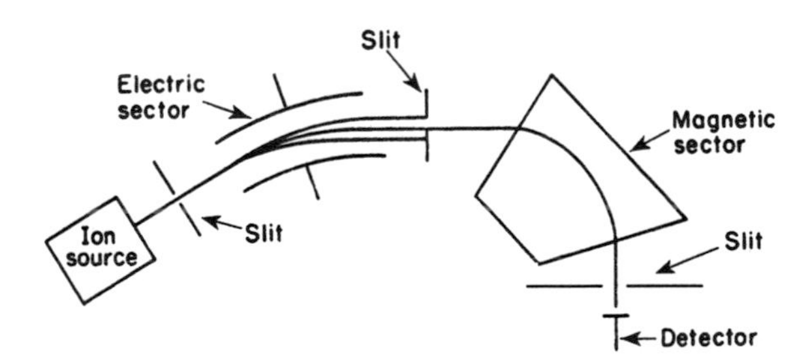


**Figure 3.3f** Schematic representation of a double-focusing mass spectrometer

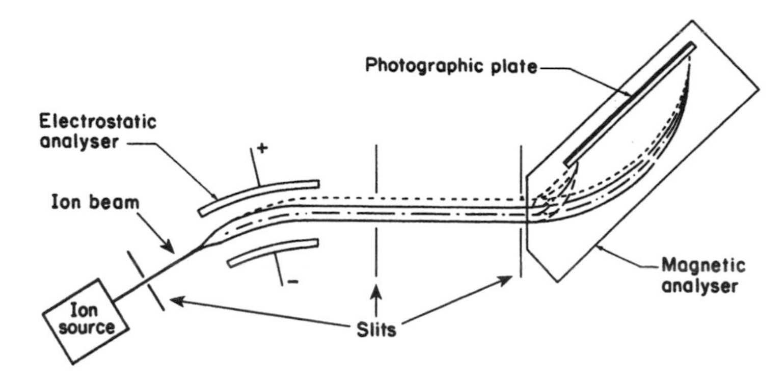


**Figure 3.3g** Schematic representation of a mass spectrometer employing Mattauch–Herzog geometry

mass analyser. With such instruments, resolving powers of the order of $10^5$ can be obtained.

The instrument shown in Figure 3.3f is said to have Nier–Johnson geometry (after the scientists who first used this design). Double-focusing mass analysers of this type were first used in the 1930s.

If the magnet polarity is reversed, then positive ions would tend to curve in the opposite direction while in the magnetic field. Such instrument usually employ photographic detectors and are normally used in conjunction with spark sources (see Chapter 9). Instruments of this kind are said to have Mattauch–Herzog geometry (see Figure 3.3g).

## 3.4 QUADRUPOLE MASS ANALYSER

Another common type of mass analyser used in mass spectrometry is the *quadrupole mass analyser*, shown in Figure 3.4a. This consists of two pairs of precisely parallel rods which are located between an ion source and a detector.

Ions from the source enter the mass analyser region as before, but in this case under a very small accelerating potential, e.g. 5–15 V. The electrical connections between the rods are shown in Figure 3.4b,

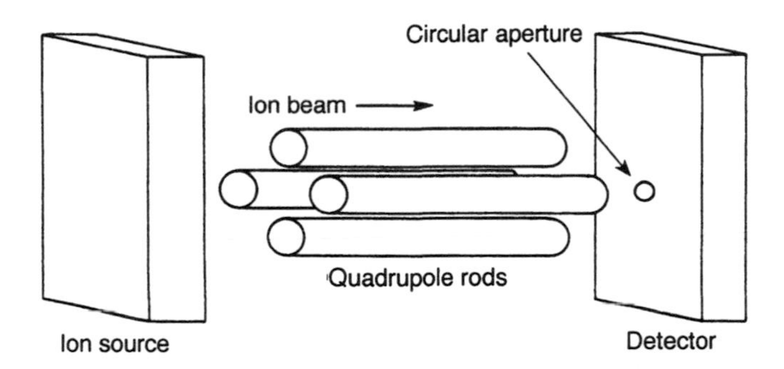


**Figure 3.4a** Arrangement of rods in a quadrupole mass analyser

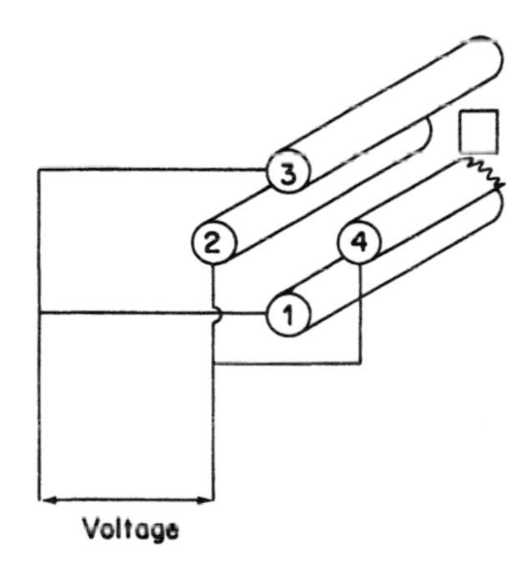


**Figure 3.4b** Electrical connections between the rods of a quadrupole mass analyser

where the rods have been numbered for clarity. A voltage made up of two components is applied to the rods; the first component is a standard DC potential, and the second an alternating radiofrequency (RF) component. Opposite rods are electrically connected.

Now let us consider how these potentials affect the ions. We will do this by first taking just one of the pairs of rods (say 2 and 3) shown in Figure 3.4b.

Π If rod 3 is at a positive potential with respect to rod 2, what will be the effect on any positive ions passing near them?

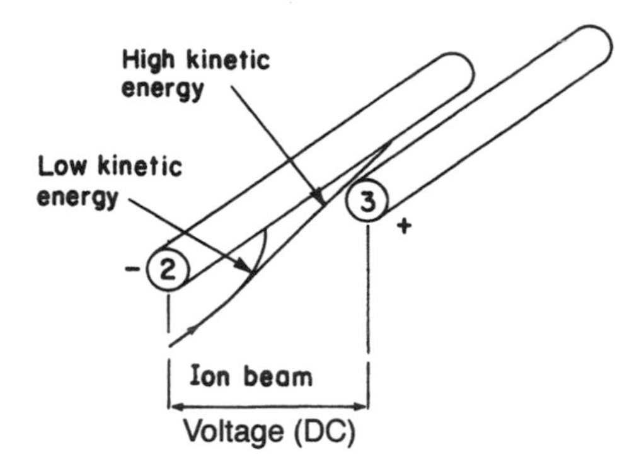


**Figure 3.4c** Deflection of positive ions between two rods set at a fixed potential difference

The positive ions will be drawn towards the negative rod (2) with those ions of lowest kinetic energy being affected the most (see Figure 3.4c).

Next, let us consider the radiofrequency component. This is effectively an AC power supply operating at a very high frequency, e.g. $10^8$ Hz.

**SAQ 3.4** What is the frequency of a normal household power supply?

Π How would the positive ions behave purely under the influence of an alternating field?

The ions would feel an attraction to rod 2 when the RF polarity is the same as in Figure 3.4c, and an attraction to rod 3 when the polarity is reversed. Figure 3.4d shows the behaviour of such ions (the frequency has been reduced for clarity).

This change in polarity is occurring at a very high rate and if we now sum together both components the ions will now follow an erratic

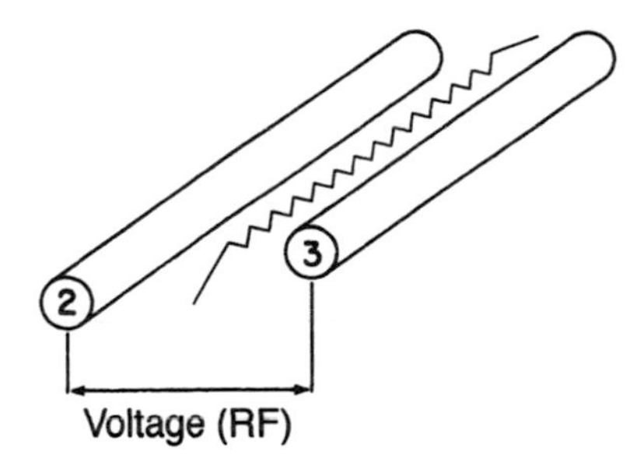


**Figure 3.4d** Deflection of positive ions between two rods at an RF potential difference

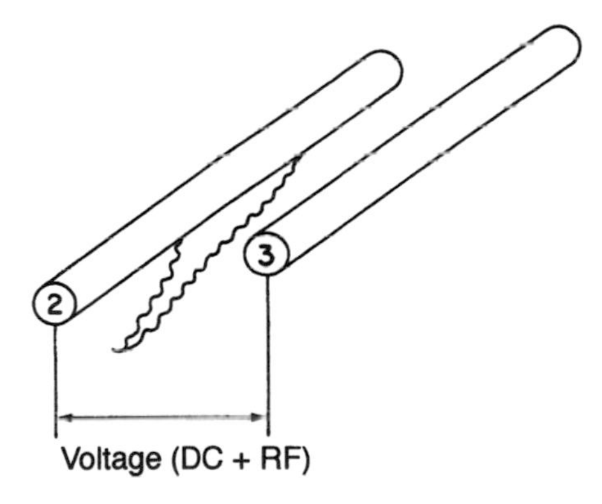


**Figure 3.4e** Deflection of ions between two rods at a variable (DC+RF) potential difference

path, which again depends on their kinetic energy, as shown in Figure 3.4e.

The same DC voltage and RF field are also applied to rods 1 and 4, simultaneously to their application to rods 2 and 3. The DC connections are such that rod 4 is at a negative potential with respect to 1, but the RF field applied to rods 1 and 4 is 180° out of phase with that applied to rods 2 and 3.

It is difficult to imagine (and represent) graphically what this means, but the ions effectively follow an oscillating trajectory between the rods. At any one pair of DC and RF values, only those ions of one particular kinetic energy pass directly between the rods and out the

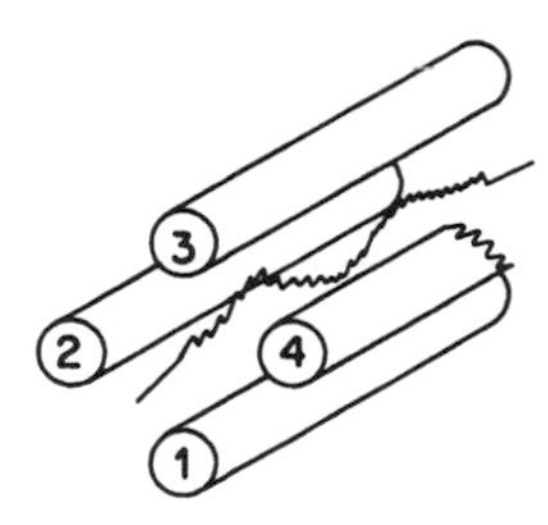


**Figure 3.4f** Deflection of positive ions in a quadrupole mass analyser

other end (as shown in Figure 3.4f). All other ions collide with the rods.

If we keep the ratio between the DC voltage and the RF field the same, but vary both parameters together, then ions of different $m/z$ values can be analysed and a mass scan can be made. Because of this separation principle, such an analyser is also called a *quadrupole mass filter*. The resolving power of this mass analyser is about the same as that of a single-focusing device, i.e. 5000.

The quadrupole analyser has certain advantages over a magnetic-sector instrument. These are as follows:

(a) it is relatively cheap to build;

(b) it is smaller and lighter;

(c) it is more robust;

(d) more accuracy is achieved in computer control of rod voltages rather than magnetic fields;

(e) the mass scale is linear, i.e. not proportional to $B^2$;

(f) scanning is very fast.

All of these factors make quadrupole instruments the most common type of analyser in use today.

## 3.5 SCAN RATE

The scan rate is the time it takes to record the spectrum. This is a very important specification and is directly relevant to mass spectrometers which are coupled to gas and liquid chromatographs, where the signal may be transient, or to studies of reaction kinetics. Scans are usually measured in number of seconds per decade of mass (s decade$^{-1}$). A decade of mass covers a range of mass which changes by a factor of 10, e.g. $m/z$ 30–300, or 50–500.

**SAQ 3.5a**

If a spectrometer is scanned at a rate of 10 s decade$^{-1}$, how long would it take to scan from $m/z$ 1 to 10 000?

This way of measuring scan rate stems from the magnetic-sector mass analysers, where the magnet follows an exponential scan law.

Scan rates with special laminated magnets can be as high as 0.1 s decade$^{-1}$. However, most scans are of the order of 1 s decade$^{-1}$ (maximum). This is because magnets need a certain amount of time to reset themselves before the next scan, otherwise technical problems such as magnet hysteresis might occur. The *cycle time* is the time a magnet needs for scanning and resetting before the next scan can take place.

The quadrupole mass analyser can scan its full mass range in a few ms, while its reset time is even less.

For this reason, and also for its ability to accurately control rod voltages, the quadrupole mass spectrometer is the instrument that is often used for a method of analysis known as *selected- (or multiple-) ion monitoring*. In this technique, the whole mass spectrum is not scanned; only a selected number of representative ions needed to characterise the compound are analysed and the instrument rapidly switches between the selected ions until the measurements are complete. Ion monitoring is often applied to mixtures of compounds where sensitive quantitative analysis is required.

**SAQ 3.5b**

Suppose that you are the head of the experimental division at NASA and you have been asked to recommend the design of a mass spectrometer to be included in an unmanned space probe to Mars (take the 1997 Pathfinder mission, for example). What type of ion source and analyser would you suggest for the instrument, bearing in mind the following important points?

(a) The spectrometer will be analysing relatively simple organic molecules of low $M_r$ values ($\leqslant 150$), although absolute identification is required.

(b) The instrument must be small and not too heavy to fit into the probe, but it must also be robust in order to withstand take-off and landing operations and it needs to be reliable.

(c) It also needs to be computer-controlled, i.e. it is an unmanned flight.

## 3.6 ION-TRAP ANALYSER

An ion trap is a device that can store ions for an extended period of time by the use of electric and/or magnetic fields. Figure 3.6a shows a cross-sectional view of a simple ion trap. This consists of a central doughnut-shaped ring electrode and a pair of end-cap electrodes. A

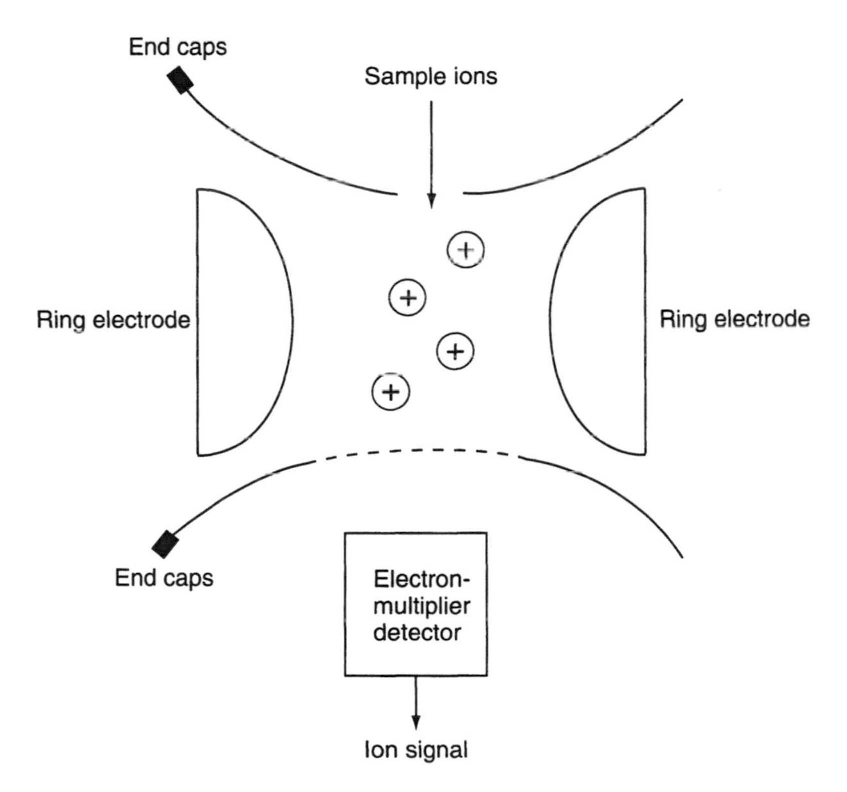


**Figure 3.6a** Schematic diagram of an ion-trap mass analyser

variable RF voltage is applied to the ring electrode, while the two end caps are connected to earth. A burst of gaseous ions from the sample under investigation is introduced through a grid in the upper end cap. The RF voltage is then scanned. Ions with an appropriate $m/z$ value circulate in a stable orbit within the ring cavity. As the voltage is increased, the orbits of the heavier ions become stabilised, while those of the lighter ions become destabilised, resulting in the latter leaving the cavity via openings in the lower end cap. The emitted ions then pass into a detector. As with the quadrupole mass analyser, the ion-trap analyser is very rugged and the resolution is of the order of 1 part in 1000.

## 3.7 TIME-OF-FLIGHT (TOF) ANALYSERS

The principle of operation of this analyser is based around the periodic production of positive analyte ions by bombardment of the sample with brief pulses of electrons, secondary ions or laser-generated photons. The frequency of these pulses is typically 10–50 kHz, with a lifetime of 0.25 μs. The positive ions that are formed are then accelerated by an electric field pulse of $10^3$–$10^4$ V. This has a similar frequency but slightly lags behind the ionisation pulse. The accelerated ions now pass into a region that contains no external field. This is known as the *drift tube* and is *ca.* 1 m in length. If we assume that all of the ions have the same kinetic energy (you will remember that this is not in fact the case), then their velocities will depend on their masses—the heavier ions arriving at the detector up to 30 μs after the lighter ones.

You may recall from equation (3.5) that the velocity of an ion can be expressed by the following:

$$v^2 = 2zV/m \qquad (3.5)$$

Therefore:

$$v = \sqrt{(2zV/m)} \qquad (3.16)$$

and, since $v = L/t$, where $L$ is the drift-region length and $t$ is the flight time:

$$t = L\surd(m/2zV) \tag{3.17}$$

The difference in transit time, $\Delta t$, for two ions of masses $m_1$ and $m_2$, can be expressed as follows:

$$\Delta t = L(m\surd_1 - \surd m_2)/\surd 2zV \tag{3.18}$$

The detector is often an electron multiplier in which the output is fed across the vertical deflection plate of an oscilloscope, while the horizontal sweep is synchronised with the accelerator pulses. Thus, we can obtain an instantaneous display of the mass spectrum. A schematic representation of a time-of-flight mass spectrometer is shown in Figure 3.7a.

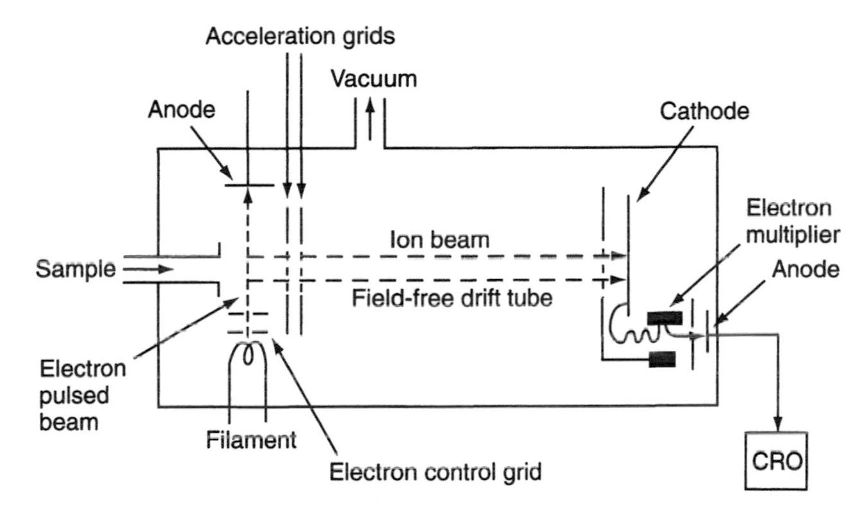


**Figure 3.7a** Schematic diagram of a time-of-flight mass analyser

**SAQ 3.7a**

How will the time resolution be changed by increasing the drift-tube length or decreasing the accelerating potential?

**SAQ 3.7b**

What can we say about the speed of the data-acquisition system that is required?

Variations in ion energies limit the resolution to about 1 part in 1000.

TOF mass spectrometers have some notable advantages, namely simplicity, virtually unlimited mass range, high scan rates ($\leqslant$ 900 amu in *ca.* 1 s), and very short ion-formation times (< 100 μs).

## 3.8 FOURIER-TRANSFORM (FT) INSTRUMENTS

Having already become standard instruments for infrared analysis, Fourier-transform devices first began to make an impact on mass spectrometry in the early 1980s. The improved signal-to-noise ratios, greater speed and higher sensitivity and resolution makes them a superior instrument in many respects.

FT spectrometers contain an ion trap (see Section 3.6) within which ions are allowed to circulate in defined orbits over extended periods of time.

The behaviour of these ions is best described by the ion-cyclotron resonance phenomenon.

When a gaseous ion moves into a strong magnetic field its motion become circular in a plane that is perpendicular to the direction of the field. The *cyclotron frequency* ($\omega_c$ or $v/r$), is defined as the angular frequency of this motion. We can rearrange equation (3.7) to make $v/r$ the subject. This gives:

$$\omega_c = v/r = Bz/m \tag{3.19}$$

Π Which parameter does the cyclotron frequency depend upon for a particular ion in a fixed field?

Only the inverse of the $m/z$ value. If we increase the ion's velocity, then its radius of rotation will also increase.

An ion that is trapped in a circular path in a magnetic field is also capable of absorbing energy from an AC electric field if the frequency of that field matches the cyclotron frequency. If absorbed, this energy can increase the ion's velocity and therefore, its radius of travel, without affecting $\omega_c$. Figure 3.8a shows the original inner circular path of an ion trapped in a magnetic field (represented by a solid line). If we now apply an AC voltage which has a frequency close to the ion's cyclotron frequency, then this field interacts with the ion, thus resulting in an increase in the velocity of the ion. The radius of the path of the ion also increases (as shown by the dashed line). If we halt the AC voltage, then the radius of the path of the ion again becomes constant. This is shown by the outer solid line in the figure. Therefore, if we have ions with similar cyclotron resonance frequencies, i.e. similar $m/z$ values, they will all be set in motion in phase with the AC field, while ions with different $m/z$ values will be unaffected.

When the frequency sweep signal is terminated, a so-called *image current* is observed which decreases exponentially with time. This current results from the circular movement of a collection of ions with the same $m/z$ value. For example, if such a group of ions approaches the upper plate in 3.8a, then induced electrons will be attracted from earth to this plate, thus causing a momentary current. As the group of ions continue around the lower plate, more electrons are induced on to the lower plate, and a current will flow in the opposite direction.

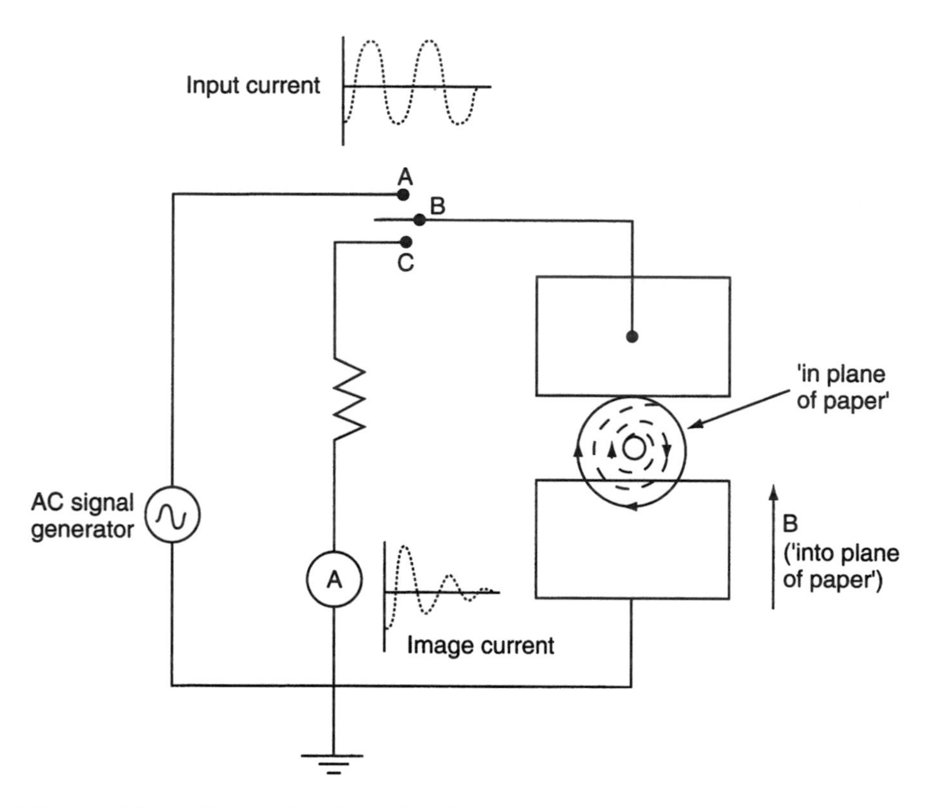


**Figure 3.8a** The path of an ion in a strong magnetic field: the inner solid line represents the original circular path of the ion; the dashed line shows the outward travelling spiral path when an AC voltage is applied (switch in position A–B). The outer solid line shows the new circular path after the AC voltage has been switched off (switch in position B–C), and the image current has been recorded

**SAQ 3.8a** What does the magnitude of this AC current depend upon?

The frequency of the current is characteristic of the $m/z$ value of the ions.

The decay of the image current is of the order of a few seconds or less.

**SAQ 3.8b**

What causes the decay of the image current?

The ion trap most frequently used on Fourier-transform instruments is the *trapped-ion analyser cell*, similar to that shown in Figure 3.8b.

In this type of cell, gaseous sample molecules are ionised at the centre by electrons and are then accelerated from the filament through the cell to a collector plate. The electron beam is periodically switched on and off via a pulsed voltage grid. A trap-plate voltage of 1–5 V holds the ions in the cell. An RF signal is applied via the transmitter plate, while the receiver plate is connected to a pre-amplifier that amplifies the image current. Storage times of a few minutes are possible when using such a set-up.

The FT instrument is operated by first generating a short electron pulse and storing the trapped ions in the cell. After a short delay, the trapped ions are subjected to a brief RF pulse (5 ms) which increases linearly in frequency during its lifetime. After the frequency sweep is discontinued, the induced image current is amplified, digitised and stored in memory. The time-domain signal that is produced is then Fourier transformed into a frequency-domain signal, and then finally into the corresponding mass spectrum (see Figure 3.8c).

**SAQ 3.8c**

By use of which equation can we convert the frequency spectrum into a mass spectrum?

FT mass spectrometers are expensive, but there are a number of advantages of this technique, namely:

(a) ion generation and mass analysis occur in the same region;

(b) all ions are detected simultaneously;

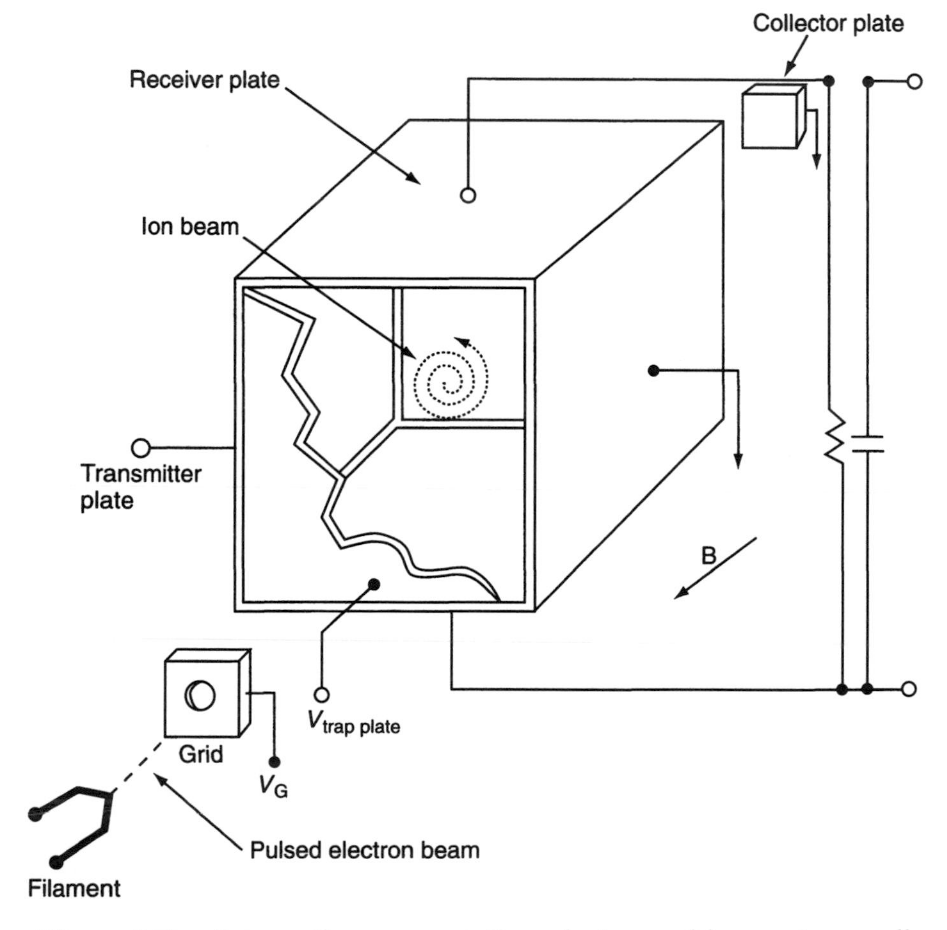


**Figure 3.8b** Schematic representation of a trapped-ion analyser cell

(c) it is very easy to switch between positive- and negative-ion spectra;

(d) extremely high resolution is possible ($>10^6$), on account of the high precision that is achievable in the frequency measurement;

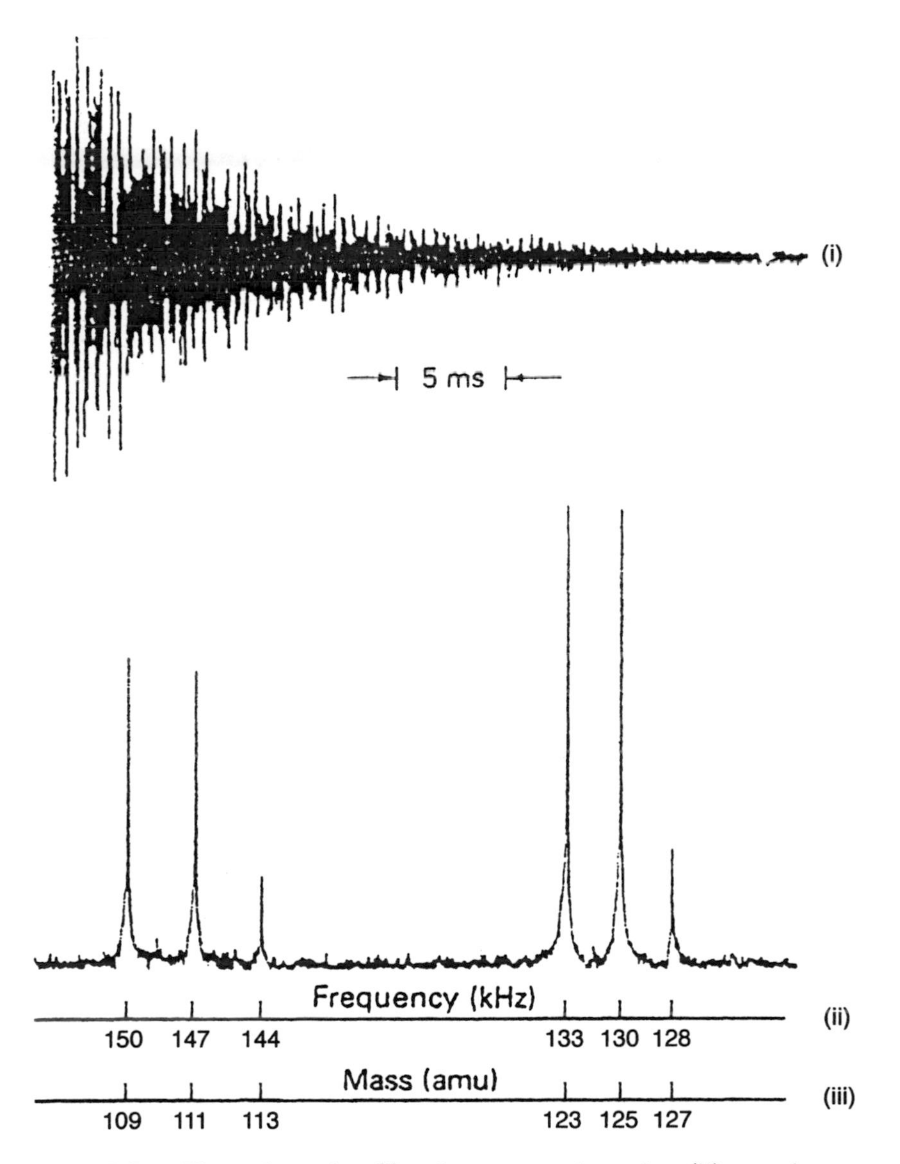


**Figure 3.8c** Time-domain (i), frequency-domain (ii), and mass-domain (iii) spectra

(e) complex ion optics and voltage are not needed;

(f) samples may be run without an internal mass standard, once the instrument is calibrated.

Recent advances have included tandem-quadrupole-Fourier-transform mass spectrometry, where ions pass first through one quadrupole, are focused by a second quadrupole, then decelerated, and finally pass into the FT–MS analyser cell. Some FT–MS instruments use superconducting magnets with fields of 2–5 T.

## 3.9 TANDEM MASS SPECTROMETRY (MS/MS)

Not to be confused with accelerator mass spectrometry (see Chapter 9) which utilises a *tandem* Van de Graaff accelerator, *tandem mass spectrometry* (MS/MS) is a technique in which a first analyser isolates a precursor ion, which then undergoes a fragmentation to yield product ions and neutral fragments:

$$m_p^+ \longrightarrow m_d^+ + m_n \qquad (3.20)$$

A second spectrometer then analyses the product ions. A typical instrument uses three quadrupoles (arranged in series with each other) as the analyser system. In this set-up the first and third of these act as spectrometers while the central one acts as an inert gas collision cell (CC), made up of a quadrupole using RF only as a means of ion focusing. Other instruments combine electric and magnetic sectors and quadrupoles.

These instruments can be scanned in several different ways (see Figure 3.9a).

The first scanning mode involves selecting an ion with a chosen mass-to-charge ratio by using the first spectrometer (MS1). This ion collides inside the central quadrupole and fragments. The reaction products are then analysed by the second mass spectrometer. This is a 'fragment-ion scan' or 'daughter scan'.

The second possibility consists in focusing the second spectrometer

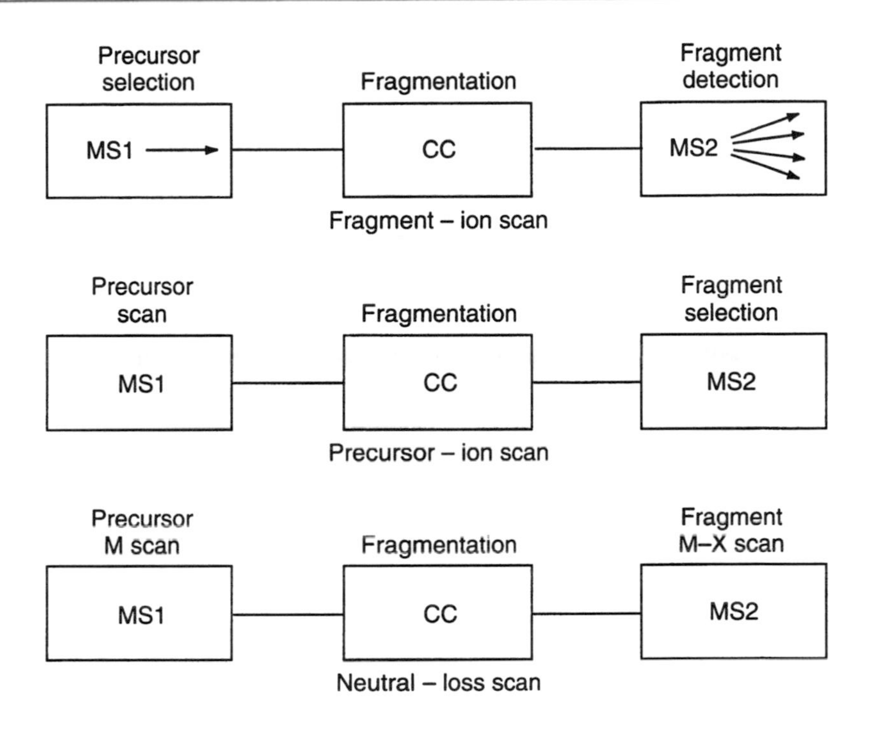


**Figure 3.9a** Examples of possible scan modes for a tandem mass spectrometer

(MS2) on a selected ion, while scanning the masses by using the first spectrometer. All of the ions that produce the ion with the selected mass through fragmentation are detected. This is a 'precursor-ion scan' or 'parent-ion scan'.

In the third common scan mode, both mass spectrometers are scanned together, but with a constant mass offset between the two. Therefore, for a mass difference of $X$, when an ion of mass $M$ goes through MS1, detection occurs if this ion has yielded a fragment ion of mass $M-X$ when it leaves the collision cell. This is called a 'neutral-loss scan', with the neutral having a mass $X$.

Collision-induced dissociations are predominant reactions with inert gases in the collision cell, but notc that collision-activated reactions can occur with reactive gases.

Π Alcohols are identified in chemical ionisation by detecting the loss of a neutral molecule. What is this molecule?

It is the water molecule, mass 18.

In theory, successive analysers can be arranged in any number. They are then labelled as $MS^n$. As the number of transmitted ions decreases at every step because of the ionic-species trajectory lengths and the presence of the collision gas, the practical maximum is three or four analysers in most cases.

Figure 3.9b illustrates how a neutral-loss scan can aid peak identification. Note how well the signal-to-noise ratio (s/n) has improved in spectrum B. The ion at $m/z$ 507 is now well detected, and the $m/z$ 491 and 523 ions, not detected in A, are now clearly visible.

Some further applications of tandem mass spectrometry are discussed in Section 8.3.

## Summary

In this chapter, we have looked at the instrumental side of mass spectrometry. We have concerned ourselves with the various techniques of mass analysis, including magnetic single- and double-focusing instruments and those using a quadrupole design. We have also looked at ion traps, time-of-flight mass spectrometers and Fourier-transform instruments, and outlined the advantages and limitations of each type. Other important features that have been discussed include resolving power, scan rates, multiple-ion monitoring and tandem mass spectrometry.

## Objectives

After studying this chapter you should be able to:

- describe the instrumental components needed in a mass spectrometer;

- describe single- and double-focusing magnetic mass analysers;
- describe the quadrupole mass analyser;
- describe time-of-flight instruments;

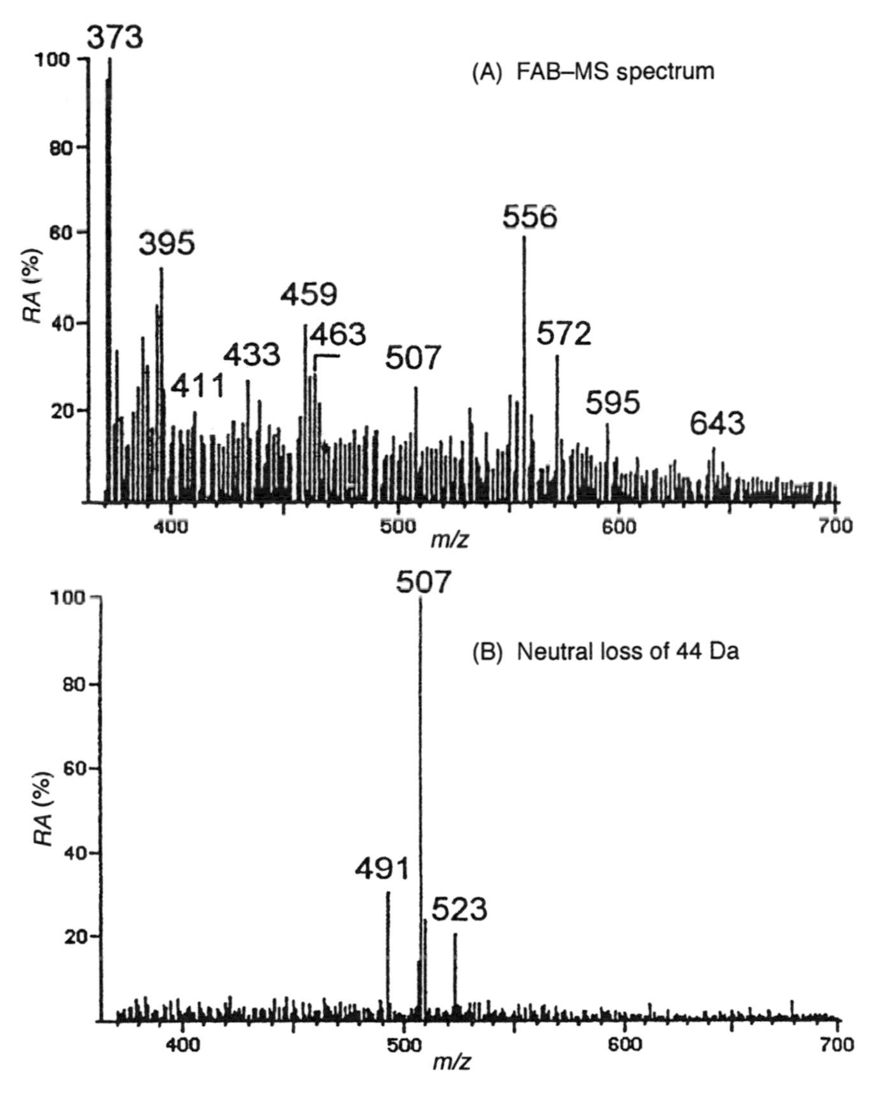


**Figure 3.9b** (A) Mass spectrum of bile acids in a urine sample; (B) neutral-loss scan of 44 Da measured immediately after spectrum A

- describe and explain the principles behind Fourier-transform mass spectrometers and ions traps;

- explain the terms, resolving power, scan rate and multiple-ion monitoring;

- explain what limits the resolving power in single-focusing mass spectrometry;

- describe the three common scanning modes used in tandem mass spectrometry.

# 4. Ion Detection, Data Recording and Processing

This chapter will conclude our look at instrumentation in organic mass spectrometry with a description of some commonly used detectors. We shall also examine different methods of data display and recording, and the use of on-line computers in data acquisition and processing.

## 4.1 ION DETECTION

### 4.1.1 Electron Multipliers

By far the most important method of ion detection employed in mass spectrometry is the use of an electron multiplier. First utilised in the 1940s, there are two common types, namely the discrete dynode electron multiplier and the continuous dynode elector multiplier. The discrete dynode system is very much like the photomultiplier detector.

**SAQ 4.1a** Where else in spectroscopy might you have seen a photomultiplier tube in operation?

The detector consists of a series of electrodes (known as dynodes) arranged close to each other, with each successive dynode being held at progressively higher voltages. The dynodes have Cu/Be surfaces and are enclosed in a vacuum jacket, which is linked to the back end of a mass spectrometer. When ions hit the first dynode, a shower of electrons is released which then strikes the second dynode. Upon impact, another larger shower of electrons is released to a third dynode, and so on. This cascading effect continues through the whole

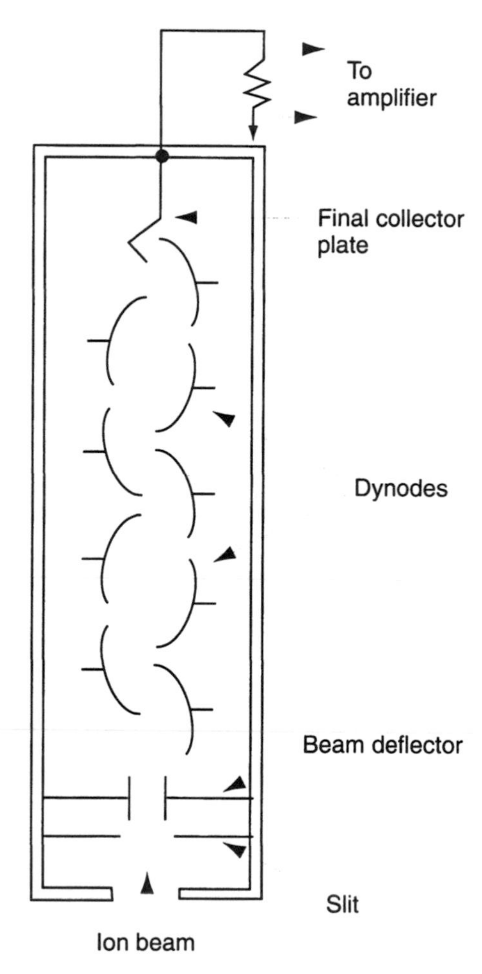


**Figure 4.1a** Schematic diagram of the discrete dynode electron multiplier

series of dynodes (usually about ten in total). Figure 4.1a shows a diagram of a discrete dynode electron multiplier.

The continuous dynode electron multiplier (CEM) is a trumpet-shaped device made of glass which has been heavily doped with lead. A potential of 1.8–2 kV is applied across the length of the detector. Ions striking the surface near the entrance eject electrons, and then move along the surface, ejecting more electrons with every impact. This type of electron multiplier has a number of advantages, notably a smaller background noise and a more compact design.

The result of the sequence of events in both electron multipliers is that the small electrical current generated when ions hit the first electrode is greatly amplified (by gains up to $10^6$) and soon becomes large enough to be passed to a recording device.

Electron multipliers are rugged and reliable devices with nanosecond response times.

**SAQ 4.1b** Why is the response time important in ion-beam measurements?

These detectors can be placed directly behind the exit slit of a magnetic-sector mass analyser, because the ions reaching the detector possess enough kinetic energy to 'knock out' electrons in the first stage of the device. When these detectors are used in tandem with quadrupole mass analysers, which have a lower-energy beam, an acceleration step (of a few thousand eV) is sometimes needed prior to the ions entering the detector.

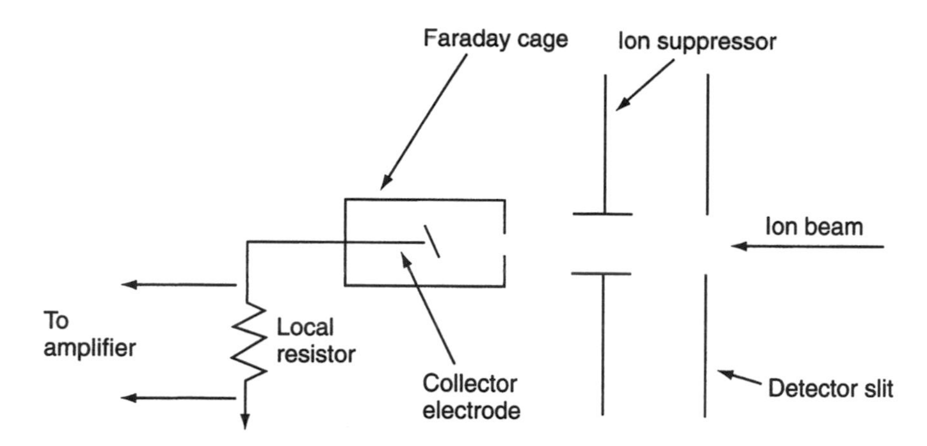


**Figure 4.1b** Schematic representation of the Faraday-cup detector

### 4.1.2 Faraday Cup

Figure 4.1b shows a schematic representation of a Faraday-cup detector. Ions exiting the analyser strike the collector electrode. The latter is surrounded by a cage which prevents the escape of reflected ions and ejected secondary electrons. The collector electrode is inclined with respect to the path of the entering ions.

Π Why is the collector electrode inclined?

This is so that particles striking or leaving the electrode are reflected away from the entrance to the cup.

Both the collector electrode and cage are connected to ground potential through a large resistor. This acts to neutralise the positive ions striking the plate. The resulting potential drop across the resistor is amplified via a high-impedance amplifier. The negatively biased ion suppressor is adjusted to minimise the unequal response of the detector to ions of different masses and energies, and to prevent secondary electrons drifting out of the chamber.

A Faraday cup is both simple and inexpensive to construct. Unfortunately, there is a limit (set by the amplifier) to the speed at which the spectrum can be scanned. There is also no internal amplification and therefore Faraday cups are less sensitive than electron multipliers.

### 4.1.3 Other Types of Detectors

It is possible to use photographic plates coated with silver bromide emulsion as detectors. These are mainly used with mass spectrometers of Mattauch–Herzog geometry (in which the ions are focused along a plane), and where it is necessary to simultaneously observe a wide range of $m/z$ values, such as a spark-source spectrometer (see Chapter 9).

Scintillation-type detectors are also sometimes used. These consist of a crystalline phosphor dispersed on a thin aluminium sheet which is mounted on the window of a photomultiplier tube. The electrons, which are produced when the positive ions strike a −40 keV cathode, impinge upon the phosphor, thus producing a scintillation which is detected by the photomultiplier tube. Advantages of this technique include low background-noise levels and an increased lifetime of the photomultiplier tube.

**SAQ 4.1c**

Where else might you have seen scintillation counting used in analysis?

## 4.2 METHODS OF DATA RECORDING

### 4.2.1 Oscilloscope and Chart Recorder

The recording device is usually either an oscilloscope, a chart recorder or a computer. The oscilloscope is useful for giving a preliminary scan of the whole sample spectrum. It allows both fine-tuning of instrument parameters and the determination of the ion signal strength before an accurate mass scan is made.

A common method of taking a permanent record of the spectrum is to use a chart recorder. This device usually employs photosensitive paper, with the image of the spectrum being developed by allowing a light beam to traverse the paper.

**SAQ 4.2a**

Why do you think a photographic chart recorder is used in preference to a pen recorder? Remember that a typical spectrometer scan speed, when using a chart recorder, is 10 s decade$^{-1}$.

We have already seen an example of the typical output from a mass scan in Figure 1.3a. The two traces are reproduced here in Figure 4.2a.

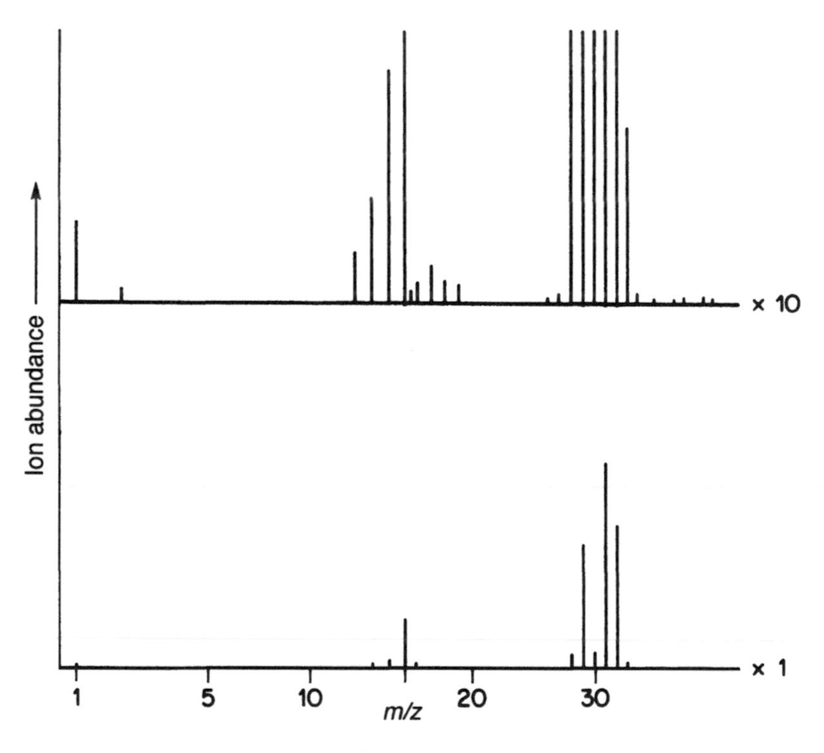


**Figure 4.2a** Mass spectra of $CH_3OH$

**SAQ 4.2b**

Can you remember why we often need more than one trace?

**SAQ 4.2c**

Can you think of any other ways of representing a mass scan?

### 4.2.2 Computer-aided Data Acquisition and Processing

Microprocessors and minicomputers are an integral part of modern mass spectrometers. There is often an abundance of information that needs to be determined, stored and displayed. A molecule with a molecular weight of 600, for example, may have over 100 fragment ions which lead to discrete spectral peaks. Therefore, because the amount of information obtained can be so great, and rapid data acquisition and processing is essential, a system capable of undertaking these tasks is needed. A computer is ideally suited for this role. The *data system*, as it is called, consists of a computer and a visual display unit (VDU), a printer/plotter and an interface to the mass spectrometer.

Π If you want to feed the electrical signal from the electron multiplier into a computer, it is necessary to carry out a signal conversion. Do you know what this is?

The electron multiplier will send out a continually varying electrical current or voltage. This is called an *analogue* signal. If it is to be handled by the computer it needs to be broken up into segments, or *digitised*. These two forms of output are illustrated in Figure 4.2b.

We achieve this digitisation by the use of an analogue-to-digital converter (ADC). The data system contains a very accurate crystal oscillator clock which defines precise regular time intervals at which the analogue voltage is sampled by the ADC. This results in a regular series of voltage pulses being fed into the computer. Figure 4.2b shows only a small portion of a mass spectrum. A real digitised spectrum might include thousands of such voltage readings. The number of readings made will have some bearing on the processing time of the computer.

Π How could we reduce the amount of information being fed into the computer, without reducing its usefulness or accuracy?

There are two obvious ways.

One is to reduce the number of voltage sampling points in our spectrum. However, we shall soon see that this will result in our peak shape being less clearly defined.

The second way is to reject voltage readings below a certain threshold value by assuming that these points arise from baseline noise in the signal. The device that sends this filtered information to the computer is called a *discriminator*. Figure 4.2c shows the results of applying such a device.

Π Can you name some sources of noise that could appear on a mass-spectral scan?

The sources of noise can be classified into two main groups:

(a) Fundamental noise; this arises from the quantisation of energy and the thermal motion of atoms at temperatures above absolute zero.

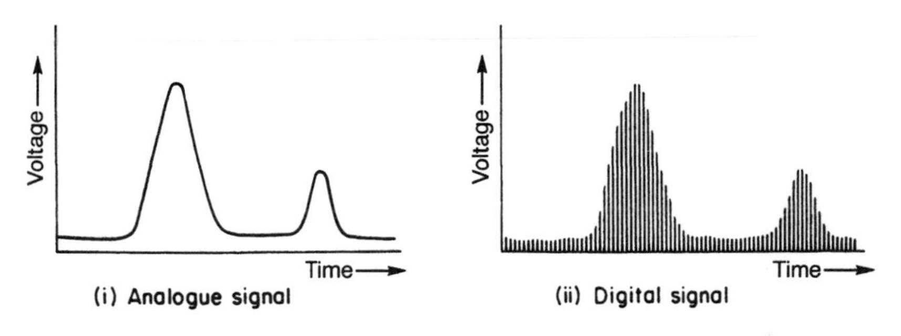


**Figure 4.2b** Analogue and digital forms of a mass-spectrometer output

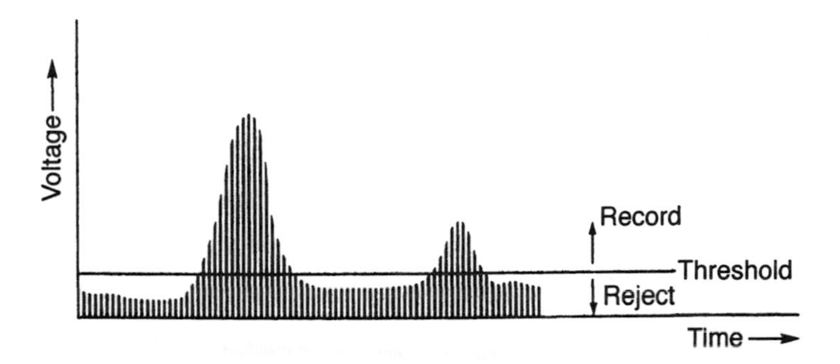


**Figure 4.2c** Digital signal from a mass-spectrometer output, showing a threshold setting

(b) Environmental noise; this results from external factors, such as build-up of charges on insulators, mechanical vibrations, pressure and temperature variations, etc.

Fundamental noise can never be fully removed, so there is a definite lower limit on ion detection.

Environmental noise can be broken up into three characteristic waveforms, namely *white* noise, *1/f* noise and *interference* noise. Graphs showing their contributions to the noise level at different frequencies are represented in Figure 4.2d.

White noise is a relatively constant background noise from a whole host of different sources superimposed on each other.

The 1/f noise is highest at lower frequencies and tails off quickly as the frequency increases.

The interference noise, which is generated by switch operations and electrical pulses, is often at a definite frequency.

A thorough discussion of data enhancement and noise reduction is beyond the scope of this present book, but improvements in the signal-to-noise ratio can be made by applying one or more of the following techniques:

(a) Electrical filtering of the analogue signal prior to digitising, in a similar manner to the discriminator action.

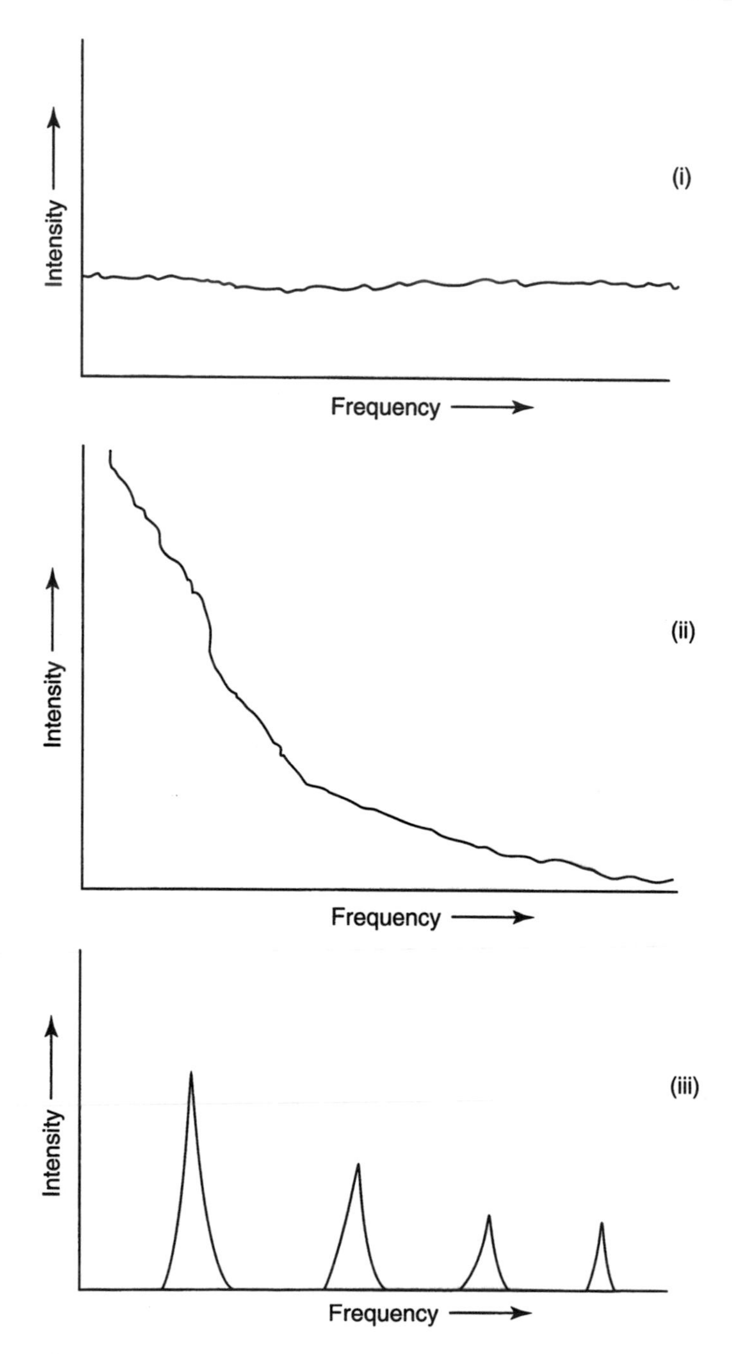


**Figure 4.2d** Intensity of environmental noise as a function of frequency: (i) white noise; (ii) $1/f$ noise; (iii) interference noise

(b) Application of mathematical deconvolution techniques to the data, such as Fourier transformation.

(c) Beam modulation by electronically or mechanically chopping the ion-beam signal prior to reaching the detector. The DC-beam signal is thus converted into an AC signal at a higher frequency, amplified and digitised. Since the 1/f noise signal is a problem only at lower frequencies, the signal and noise can be discriminated against.

(d) Time-based signal averaging can cut down on white noise. In this case, the noisy signal is repeatedly sampled and digitised at fixed time intervals and stored in the computer's memory. Algebraic summing at each sampling point results in the coherent data signal increasing with the number of sampling cycles, while the non-coherent noise sums to a smaller value (see Figure 4.2e).

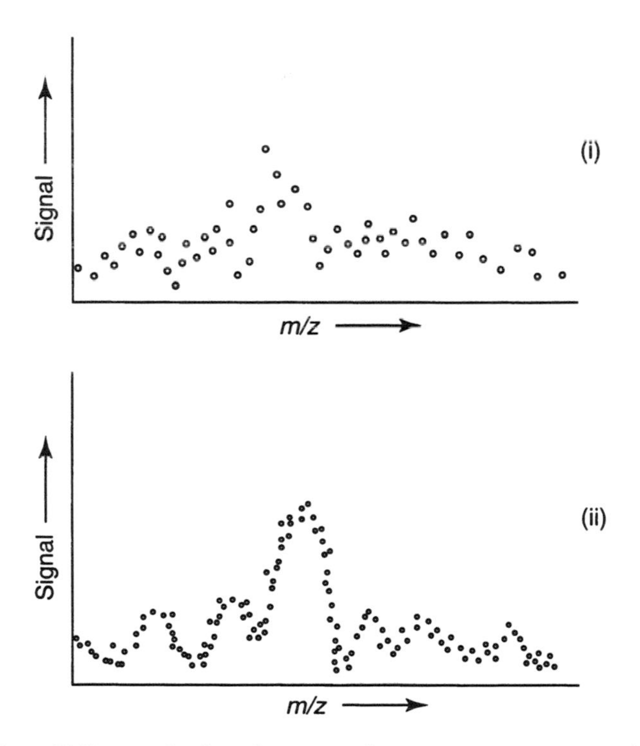


**Figure 4.2e** Effect of signal averaging on mass-spectral data: (i) normal scan; (ii) time-based, signal-averaged data

The signal-to-noise ratio increases at approximately the square root of the number of scans made.

**SAQ 4.2d**

For what general class(es) of measurement would the time-based, signal-average technique be useful?

(e) Derivative mass spectrometry. The digitised signal is derivatised (or doubly derivatised) with respect to $m/z$. Figure 4.2f shows a symmetrical peak and an obscured doublet that have been derivatised.

Π What can you say about the resolution of the derivatised doublet?

By using the derivatives of the signal, the two peaks are much better resolved.

Now let us return to the problems of peak shape and number of sampling points. Figure 4.2g shows analogue and digital peaks from a mass spectrum obtained using different sampling rates.

Π What can you say about the peak shape and resolution when the sampling rate is low (peaks iii to vi)?

The singlet peak appears to be distorted and the doublet appears to be a grossly distorted singlet. Therefore, too low a sampling rate must be avoided. Rates of 10–20 points per amu ensure adequate mass measurements. It is also preferable to have a constant signal for each digitised time point. A 'sample-and-hold' amplifier preceding the ADC enables this task to be carried out quite simply.

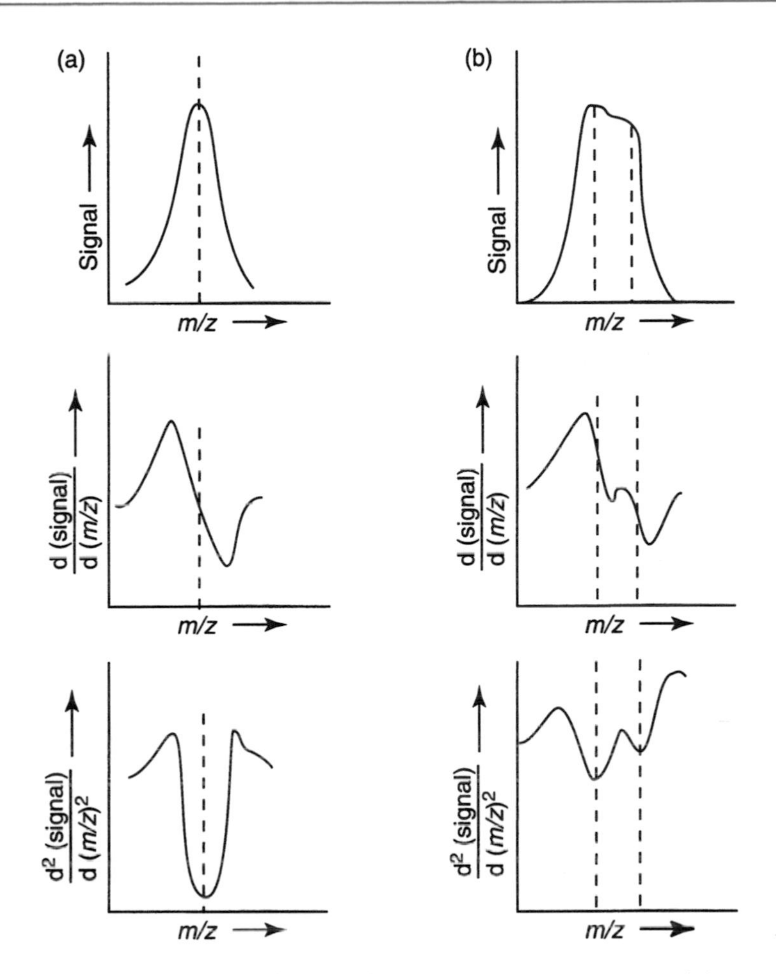


**Figure 4.2f** Representations of a symmetrical peak (a) and an obscured doublet (b) that have been derivatised and doubly derivatised

The computer now calculates the *centroid* of each peak. The centroid should be directly below the peak maximum if the peak is symmetrical; this is shown in Figure 4.2h using similar peaks to those shown in the previous figure.

Π What happens to the centroid position in peaks v and vi?

The centroid position has clearly been displaced from its true value.

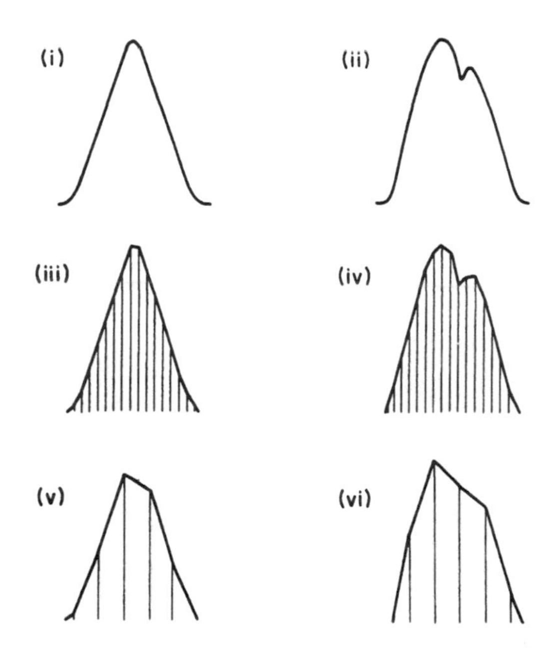


**Figure 4.2g** Peaks from a mass spectrum in analogue (i, ii) and digital (iii–vi) forms, obtained by using different sampling rates

The computer uses these centroids to define the peak at one time value. It does this by essentially dropping a line down from the centroid to the $x$-axis (time axis) and then reading off the time value.

**SAQ 4.2e**

What can we say about the time value obtained and the true time value from peak (vi) in Figure 4.2h?

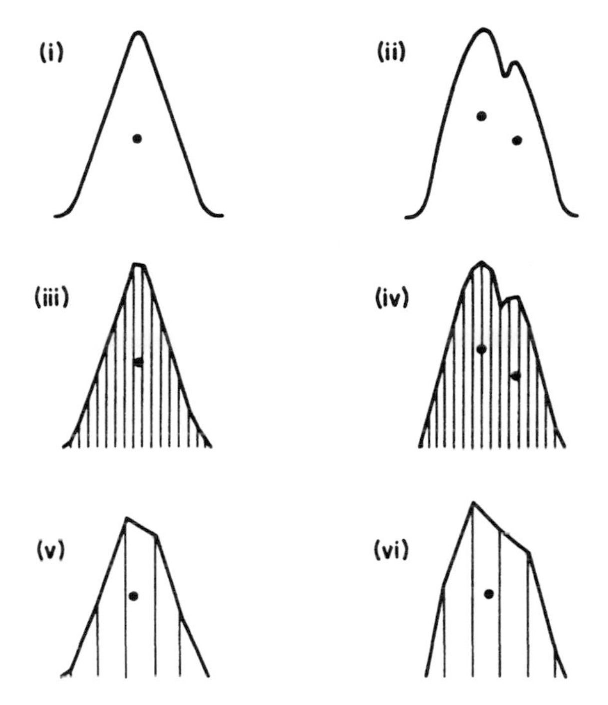


**Figure 4.2h** Peaks from a mass spectrum in analogue (i, ii) and digital (iii–vi) forms, showing the position of the centroid of each peak (•)

Once a peak has been defined, the computer sums up all the voltage readings under this peak, i.e. the peak area. The computer can therefore obtain a measure of the peak's relative intensity by comparing other peaks in the spectrum. These data can be tabulated, or put on a relative abundance plot of peak heights which may be normalised against the largest peak, i.e. the base peak.

**SAQ 4.2f**

List some of the advantages of normalising peaks on the base peak.

The base peak may be arbitrarily given a peak height of e.g. 100 or 1000. Table 4.2a is an example of the printout that can be obtained from a computerised mass spectrometer. The compound being analysed in this case is benzoic acid. The left-hand columns containing

**Table 4.2a** A computer printout of the mass spectra data obtained for benzoic acid, $C_6H_5COOH$

```
Sample No:  001
Operator:    JB
Background:   0
Substrate:    0

37  14      75   7      122  100
39  18      76  14      123    1
50  42      77  62
51  53      78  15
52  12     105  80
74  16     106   5
```

numbers in increasing order are $m/z$ values while the right-hand columns give the corresponding intensities of these ions. The highest peak is obtained for the base peak, i.e. $m/z$ 122. Here, this is given a number of 100; all of the other peaks are relative to this one. Therefore, the peak at $m/z$ 51 has a corresponding intensity of 53, i.e. it is 53% as strong as the intensity of the base peak.

**SAQ 4.2g**

Draw a bar diagram corresponding to the data recorded in Table 4.2a.

Π However, there is still one vital piece of information missing. Which is this?

So far, we have obtained only a time value for each peak. If the spectrum is to make any sense, we need to convert these points into $m/z$ values. We can do this by scanning a reference compound, where the peaks have known $m/z$ values. Perfluorotri-$n$-butylamine (PFTBA) or perfluorokerosene are often used for this purpose. Therefore, the time values can be converted into $m/z$ values for unknown compounds by drawing a curve such as that shown in Figure 4.2i where the peak of the unknown compound is superimposed on the calibration curve. Its time value is expressed as $T_x$ and its $m/z$ value as $(m/z)_x$. This conversion curve is sometimes referred to as the *scan-law* curve. For high-resolution work where there is little peak overlap, the standard will be admitted with the sample.

The scan-law curve will, of course, be specific for the particular spectrometer in question and the scan rate that is being used. You will probably never see the scan-law curve appear on your output. The computer will have already applied an equation to the calibration curve's best-fit line and stored it in its memory. The sample data is simply converted by using this equation and you are presented with the $m/z$ output directly. You are therefore given the data obtained from calculated $m/z$ values and intensities in tabular form or as a bar diagram.

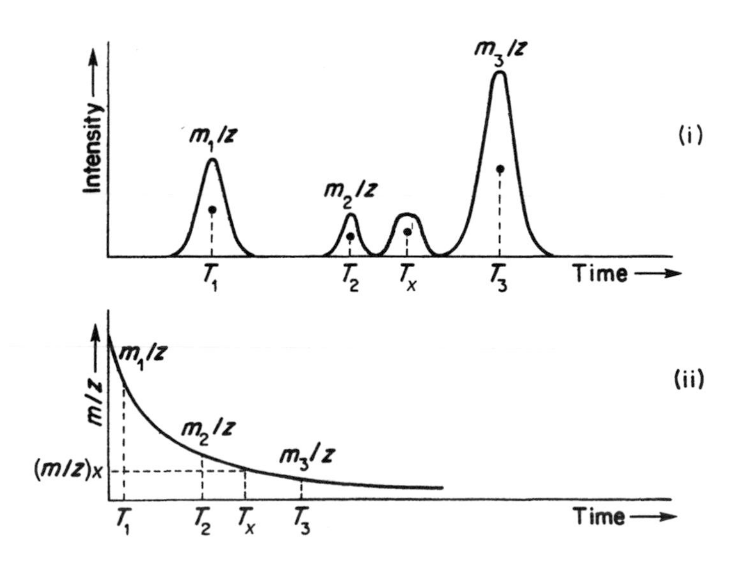


**Figure 4.2i** Representations of mass-spectral data: (i) in the form of intensity against time; (ii) conversion curve, $m/z$ against time (scan-law curve)

Sometimes a technique called *peak matching* is employed to calculate an unknown mass. Here, the exact mass ratio between *m/z* of the calibrant and the unknown is determined by precise variation of the acceleration voltage *V* in a magnetic-sector instrument. Since the variation of *m/z* at the focal point is linear with voltage, the unknown mass is precisely determined by the product of the known mass and the ratio of the voltages.

Computers can also be used for fine-tuning the ion source and controlling the mass analyser. Several instrumental variables may need to be accurately tuned in a short space of time. A human operator is obviously much less effective at this task than a computer.

Figure 4.2i shows a block diagram of the computerised control and data-acquisition and processing systems of a quadrupole mass spectrometer. This figure shows two features that will be encountered on any instrument.

First, the computer serves as the main instrument controller. The operator communicates with the spectrometer via interactive

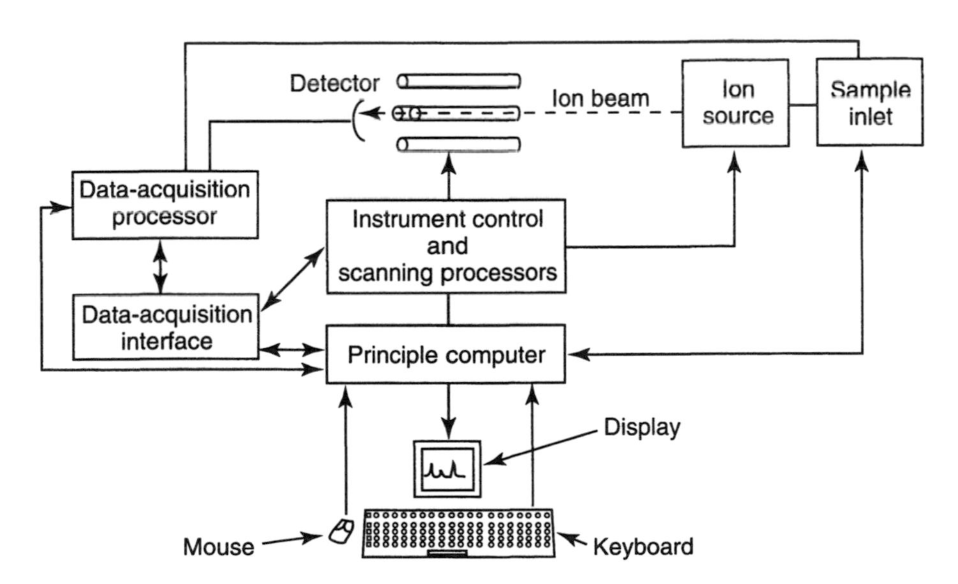


**Figure 4.2j** Schematic diagram of the instrument control and data-acquisition and processing systems of a quadrupole mass spectrometer

software. This enables the selection of specific operating conditions and machine parameters for the analysis. The computer also controls the programs responsible for manipulations of the data and the output.

Secondly, most instruments possess a series of microprocessors (about six in total) which are responsible for specific aspects of instrument control and information transmission between the computer and the spectrometer. There are a number of instrumental variables that the computer can control.

Π Can you name any of these?

Source temperature, accelerating voltage, scan rate, and magnetic field strength or quadrupole voltage are some of the possibilities.

With most systems, the computer stores all of the spectra and related information on disk. These may be plotted or printed directly, or the spectroscopist may employ certain data-reduction software to clarify the pictures or extract specific information.

With regard to data processing, mass spectrometry is similar to infrared and nuclear magnetic resonance spectroscopies, in that large libraries (>200 000 entries) of mass spectra are available in computer databases. Most on-line computers are able to quickly search these files for spectra that match or closely match the spectrum of the sample being investigated.

**SAQ 4.2h**

What are the two most important pieces of information that the computer obtains from an initial scan of an unknown sample?

**SAQ 4.2i**

> What factors do you think determine the sensitivity of measurement for a particular ion in a mass spectrometer?

**Summary**

We have now completed our coverage of the main instrumental aspects of organic mass spectrometry. You should now know how ions can be detected and their spectra processed, displayed and recorded, by using photographic plates, scintillation devices, oscilloscopes, chart recorders and computers.

**Objectives**

After studying this chapter you should be able to:

- describe how discrete and continuous dynode electron multipliers function;
- describe how a Faraday cup works;
- describe how ions may be detected on a photographic plate and with a scintillation-type collector;

- describe the recording of ions on an oscilloscope and chart recorder;

- describe and explain how a computer can be used to acquire, process and record mass-spectral data.

# 5. Use of Isotopes in Mass Spectrometry

This part of the Unit will introduce you to calculating abundances of isotopes from mass spectral data. Armed with this information, you will then be able to work out molecular formulae.

## 5.1 INTRODUCTION

You may remember that in the last chapter we saw a typical printout from the mass spectrum of benzoic acid (Table 4.2a). In this spectrum, the most abundant peak, i.e. the base peak, was given a value of 100, and all of the other peaks in the spectrum were given values relative to this peak height. In SAQ 4.2g you were asked to plot the spectrum you would observe as a result of this tabulated printout. You may have noticed that the values given in this table were only rounded up to the nearest per cent value.

Π Why do you think that these relative abundance values are not expressed more precisely?

This is simply because the peak heights of fragment ions can vary by a few per cent as a result of both temperature and pressure variations in the instrument during each scan. In addition, when comparing spectra using instruments with a differing beam geometry, resolution and efficiency, it would be highly unlikely to obtain an exact similar fragmentation pattern, even if the same ion source and machine operating parameters were used.

It is often the practice to identify the *m/z* value of the base peak by putting the value in brackets, by underlining or by using italics.

## 5.2 ISOTOPE ABUNDANCES: NON-HALOGEN ISOTOPES

Mass spectroscopists have obtained and published very accurate values of the atomic weights and relative abundance of all of the naturally occurring isotopes of the elements. These are collected together in tables to be found in many different reference books, such as the well known *Handbook of Chemistry and Physics*, which is published annually by CRC Press.

Π How do isotopes of elements differ? Is it their atomic weight or atomic number that varies?

It is the atomic weight that varies. The isotopes differ in only the number of neutrons in their nuclei. Their chemical properties can be assumed to be identical, since their atomic numbers, and therefore their number of electrons, are identical.

Sometimes you will meet a mass spectrum which shows an unfamiliar cluster of ions. When these are clearly the molecular ions or closely related species you will then have to scour the reference sources to find out which element (or elements) give rise to the observed patterns. This sounds a formidable task, bearing in mind the ninety plus elements that are available, but in practice you will probably have a good idea from the origin of the sample as to which elements are likely to be present. This will narrow your search down considerably.

**SAQ 5.2a**

You have obtained a soil sample from a contaminated land site previously used by local industry. Name some elements that you *would not* expect to be present in the sample, by considering their abundances in nature.

Depending on the industrial use of the site, you could probably eliminate many more elements which have limited industrial applications.

Π A water sample has been analysed by mass spectrometry (using inductively coupled plasma mass spectrometry—see Chapter 9) and the spectrum shown in Figure 5.2a has been obtained for the given mass region. If we assume that all of the ions formed in this region are purely elemental ones, what element is present? Use the table of isotopic abundances of some elements in this mass region (Table 5.2a) to help in your identification.

**Table 5.2a** Natural isotope abundances of various elements in a specified mass region

| Element | Atomic number | Atomic mass | Natural abundance[a] (%) |
|---|---|---|---|
| Ti | 22 | 46 | 8.0 |
| | | 47 | 7.5 |
| | | 48 | 73.7 |
| | | 49 | 5.5 |
| | | 50 | 5.3 |
| V | 23 | 50 | 0.3 |
| | | 51 | 99.7 |
| Cr | 24 | 50 | 4.3 |
| | | 52 | 83.8 |
| | | 53 | 9.5 |
| | | 54 | 2.4 |

[a]To 3-figure accuracy

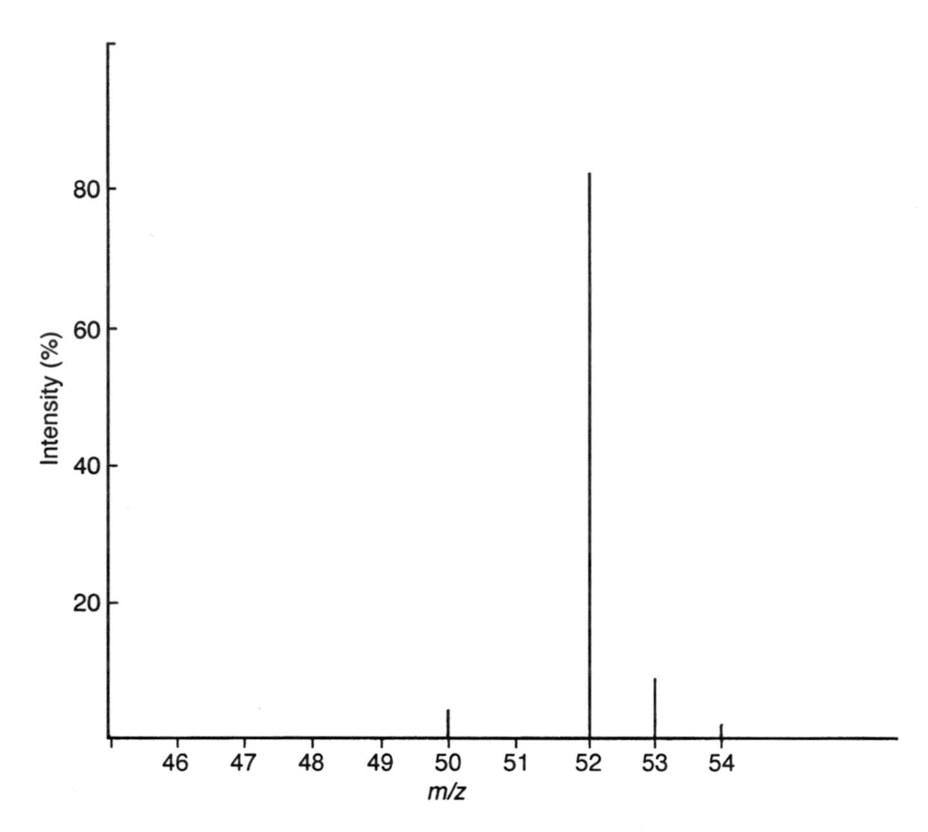


**Figure 5.2a** Mass spectrum of a water sample

The correct answer is chromium (Cr). The spectrum cannot be that of titanium (Ti), since there are no peaks in the mass region 46–49, nor can it be vanadium, since there is no peak at 51.

Of course, in mass spectrometry in general, we don't always obtain mass spectra of simple elemental ions. These ions are often combined with carbon and hydrogen (as seen in the next example), a situation which is not as simple as it may at first seem.

Π Figure 5.2b shows the molecular-ion ($M^{+\bullet}$) cluster of an organic compound containing an unknown heavy metal. From previous knowledge, the search has been narrowed down to three possibilities for the metal, i.e. mercury (Hg), cadmium (Cd), or lead (Pb). By looking at the table of isotopic abundances of these elements (Table 5.2b), can you identify the element present in the sample?

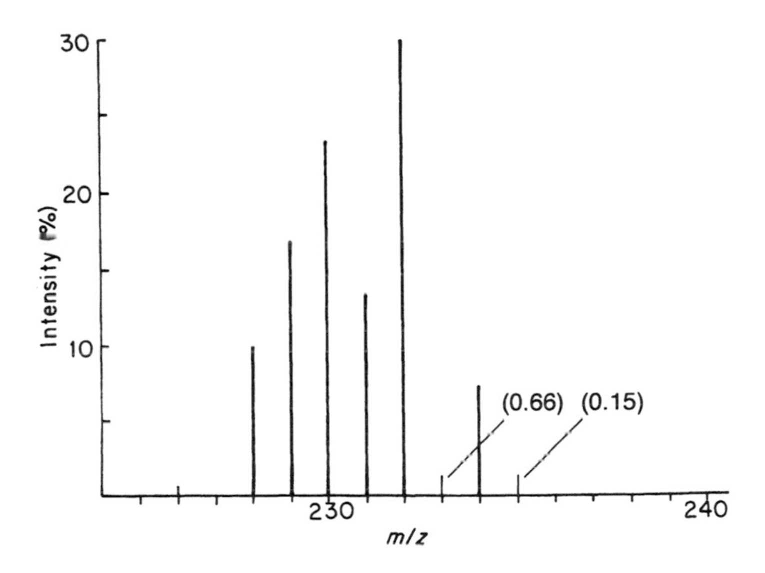


**Figure 5.2b** $M^{+\bullet}$ cluster of a heavy metal compound; the relative heights of the two most abundant peaks are 1/1.29

The answer is mercury. Did you get it correct? Mercury is used in the chloralkali industry and is a common environmental pollutant which needs regular monitoring in both soil and water samples. The molecular ion present is that of $(CH_3)_2Hg$ (dimethyl mercury) which is a relatively volatile compound, and is often found in contaminated soils and water, being quite easy to analyse by mass spectrometry.

How did you go about tackling this problem? You can tell straight away that Pb is not involved because its molecular ion would show only four peaks, with the most intense at the highest mass number, and this is clearly not the case. There are six major peaks and three minor peaks in the spectrum. Cd has eight isotopes and Hg just seven. So, which element is it? Let us ignore the minor peaks right now, since they can often be misleading, and look at which peaks are missing and which are the largest. The $^{109}$CD, $^{115}$Cd, $^{197}$Hg and $^{203}$Hg isotopes should not be there in our spectrum, while intense peaks should occur for Cd at the fifth and seventh isotopes ($^{112}$Cd and $^{114}$Cd), and for Hg at the

**Table 5.2b** Natural isotopc abundances of cadmium, mercury and lead

| Element | Atomic number | Atomic mass | Natural abundance[a] (%) |
|---|---|---|---|
| Cd | 48 | 106 | 1.2 |
| | | 108 | 0.9 |
| | | 110 | 12.4 |
| | | 111 | 12.8 |
| | | 112 | 24.1 |
| | | 113 | 12.3 |
| | | 114 | 28.9 |
| | | 116 | 7.6 |
| Hg | 80 | 196 | 0.2 |
| | | 198 | 10.0 |
| | | 199 | 16.8 |
| | | 200 | 23.1 |
| | | 201 | 13.2 |
| | | 202 | 29.8 |
| | | 204 | 6.9 |
| Pb | 82 | 204 | 1.4 |
| | | 206 | 25.2 |
| | | 207 | 21.7 |
| | | 208 | 51.7 |

[a]To 3-figure accuracy

fourth and sixth isotopes ($^{200}$Hg and $^{202}$Hg). Unfortunately, the matter is still not completely resolved since the two largest peaks are spread out over three mass units in each case and there only appears to be one mass gap in our given spectrum, not two as we might expect from the abundance table.

Π If all else fails, what should we do?

The answer is to try and obtain estimates of the relative peak heights of our two most abundant peaks from the data given and compare these values with the tabulated natural abundances. If we do this, we soon find that the ratios of the two largest peaks are 28.9/24.1 = 1/1.20 (for Cd), and 29.8/23.1 = 1/1.29 (for Hg). The spectral data given indicated that the ratio of the two most intense peaks was 1/1.129. This confirms that the compound does include Hg, and not Cd. It is now a good idea to check that the abundances of the other major peaks, relative to the largest peak, agree with the accepted values given in the table. From a visual inspection, it is easy to see that the abundances of the third and fourth most abundant isotopes differ by only a few per cent. This further rules out Cd, since its third and fourth most abundant peaks are approximately the same intensity.

Π What sort of elements would it be very difficult to recognise in compounds by using the procedures given above?

These are the elements that have only one isotope; such elements are said to be monoisotopic. Only one peak may be seen in any particular molecular-ion cluster as a direct result of these elements' masses. Therefore, the element would be very difficult to characterise unless it gave an ionised atom at an unambiguous mass number.

**SAQ 5.2b**

Can you think of some examples of monoisotopic elements? (If you have access to a 'chart of the nuclides' or a suitable reference book, then this will help you out, but don't worry if you can't think of many.)

Some of the more common monoisotopic elements that you might meet in organic mass spectrometry are fluorine, phosphorous and iodine.

Π What were the very small peaks at $m/z$ 233 and 235 in Figure 5.2b?

Well, they could be metastable peaks, peaks due to $X^{2+}$ charged interferent ions (unlikely at this high mass level), or $(M+1)^+$ peaks from hydrogen-transfer reactions, such as those present in the CI ion source. They are, in fact, peaks due to the isotopes of carbon and hydrogen and mainly result from $^{13}C$ (abundance 1.08%) in the molecular ion $(CH_3)_2Hg^{+\bullet}$, i.e. $m/z$ 233 $((^{13}C^1H_3)_2{}^{202}Hg^{+\bullet})$, and $m/z$ 235 $((^{13}C^1H_3)_2{}^{205}Hg^{+\bullet})$. Note that we would also see peaks due to the other isotopes of Hg, but these are obscured by other peaks, or in the case of $^{106}Hg$, they are too small to observe at $m/z$ 277.

Table 5.2c shows the normalised natural abundances of some important isotopes.

Π What is the probability of seeing a $^{13}C$ and a $^2H$ atom in the above methyl mercury ion?

The answer is 2 × 1.08 = 2.2%, for $^{13}C$, and 6 × 0.016 = 0.1%, for $^2H$.

**Table 5.2c** Normalised relative abundances of the isotopes of some common elements

| Isotope | $A_r^a$ | Abundance (%) | $A_r^a$ | Abundance (%) | $A_r^a$ | Abundance (%) |
|---|---|---|---|---|---|---|
| H | 1 | 100 | 2 | 0.016 | | |
| C | 12 | 100 | 13 | 1.08 | | |
| N | 14 | 100 | 15 | 0.36 | | |
| O | 16 | 100 | 17 | 0.04 | 18 | 0.20 |
| F | 19 | 100 | | | | |
| Si | 28 | 100 | 29 | 5.07 | 30 | 3.31 |
| P | 31 | 100 | | | | |
| S | 32 | 100 | 33 | 0.78 | 34 | 4.39 |
| Cl | 35 | 100 | | | 37 | 32.4 |
| Br | 79 | 100 | | | 81 | 97.5 |
| I | 127 | 100 | | | | |

[a]Relative atomic mass

In both cases, the probability is simply the relative abundance of the isotope multiplied by the number of atoms in the molecule. This shows that even minor isotopes can give rise to small but significant peaks in mass spectra. The more atoms of such elements that you have present, then the more chance there is that you will observe an M+1 or M+2, etc., peak in the spectrum. It can therefore generally be said that:

**'Molecular ions are rarely single peaks'**

**SAQ 5.2c**

Using the data given in Table 5.2b, calculate and plot the normalised spectrum (with reference to the base peak) for the $Pb^{+\bullet}$ ion cluster.

**SAQ 5.2d**

A normalised spectrum of a molecular-ion cluster is shown in Figure 5.2c. Calculate the isotopic abundance of the element concerned. Given that the species has the formula $(CH_3)_2X$, what is this element? Use mass tables is you are unsure.

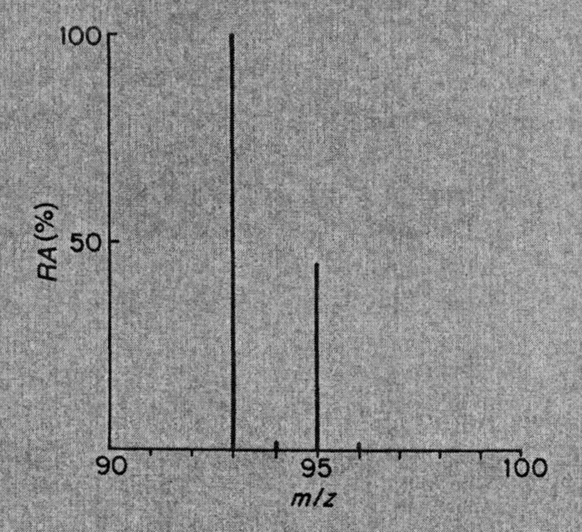


**Figure 5.2c** Mass spectrum of the molecular ion of $(CH_3)_2X$

Π You may have noticed small peaks at $m/z$ 94 and 96 in Figure 5.2c. What are these due to?

Again, these are the result of the presence of $^{13}C$ in our molecular-ion cluster (from the peaks at $m/z$ 93 and 95).

How can we calculate the abundances of these peaks? This can be done by application of the following simple formula:

$$\text{Isotope intensity} = (I \times n \times RA)\% \qquad (5.1)$$

where $I$ is the normalised relative abundance factor of the peak to the base peak.

Therefore, for a peak of 77% normalised relative abundance, we would insert a value of $I$ of 0.77.

In addition, $n$ is the number of carbon atoms present, and $RA$ is the normalised relative abundance (%) of the isotope in question.

Isotope intensity is expressed in percentage terms of the base-peak height.

Therefore, in Figure 5.2c, we would expect $^{13}C$ (at $m/z$ 94) to reach an intensity of:

$$1.0 \times 2 \times 1.08 = \mathbf{2.16\%} \text{ of the base-peak height}$$

(where $I = 1.0$, $n = 2$, and $RA = 1.08$)

You may find it easier to construct a mnemonic (memory-aid) around this formula. A possibility might be:

'*I* never *R*ang *A*nn'.

Π What would you predict the $m/z$ 96 ion intensity in Figure 5.2c to be?

Applying the 'I never Rang Ann' mnemonic, and having an approximate value for the normalised elemental abundance for $^{65}Cu$ of 44.5% (from SAQ 5.2d), gives the following:

$$I = 0.445, n = 2, RA = 1.08\%.$$

Therefore, the isotope intensity at $m/z$ 96 is:

$$0.445 \times 2 \times 1.08 = \mathbf{0.96\%}$$

Π We have ignored deuterium ($^{2}H$) so far in our calculations. Is this acceptable?

Applying the mnemonic for $m/z$ 94 gives:

$$I = 1.0, n = 6, \text{and } RA = 0.016.$$

Therefore, the isotope intensity = $1.0 \times 6 \times 0.016 = 0.096\%$, i.e. *ca.* 0.10%.

This is a negligible intensity compared to 2.16% from $^{13}C$ in Figure 5.2c, so we were correct to ignore it. The $m/z$ 96 peak would be even smaller!

However, as the number of carbon and hydrogen atoms increases in larger organic molecules, account does then need to be taken of their contributions to the (M+1) ions. Fortunately, this can be achieved in a relatively straightforward way, when there are just carbon, hydrogen, nitrogen, and oxygen atoms present:

$$(M+1)/M = (1.08 \times nC)\% + (0.016 \times nH)\% + (0.36 \times nN)\% + (0.04 \times nO) \quad (5.2)$$

where $nC$, $nH$ and $nN$ are the number of C, H and N atoms, respectively.

Let us apply this formula to our molecular ion, $(CH_3)_2Cu^{+\bullet}$, in Figure 5.2d where $n$C = 2, $n$H = 6, $n$N = 0, and $n$O = 0.

Therefore:

$$(M+1)/M = (1.08 \times 2) + (0.016 \times 6) + 0 + 0 = 2.26\%.$$

This is the same result as the sum of 2.16% and 0.1% obtained from the 'I never Rang Ann' nmemonic for C and H in the previous exercises. Therefore, our formula works properly. To a first approximation, we can still ignore the H term for this ion.

Π What modifications would we have to make to equation (5.2) in order to take into account ions that also contain, F, P, Cl, Br and I?

The answer is, none at all. All of these isotopes are monoisotopic.

Therefore, it can be said that unless the compound contains silicon or sulphur, the above formula applies to most compounds that are commonly analysed in organic mass spectrometry.

When we deal with Si or S, we just need to add $(5.07 \times n\text{Si})\%$ or $(0.78 \times n\text{S})\%$, respectively, to our other terms.

The relative abundances of (M+2) peaks in organic compounds, due to the presence of two $^{13}C$ and/or $^{18}O$ atoms, can also be calculated from a simple formula. Unfortunately, there are a number of other common elements which have significant (M+2) isotopes, namely Si, S, Cl, and Br, and so the formula is applicable to a smaller number of atoms.

The formula that we use is as follows:

$$(M+2)/M = \frac{(1.1 + n\text{C})^2}{200}\% + (0.2 \times n\text{O})\% \qquad (5.3)$$

Π Why are there no terms for H or N?

These two elements do not have stable (M+2) isotopes.

**SAQ 5.2e**

(a) Calculate the abundances of the (M+1) peaks of CO, $N_2$ and $CH_2=CH_2$ ($M^{+\bullet} = 28$).

(b) Calculate the abundance of the (M+1) peak of naphthalene, $C_{10}H_8$ ($M^{+\bullet} = 128$). Would you expect to be able to distinguish naphthalene from cyclohexanecarboxylic acid, $C_7H_{12}O_2$ (also $M_r$ 128), from the height of its (M+1) peak alone?

(c) Calculate the abundances of the (M+1) and (M+2) peaks in the mass spectrum of cholesterol, $C_{27}H_{45}OH$.

(d) Calculate the abundance of the (M+2) peak in the mass spectrum of trimethyl phosphate $(CH_3O)_3PO$.

(e) At what carbon number (approximately), would you expect the (M+2) peak of an organic compound, containing C, H, N, F, P, and/or I, and four oxygen atoms, to appear above a general background level of 2% (relative to $M^{+\bullet} = 100\%$)?

Π If we have a cluster of molecular-ion peaks as a result of the different isotopes contained in a compound, how do we define the molecular ion?

The molecular ion is defined as *that single species in which all of the elements present are found in their most abundant forms*.

Any other isotopic species in the cluster are then defined as (M−1), (M+1), (M+2), etc., relative to the most abundant isotope species.

**SAQ 5.2f**

What are the *m/z* values of the molecular ions of the following compounds:

(a) 1-chloronaphthalene, $C_{10}H_7Cl$;

(b) diethyl mercury, $(C_2H_5)_2Hg$;

(c) tetraethyl lead, $(C_2H_5)_4Pb$;

(d) ferrocene (dicyclopentadienyliron), $(C_5H_5)_2Fe$?

## 5.3 ISOTOPE ABUNDANCES: HALOGEN ISOTOPES

### 5.3.1 Introduction

Do you remember that we pointed out that some elements, e.g. chlorine and bromine, have significant (M+2) isotopes? The isotopes, $^{37}$Cl and $^{81}$Br, have normalised abundances of 32.4 and 97.5%, respectively. Therefore, if you see a mass spectrum with a cluster of intense peaks differing by 2 mass units, there is a good possibility that Br or Cl will be present.

One way in which the monoisotopic elements, fluorine and iodine, could be identified is with the use of a high-resolution mass spectrometer. They are both *mass-deficient* atoms, i.e. the atomic weight of $^{127}$I is less than its mass number ($A_r$ = 126.904470), as is the atomic weight of $^{19}$F ($A_r$ = 18.998495). Two of the other common atoms found in organic mass spectrometry, i.e. $^{14}$N and $^{1}$H, are both *mass proficient*; their mass numbers are greater than their precise atomic weights, which are 14.003074 and 1.007825, respectively. Therefore, you should be alerted when you see a single peak with an accurately measured mass which is slightly less than the nominal value given. However, be careful, as $^{31}$P is also monoisotopic and mass deficient, while $^{16}$O is mass deficient and has low-abundance isotopes.

Π What are all of these atomic-weight values based upon?

They are based upon the mass of the $^{12}$C isotope, with an atomic weight of 12.000000.

One way in which this mass deficiency can be utilised is that polyfluoro calibrants, such as poly(fluoro kerosene), $CF_3(CF_2)_nCF_3$, can sometimes be used in high-resolution studies. Ions from these polymeric mixtures appear in regular series, such as $CF_3^+$,$CF_3CF_2^+$,$CF_3(CF_2^+)_2^+$, with *m/z* values of 69, 119, and 160 respectively. They are significantly mass deficient (see Table 5.3a, and therefore enable clear, definitive calibration points to be established.

If only low-mass-resolution spectra are available, the presence of F or I can be observed by the presence of two features:

(a) mass increments of 18, 36, 54, 72, etc., for each F present, and 126, 252, 378, 504, etc., for each I present;

(b) successive losses of 127 ($I^\bullet$), or 128 (HI) amu from the $M^{+\bullet}$ of an iodo compound.

Π Why do F and I have increment values of 18 and 126 amu, respectively, when their corresponding $A_r$ values are 19 and 127?

The reason is that in order for a fluorine or iodine atom to be substituted into an organic molecule, a hydrogen atom must be replaced. Hence, the net increase is 1 amu less than the $A_r$ value. As an example of this, look at Figures 5.3a and 5.3b, which show the mass spectra of 1,3-dinitrobenzene, and a derivative of this compound, respectively.

**Table 5.3a** Some of the ions produced from poly(fluoro kerosene)s when used as calibrants in high-resolution studies

| Ion | Accurate mass[a] |
|---|---|
| $CF_3^+$ | 68.99549 |
| $CF_3CF_2^+$ | 118.99248 |
| $CF_3(CF_2)_2^+$ | 168.98947 |

[a]Based upon $^{12}C = 12.0000$

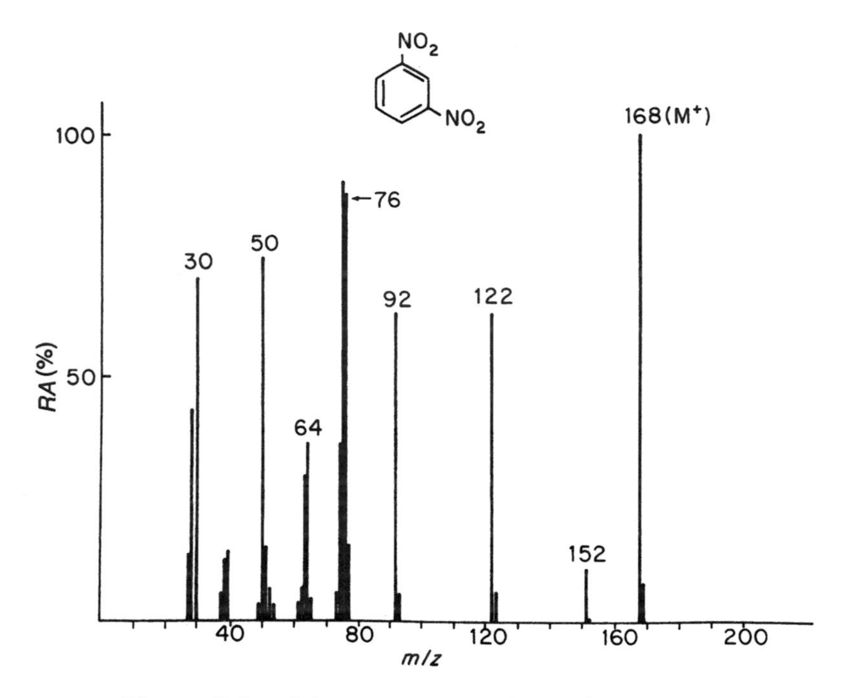


**Figure 5.3a** Mass spectrum of 1,3-dinitrobenzene

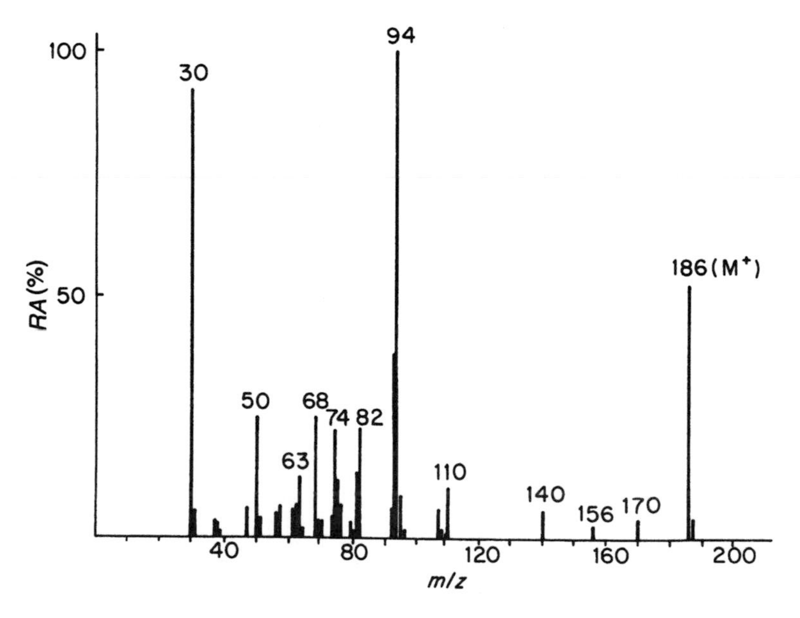


**Figure 5.3b** Mass spectrum of a derivative of 1,3-dinitrobenzene

You will see that the spectra are quite different, but if we simply concentrate on the molecular ion in both cases, the $M_r$ value has risen from 168 to 186, an increment of 18 amu. Hence, Figure 5.3b must be the spectrum of dinitrofluorobenzene (in fact, it is the 2,4-dinitro isomer):

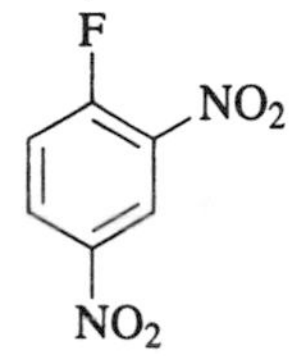

F
$NO_2$
$NO_2$

Note also that there isn't any apparent loss of F, i.e. no peaks 19 amu lower than the molecular ion or from any of the fragment ions. The characteristic and unique increase of 18 amu is the only clue to the presence of a fluorine atom.

Π Can you think of any reason why the fluorine atom is not lost?

The reason lies in the very high bond energy of the C—F bond, i.e. *ca.* 423 kJ mol$^{-1}$. This bond is much stronger than some of the other bonds in the molecular ions which fragment first.

### 5.3.2 Calculation of Chlorine and Bromine Isotope Ratios

When one Cl atom is present in a molecule, any ion derived from it which still contains the halogen will show a $^{35}Cl/^{37}Cl$ ratio of 100/32.4 (Table 5.2c; see also Figure 5.3c(i)). Similarly, a single bromine-containing ion will show a $^{79}Br/^{81}Br$ ratio of 100/97.5 (Table 5.2c; see also Figure 5.3d(i).

But how do we account for the three peaks seen in Figure 5.3c(ii) from an ion containing two Cl atoms, or the four peaks seen in Figure 5.3c(iii), which arise from an ion containing three Cl atoms?

Notice first of all that there are $(n+1)$ isotope peaks, where $n$ is the number of halogen atoms that are present. The intensities of the

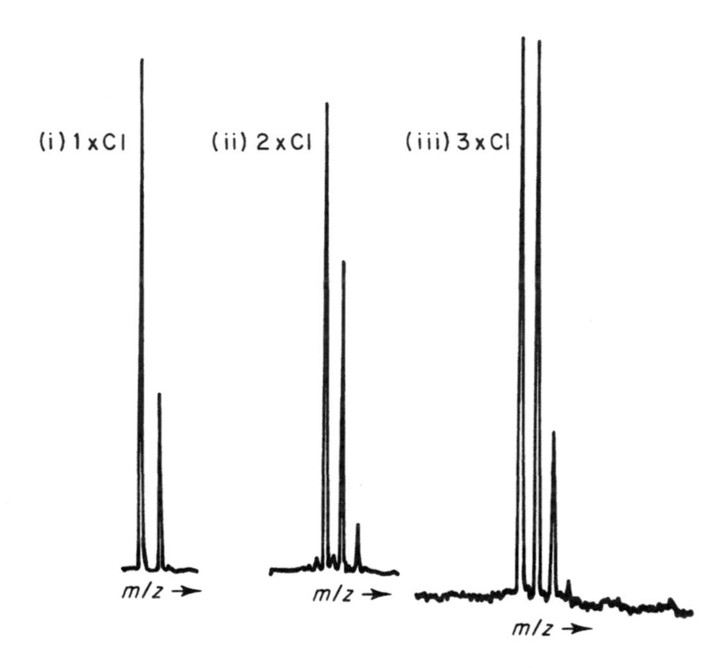


**Figure 5.3c** Patterns produced by elements with abundant stable isotopes, showing examples of ions containing (i) one, (ii) two, and (iii) three chlorine atoms

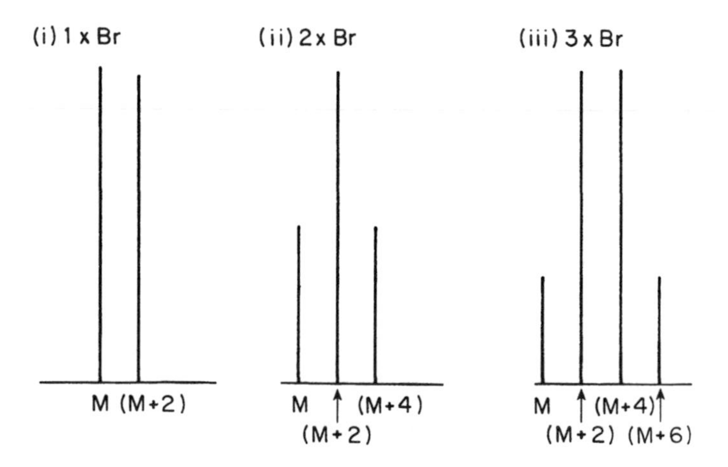


**Figure 5.3d** Patterns (not normalised) produced by elements with abundant stable isotopes, showing examples of ions containing (i) one, (ii) two, and (iii) three bromine atoms

isotopic species can be found by applying the following formula:

$$(a + b)^n \tag{5.4}$$

where $a$ is the relative abundance of the light isotope, $b$ is the relative abundance of the heavy isotope, and $n$ is the number of halogen atoms present.

Abundances $a$ and $b$ can be to any desired accuracy, but for our purposes, and also for simplicity, we will use whole number ratios:

e.g. $^{35}Cl = 3$ (almost) and $^{37}Cl = 1$

$^{79}Br = 1$ and $^{81}Br = 1$ (almost).

Let us first take the simple case where $n = 1$ for both Cl and Br. Equation (5.4) gives the relative abundances of the isotope peaks as 3/1 and 1/1 (as it should, of course).

Now consider $n = 2$. For $Cl_2$, equation (5.4) becomes:

$$(a+b)^2 = (a+b)(a+b) = a^2 + 2ab + b^2$$

If we now substitute '3' everywhere we have $a$, and '1' everywhere we have $b$, we obtain the following:

$$3^2 + (2 \times 3 \times 1) + 1^2$$

and by replacing the addition signs by proportionality signs, this becomes:

9/6/1

Π Now, check Figure 5.3c(ii) to see if our calculation is correct (note that the ion intensities have not been normalised).

Hopefully, you found that the ratios were as expected, give or take 5% or so.

Π Why might we expect relative abundance determined by a mass scan to not be very accurate?

There are random and systematic errors from the source, analyser, multiplier, amplifiers, data systems and plotter, all of which could give rise to inaccuracies in the measurement. For really accurate measurements, ion monitoring, using a technique known as accelerator mass spectrometry, is a better choice (see Chapter 9).

**SAQ 5.3a**

Use equation (5.5) to calculate the ratios of the isotope peaks expected for a $Br_2$-containing ion. Compare your predictions with Figure 5.3d (ii).

Let us continue now with the case where $n = 3$.

$$(a+b)^3 = (a+b)(a^2+2ab+b^2) = a^3 + 3a^2b + 3\,ab^2 + b^3 \qquad (5.6)$$

For Cl, $a^3 = 27$, $3a^2b = 27$, $3ab^2 = 9$, and $b^3 = 1$, so the relative abundances for the ions are 27/27/9/1 (approximately).

**SAQ 5.3b**

Use equation (5.6) to calculate the ratios of the isotope peaks expected for a $Br_3$-containing ion. Compare your predictions with Figure 5.3d(iii).

The series continues with each additional halogen atom. For $Cl_4$, the relative abundances are 81/108/54/12/1 for the $M^{+\bullet}$, $(M+2)^{+\bullet}$, $(M+4)^{+\bullet}$, $(M+6)^{+\bullet}$ and $(M+8)^{+\bullet}$ ions, respectively.

An interesting point to note is that when $n \geqslant 3$, the $(M+2)^{+\bullet}$ ion in the $Cl_n$ cluster is as, or even more, abundant than the $M^{+\bullet}$ ion. As $n$ increases, this phenomenon becomes more marked. In addition, the ion of highest mass (where the atoms are all $^{37}Cl$), becomes relatively small, and essentially negligible for $n > 5$. It is important not to miss this peak and wrongly identify the $Cl_{n-1}$ peak by mistake. By identifying the peak of highest mass in this cluster, you will be able to decide what $n$ is, since the mass of this ion is $(M+2n)^{+\bullet}$.

All of the bromine clusters have symmetrical abundances. For example, $Br_2$ shows the ratio 1/2/1, $Br_3$, 1/3/3/1, and $Br_4$, 1/4/6/4/1.

Π Where might you have seen this symmetrical abundance pattern before?

NMR spectrum intensities, perhaps?

Let us now move on to consider how we can calculate the relative abundances of clusters of ions which contain *both* chlorine and bromine.

We simply extend equation (5.4) to include the isotopes of Br. This then becomes:

$$(a+b)^n \times (c+d)^m \tag{5.7}$$

where $a$ and $b$ are the relative abundances of $^{35}Cl$ and $^{37}Cl$, respectively, $c$ and $d$ are the relative abundances of $^{79}Br$ and $^{81}Br$, respectively, and $n$ and $m$ are, respectively, the number of chlorine and bromine atoms present in the molecule.

If we take the simplest case first where we have one molecule of bromine and one molecule of chlorine, and use equation (5.7):

$$n = m = 1$$

and $(a+b)(c+d)$ then expands to give:

$$ac + ad + bc + bd$$

Here though, we need to remember that $ad$ and $bc$ will in practice have the same mass, i.e. $(M+2)^{+\bullet}$, so we should group them together to give the following:

$$ac + (ad + bc) + bd \qquad (5.8)$$

If we substitute $a = 3$, $b = 1$, $c = 1$, and $d = 1$ into equation (5.8), we find that the ratio $M^{+\bullet}/(M+2)^{+\bullet}/(M+4)^{+\bullet}$ is predicted to be 3/4/1 (to the same accuracy as that used previously).

**SAQ 5.3c**

Figure 5.3e shows the unnormalised mass spectrum of 4-bromochlorobenzene. Are the molecular ions present in the predicted 3/4/1 ratio? What other ions containing halogens are present in the spectrum?

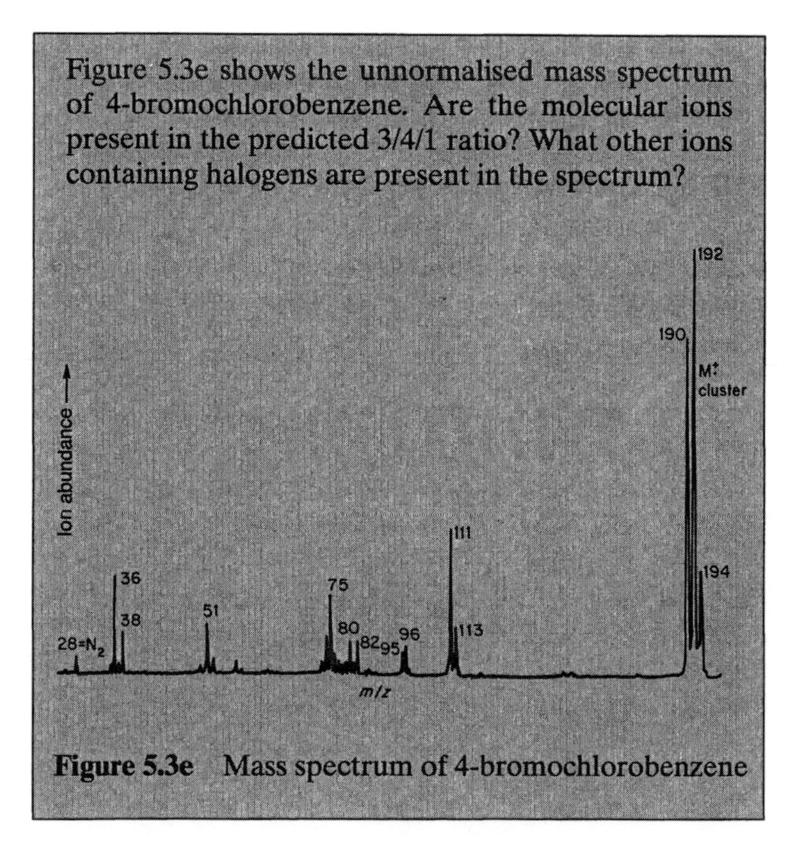


**Figure 5.3e** Mass spectrum of 4-bromochlorobenzene

Unfortunately, as the number of atoms of Cl and Br gets bigger, so does the amount of work needed to come to the correct answer! However, there is a simplified method that will allow you to come to the same result (for non-complex cases only).

Let us take the ClBr ion again. We know that the relative abundances for the Cl isotopes ($^{35}Cl/^{37}Cl$) can be approximated as 3/1 and for the Br isotopes ($^{79}Br/^{81}Br$) as 1/1.

We can 'multiply' out these terms (3/1)(1/1), and after each multiplication move the following terms one column to the right: if we then add up all of the columns we get the following:

| | | | |
|---|---|---|---|
| (3/1) × 1 = | 3 | 1 | |
| (3/1) × 1 = | | 3 | 1 |
| Total | 3/ | 4/ | 1 |

These are the relative abundances we have just worked out from equation (5.7) for the $M^{+\bullet}/(M+2)^{+\bullet}/(M+4)^{+\bullet}$ ions.

Now, let us consider $Cl_3Br$, as a further example. What are the relative abundances of the ions?

You could now work through the method outlined above, putting numbers into equation (5.7) and ending up with the following:

$$(a+b)^3(c+d)$$

which can be expanded, sorted and substituted, as before.

Alternatively, you can follow the simplified method.

For $Cl_3$, the abundances of the isotopes given by equation (5.6) are 27/27/9/1.

For Br, the relative abundances are 1/1.

Combining these ratios, i.e. (27/27/9/1) and (1/1), and placing them in the order shown below, gives:

| | |
|---|---|
| (27/27/9/1) × 1 = | 27/27/ 9/ 1 |
| (27/27/9/1) × 1 = | 27/27/ 9/1 |
| Total | 27/54/36/10/1 |

These are the relative abundances for the $M^{+\bullet}/(M+2)^{+\bullet}/(M+4)^{+\bullet}/(M+6)^{+\bullet}/(M+8)^{+\bullet}$ ions.

Table 5.3b lists the relative abundances of the ions in the halogen clusters that we have been considering in this section. Note that most of the clusters are distinctly different, thus enabling straightforward identification of the number of halogen atoms to be made in most cases.

This simplified method does not just apply to halogen mixtures. It will also work perfectly well for other simple combinations, e.g. HgBr.

**SAQ 5.3d**

Calculate the relative abundances, using the simplified method, of the major isotope peaks in clusters containing:

(a) $CuBr_2$;

(b) SClBr.

**Table 5.3b** Relative abundances of ions in clusters containing chlorine and/or bromine

| Composition | M | (M + 2) | (M + 4) | (M + 6) | (M + 8) | (M + 10) | (M + 12) |
|---|---|---|---|---|---|---|---|
| $Cl_2$ | 9 | 6 | 1 | | | | |
| ClBr | 3 | 4 | 1 | | | | |
| $Br_2$ | 1 | 2 | 1 | | | | |
| $Cl_3$ | 27 | 27 | 9 | 1 | | | |
| $Cl_2Br$ | 9 | 15 | 7 | 1 | | | |
| $ClBr_2$ | 3 | 7 | 5 | 1 | | | |
| $Br_3$ | 1 | 3 | 3 | 1 | | | |
| $Cl_4$ | 81 | 108 | 54 | 12 | 1 | | |
| $Cl_3Br$ | 27 | 54 | 36 | 10 | 1 | | |
| $Cl_2Br_2$ | 9 | 24 | 22 | 8 | 1 | | |
| $ClBr_3$ | 3 | 10 | 12 | 6 | 1 | | |
| $Br_4$ | 1 | 4 | 6 | 4 | 1 | | |
| $Cl_5$ | 243 | 405 | 270 | 90 | 15 | 1 | |
| $Br_5$ | 1 | 5 | 10 | 10 | 5 | 1 | |
| $Cl_6$ | 729 | 1458 | 405 | 540 | 135 | 128 | 1 |
| $Br_6$ | 1 | 6 | 15 | 20 | 15 | 6 | 1 |

## 5.4 AN APPROXIMATE METHOD FOR CALCULATING RELATIVE ABUNDANCES OF CLUSTERS DUE TO MINOR ISOTOPES

The methods so far described can be used to calculate the relative abundances of any isotope, whatever its abundance.

We will now present an approximate method which can be used for isotopes which have a low, but nevertheless significant, natural abundance, such as $^{29}Si$ or $^{34}S$.

Π Do you remember the expression describing the relative abundances of ions in a cluster, i.e. equation (5.4)?

It is $(a+b)^n$, where $a$ and $b$ are isotopes of a particular element and $n$ is the total number of atoms present.

If $a >> b$, it turns out that terms beyond the fourth, i.e. $(M+6)^{+\bullet}$ ion, can often be ignored because they will become relatively small compared to the other terms when the isotopes have a mass difference of two or more. When the isotopes have a mass difference of one, terms beyond $(M+2)^{+\bullet}$ can be ignored.

These assumptions lead to a general equation (derived from the binomial theorem) which describes the relative abundances of the first four ions in any cluster. This equation can be used for minor isotopes that have up to 10% relative abundance (to the major isotope) and where there are four or more atoms present in the molecule.

The relative abundances of the ions are given by the following expression:

$$a^3 / na^2b / \frac{n(n-1)}{2}ab^2 / \frac{n(n-1)(n-2)}{6}b^3 \qquad (5.9)$$

where $a$ is the relative amount of the most abundant isotope, $b$ is the relative amount of the least abundant isotope, and $n$ is the number of isotopes in the molecule.

**SAQ 5.4a**

Reference to Table 5.2c shows that $^{33}S$ has a natural abundance of 0.78% relative to $^{32}S$ (100%).

(a) What is the relative abundance of $(M+1)^{+\bullet}$ in the $S_8^{+\bullet}$ cluster? (Hint—use the nmemonic.)

(b) What is the contribution made by species containing two $^{33}S$ to the $(M+2)^{+\bullet}$ ion?

(c) Would $(M+3)^{+\bullet}$ be observable in the $S_8^{+\bullet}$ cluster?

In attempting parts (b) and (c) you may find the binomial approximation more helpful (see equation (5.9)).

**SAQ 5.4b**

A common method for analysing polyhydroxy compounds, such as glucose, is to convert each —OH group into a —OSi($CH_3$)$_3$ (trimethylsilyl, TMS) group; this is then used in GC/MS analysis. How intense would you expect the $(M+1)^{+\bullet}$ ions to be for the penta-TMS derivative of glucose ($C_6H_{12}O_6$), which has the formula, $C_{21}H_{47}O_6Si_5$?

**SAQ 5.4c**

Figure 5.4a shows the mass spectrum of the pesticide DDT, which contains a number of atoms of a certain halogen. Which halogen is present, and how do you think the cluster *m/z* 235/237/239 has been formed from $M^{+\bullet}$?

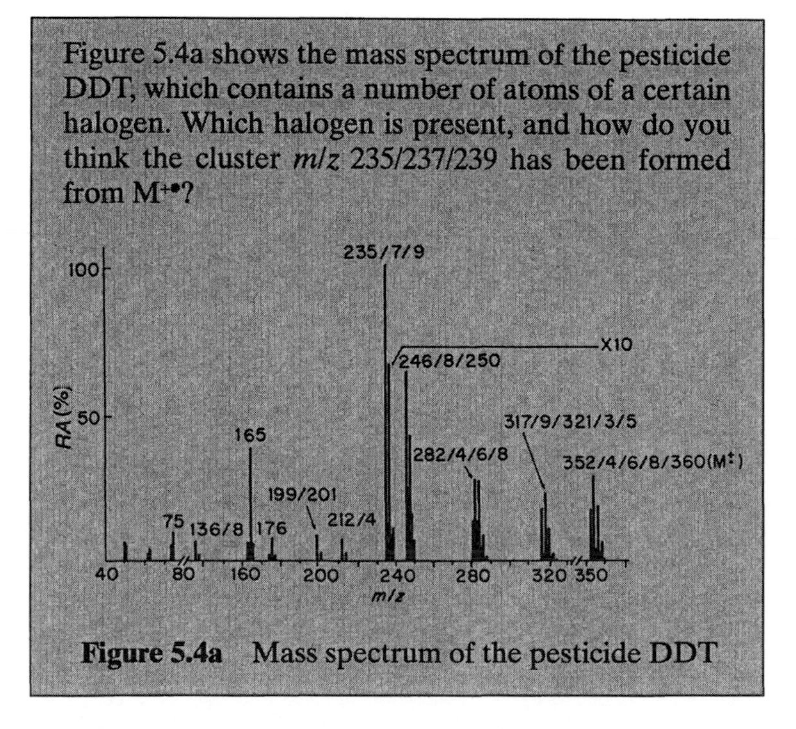


**Figure 5.4a** Mass spectrum of the pesticide DDT

Π Figure 5.4a shows two features commonly found in mass spectra which can be used to aid compound elucidation. What are they?

First, the *m/z* axis has discontinuities between *m/z* 75 and 136, and *m/z* 325 and 352. This is in order to keep the figure reasonably compact and to omit ion peaks that are of no interest. The discontinuities are indicated by a // sign. Secondly, when important ions have a very low relative abundance, they are sometimes shown with deliberately enhanced abundances. This is particularly important when it is necessary to show complex isotope patterns. Usually, such peaks are enhanced by a constant factor (e.g. × 5 or × 10, etc.). In Figure 5.4a all of the ions from *m/z* 246 onwards are enhanced by a factor of 10.

## 5.5 ISOTOPE-DILUTION MASS SPECTROMETRY

*Isotope-dilution mass spectrometry* (IDMS) is often used to establish standard concentrations. It is, in fact, a special use of the method of *internal standards*, in which the intensities of signals corresponding to the product need to be quantified with those of a reference standard. It is often a deuterated derivative of the compound being examined that is used as an internal standard. Therefore, the physical and chemical properties of the two compounds should be virtually identical. The method consists of examining the mass spectrum of the compound that needs to be determined in order to select an intense characteristic peak, which is then used to measure the analyte. A known, exact quantity of labelled internal standard is added to the sample of unknown concentration. This has the effect of moving the signal corresponding to the characteristic peak to a different position in the spectrum, according to the number and nature of the atoms that were used in the labelling. The ratio of these two signal intensities is then used to measure their relative proportion.

Suppose that there are $N$ atoms or molecules in the mixture that yield a peak which is characteristic of the mass $m$, and there are $M$ atoms or molecules of the labelled compound in the substance which yield a characteristic peak for the mass $m + n$ (= $o$), where $n$ corresponds to the mass displacement caused by the introduction of isotopes into the molecule.

The ratio $R_{mo}$ of the ion intensities over the masses $m$ and $o$ is given by the following expression:

$$R_{mo} = \frac{NP_m + MQ_m}{NP_o + MQ_o} \tag{5.10}$$

where $P_m$, $P_o$, $Q_m$, and $Q_o$, represent the relative isotopic abundances normalised over the isotopes, for the natural product and the labelled product with masses $m$ and $o$, respectively. Since $P_m$, $P_o$, $Q_m$, $Q_o$, and $M$ are known, and $R_{mo}$ can be deduced from the spectrum of the mixture containing the natural product and the labelled compound, then the value of $N$ can be accurately calculated. This process is illustrated in Figure 5.5a.

If $A$ represents the Avogadro constant, $x$ and $y$ the quantities of natural product and labelled product added to the sample, respectively, and $E$ and $F$ the molecular masses of natural and labelled products, respectively, then the equation becomes:

$$R_{mo} = \frac{(Ax/E)P_m + (Ay/F)Q_m}{(Ax/E)P_o + (Ay/F)Q_o} \tag{5.11}$$

which can be rearranged to give:

$$R_{mo} = \frac{(x/y \times P_m/E) + Q_m/F}{(x/y \times P_o/E) + Q_o/F} \tag{5.12}$$

Now, if $P_o$ and $Q_m$ are zero (there is no interference between the

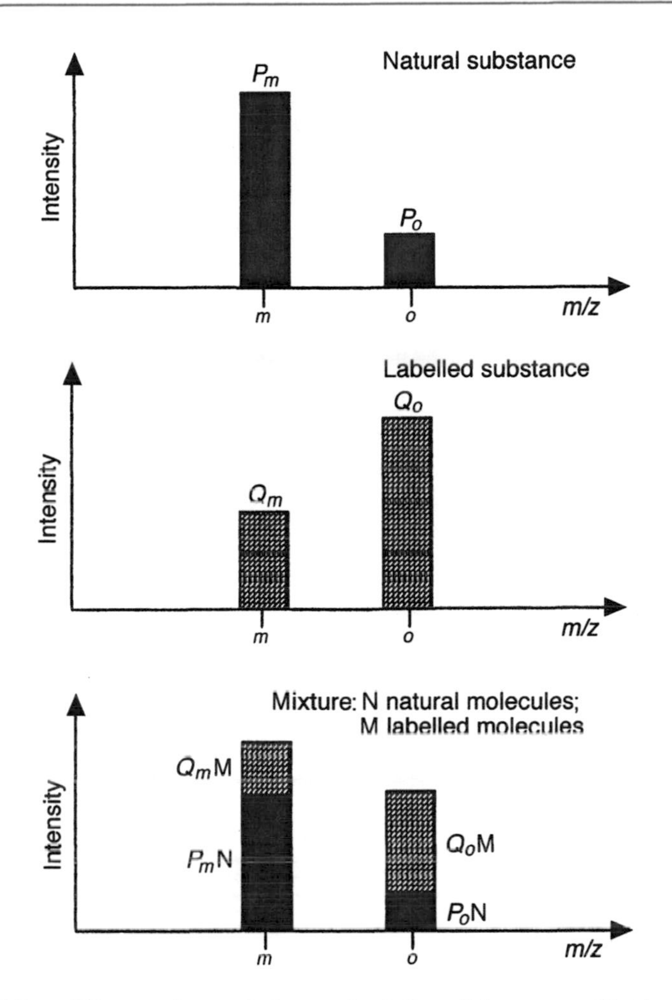


**Figure 5.5a** Illustration of the principle of isotope dilution using a molecule containing two isotopes with masses *m* and *o*

natural compound at high mass *o* and the labelled compound at low mass *m*), then the equation becomes:

$$R_{mo} = \frac{x}{y} \times \frac{P_m}{Q_o} \times \frac{F}{E} \tag{5.13}$$

Π What sort of curve does equation (5.13) represent?

This is the equation of a straight line passing through the origin.

Therefore, in theory, the calibration curve for a particular analysis can be calculated without any reference to experiments on samples containing increasing quantities of the compound to be measured. If the isotopic abundances are known with precision, this method allows the determination of the exact quantity of the substance to be measured without having to set up a calibration curve.

**Summary**

We have now seen a number of ways in which relative isotope intensities can be calculated. With this knowledge, compounds of similar $M_r$ values, but containing different constituent atoms, may be distinguished by comparing the relative abundances of their (M+1) peaks, etc., with the calculated theoretical values. Fluorine and iodine may be distinguished using high-resolution mass spectrometers by the fact that they are mass-deficient, and from low-resolution mass spectra by noting mass increments of 18 and 126, respectively, for each hydrogen atom that is replaced. Combinations of bromine and chlorine atoms in various compounds were also discussed.

In addition, the principles of isotope dilution mass spectrometry have been outlined.

**Objectives**

Now that you have completed this chapter you should be able to:

- recognise that most elements have distinctive isotopic intensity patterns;
- recognise the patterns produced by common elements such as C, H, O, N, Si, S, Cl, Br, Cu, Cd, Hg, Pb, Cr, V and Ti;
- identify unknown elements from their isotopic-abundance patterns;
- apply simple formulae to calculate the isotopic intensities of (M+1) and (M+2) peaks for the ions produced in the mass spectra of organic molecules containing C, H, N, O, F, P, Cl, Br, H and I, and C, H, N, O, F, P and I, respectively;

- adapt these formulae to take account of the presence of other elements with (M+1) isotopes, such as $^{29}Si$ and $^{32}S$;
- define what is meant by the molecular ion of a compound and calculate the relative mass $M_r$ of the molecular ion of any given compound;
- recognise that fluorine and iodine are mass-deficient and may be distinguished by high-resolution mass spectral measurements;
- recognise that fluorine and iodine can also be detected in low-resolution mass spectra by noting mass increments of 18 and 126, respectively, for each hydrogen atom replaced;
- explain why polyfluoro compounds are used for mass calibration in mass spectrometry;
- apply the formula $(a+b)^n$ to calculate the relative abundances of clusters containing C1 and Br, and recognise the typical patterns produced by $C1_n$ where $n \geqslant 6$;
- apply the formula $(a+b)^n(c+d)^m$ to calculate the relative abundances of clusters containing Cl and Br for all combinations of C1 and Br up to $Cl_4$ and $Br_4$;
- recognise that $Br_n$ clusters are typically symmetrical, and distinguish patterns up to $Br_6$;
- apply a simplified method for the calculation of the relative abundances of ions in clusters containing Cl and Br combined with other isotopic elements, such as Si, S, Cu and Hg;
- apply a binomial-expansion approximation to the expression $(a+b)^n$ when the relative abundance of an isotope is less than *ca.* 10% of the major isotope, to give the relative abundance of the first four peaks in a cluster;
- understand the principles of isotope dilution.

# 6. Metastable Ions and Modes of Fragmentation

In the first three sections of this chapter, we will concentrate on an entirely new type of ion that often leads to confusion, i.e. the *metastable ion*, and examine how such ions can be used to elucidate fragmentation pathways.

We will then go on to look at the ways in which the energy given to molecular ions is 'moved around' and used to bring about fragmentations, how molecular ions and their daughter ions can be stabilised, and the mechanisms of the fragmentations themselves in terms of how the bonds actually break. We will continue by introducing the *Quasi-Equilibrium Theory* and look at how positive ions can be stabilised by inductive and mesomeric effects. Finally, we shall examine the important mechanisms of bond cleavage in mass spectrometry.

If you have studied organic chemistry before, you may find some of the parts on stability and mechanisms quite familiar. If this is the case, then read over the material briefly and attempt a few of the exercises and SAQs.

## 6.1 ORIGIN OF METASTABLES

Figure 6.1a shows a photographic trace of a mass spectrum. There are a number of sharp peaks at integer *m/z* values and also some much

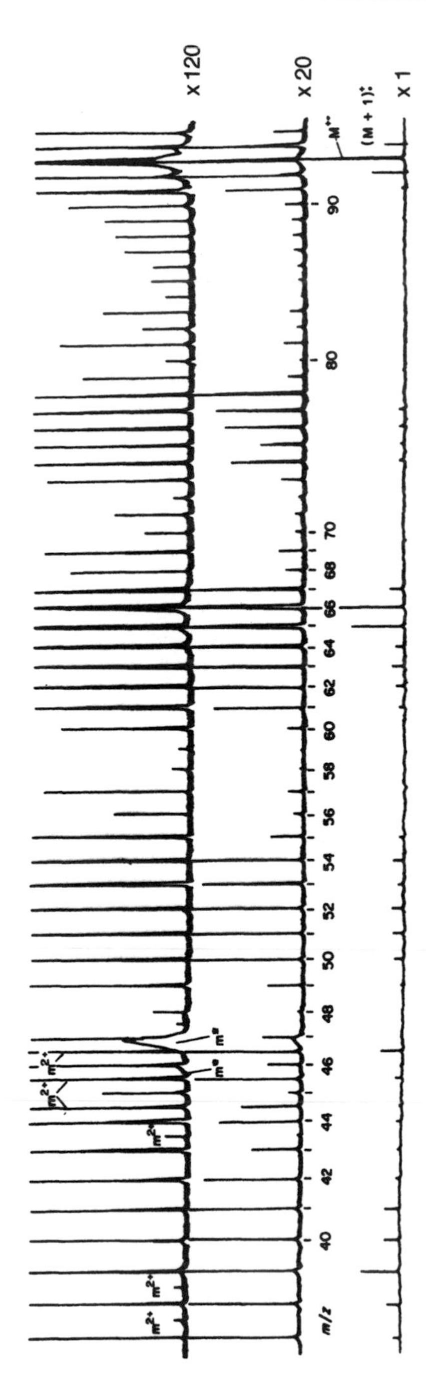

**Figure 6.1a** Photographic trace of a mass spectrum

weaker, broader peaks occurring at non-integer *m/z* values (marked as *m**). These *m** peaks are known as metastable ions and although they might appear at first to complicate the spectrum, they can be used to aid identification of fragment ions and pathways.

Π Do you remember from Section 1.2 which factors determine whether an ion fragments or not?

Fragmentation depends upon a whole host of factors, including the strength of the bonds to be broken, the stability of the products, the time interval between ion formation and detection, and on the amount of energy that the ion carries the instant it has been ionised. Those molecular ions without the necessary (threshold) energy to decompose remain as molecular ions and are recorded as $M^{+\bullet}$ ions. However, those ions with internal energies above the threshold value can decompose in the ion source and are analysed and recorded as fragment ions.

In general, processes that involve both bond making and bond breaking are slower than processes involving cleavage of only one bond.

Π Why is this the case?

Bond breaking and bond forming in the same process is slow since the ion needs to achieve the correct transition state for the reactions to take place. Only a small proportion of molecules will achieve the correct geometry and possess the correct energy for such reactions. Therefore, the number of reactions per second will be relatively small compared with simple bond cleavages, where molecules only have to achieve the correct energy.

In the mass spectrometer, collisions between ions are much rarer than in the liquid phase since the atoms are much further apart and the ion source is kept under high vacuum. Therefore, gaseous ions have a long mean free path and there is no equilibration of energy among them, as a result of collisions in the ion source. Therefore, whether an ion decomposes or not depends upon the internal energy gained at its formation. A hypothetical internal energy distribution is shown in Figure 6.1b.

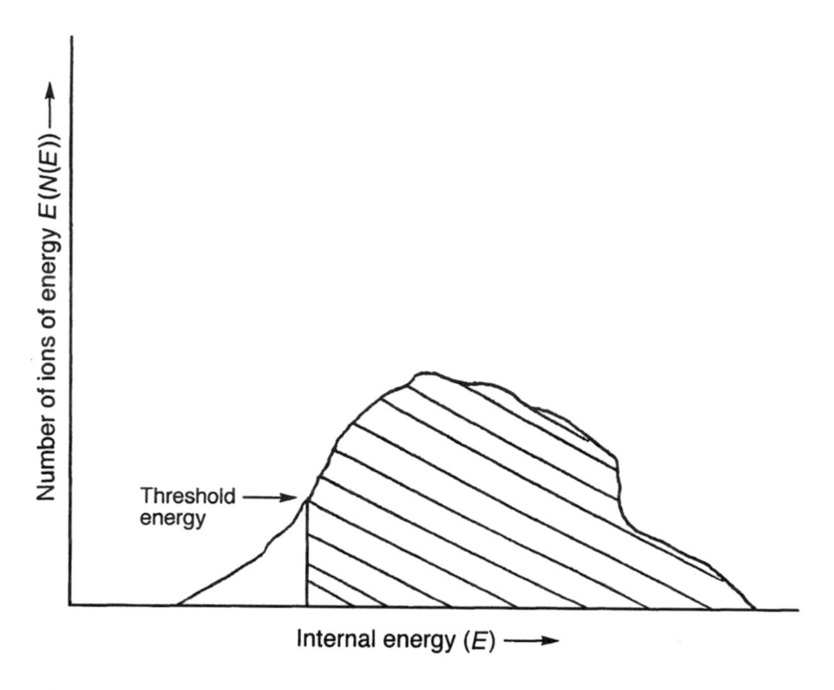


**Figure 6.1b** Hypothetical internal energy distribution of ions

In this figure, the proportion of ions in the unshaded area have insufficient internal energy to fragment within the time scale of the mass spectrometer and are thus detected as molecular ions. In contrast, the proportion of ions in the shaded area have sufficient energy to fragment and, depending on the other factors listed above, will therefore form fragment ions.

Π Consider those ions that have intermediate energy values, i.e. energies around the threshold value. What will happen to them?

These molecular ions would have lifetimes which are sufficiently long for them to leave the ion source, but they would then decompose during flight, before detection. This means their lifetimes would be less than ca. $10^{-5}$ s, but greater than ca. $10^{-6}$ s.

We have so far only referred to molecular ions decomposing to give fragment ions, but our comments could also be applied to the decomposition of unstable fragment ions.

Metastable ions can be formed in any part of the mass spectrometer between the exit from the ion source and the detector, but we shall only concern ourselves here with the ions formed in the small region between the source exit and the magnet (for magnetic-sector instruments). This is known as the field-free region because it is situated between the regions of the accelerating field and the magnetic field.

Let us now consider what happens to an ion ($M_1^{+}$), with mass $m_1$, which decomposes in the field-free region to give a fragment ion ($M_2^{+}$), of mass $m_2$, and a neutral molecule ($M_1$–$M_2$); the fragmentation can be represented by the following:

$$M_1^{+} \longrightarrow M_2^{+} + (M_1\text{–}M_2) \tag{6.1}$$

The ion $M_1^{+}$ is accelerated out of the ion source and attains a velocity $v_1$, so that it has a kinetic energy of $\frac{1}{2}m_1v_1^2$. This gain in kinetic energy is proportional to the potential difference through which the ions pass.

Π Can you recall the above relationship? (See equation (3.4) if you have difficulty remembering.)

The expression is as follows:

$$\tfrac{1}{2}m_1v_1^2 = zV \tag{6.2}$$

where $z$ is the charge on the ion and $V$ the potential difference.

If the ion $M_1^{+}$ has a momentum $m_1v_1$, the 'law of conservation of momentum' tells us that this momentum is shared out between the products of decomposition, i.e. $M_2^{+}$ and ($M_1$–$M_2$).

Π If we assume that both product ions have the same velocity when they fragment, can you write down an equation which describes this conservation of momentum?

The appropriate equation is:

$$m_1v_1 = m_2v_2 + (m_1\text{–}\mathrm{m}_2)v_2 \tag{6.3}$$

**SAQ 6.1a**

Expand and rearrange equation (6.3) to show that $v_1 = v_2$.

From now on, we shall refer to this common velocity as $v$.

The ion decomposes in the field-free region, so the motion of $M_2^+$ will be governed by the magnet.

Π Can you write down the equation for the forces acting on this ion, i.e. equation (3.3) from earlier?

The equation is:

$$m_2 v^2 r = Bzv \quad (6.4)$$

where $r$ is the radius of curvature of the magnet and B is the magnetic field strength.

If we take equations (6.3) and (6.4) and perform a series of algebraic manipulations, we can end up with a relationship which is similar to that given previously as equation (3.9):

$$B^2z^2r^2/m_2^2 = 2zV/m_1 \quad (6.5)$$

(Remember that the velocities of all of the ions are assumed to be equal.)

If you have difficulty with this operation, refer back to Section 3.1 and perhaps work through the exercise again.

Rearranging equation (6.5) gives:

$$m_2^2/m_1z = B^2r^2/2V \quad (6.6)$$

and if:

$$m^* = m_2^2/m_1 \quad (6.7)$$

we find:

$$m^*/z = B^2r^2/2V \quad (6.8)$$

This means that the mass of the metastable ion ($m^*$) will always be less than $m_2$.

**SAQ 6.1b**

Suppose that an ion of $m/z$ 60 decomposes by loss of a hydrogen atom to $m/z$ 59. What would be the corresponding $m^*$ value?

The actual masses of metastable ions are usually only quoted to single decimal places. This is because of the reduced ability of the magnet to focus metastable ions with a broad spread of kinetic energies. The result of this is a rather broad ion peak, whose $m/z$ value covers a small range.

This energy distribution arises out of the fragmentation process itself. Fragments formed will shoot off in all directions in the field-free region. However, the general direction for many of the ions will be forward as a result of the high accelerating voltage that was imparted upon them in the ion source. As a consequence, those ions that were fragmented in a forward direction will end up a little way ahead, while those fragmented backwards will lag a bit behind.

A mass spectrum based on this theoretical fragmentation might look something like the one shown in Figure 6.1c.

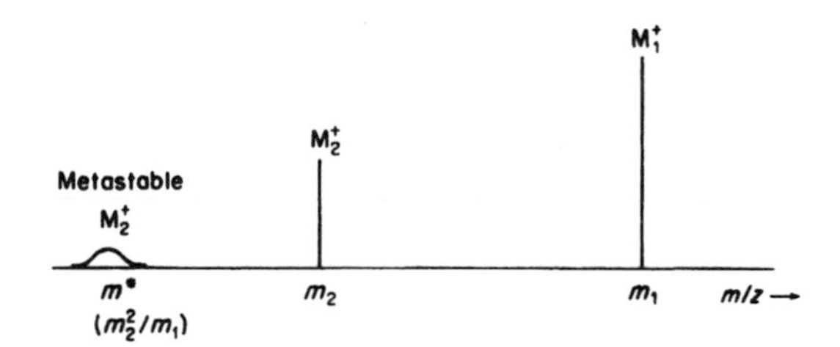


**Figure 6.1c** Partial mass spectrum obtained for the fragmentation $(M_1^+ \longrightarrow M_2^+ + (M_1{-}M_2))$

## 6.2 THE USE OF METASTABLE IONS

Metastable ions can be useful in helping to establish fragmentation routes, since if one sees all three ions, i.e. $M_1^+$, $M_2^+$ and $(M^*)^+$ in the mass spectrum, then at least a proportion of the $M_2^+$ ions, i.e. those formed in metastable decompositions, must be formed directly from

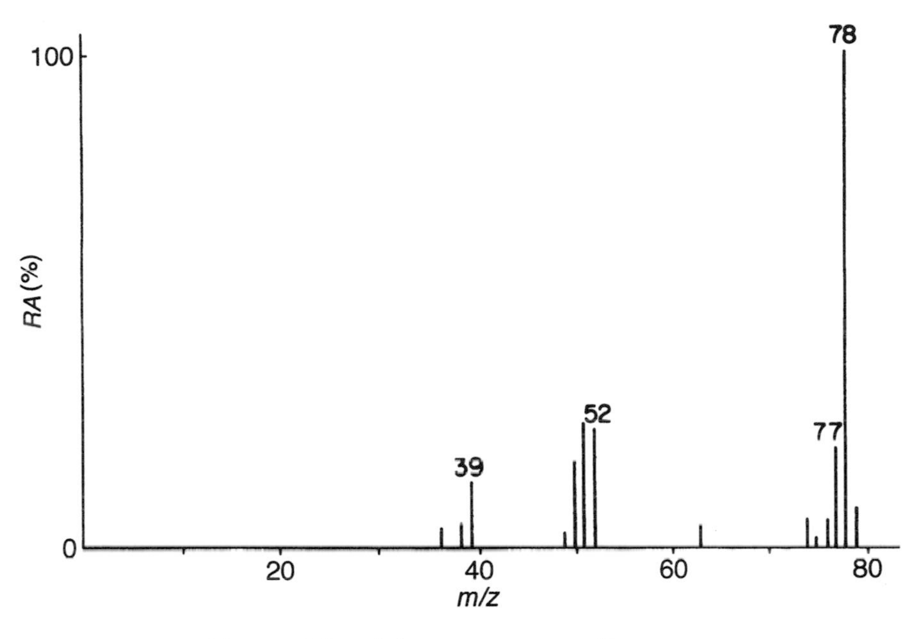


**Figure 6.2a** Mass spectrum of benzene

$M_1^+$. Figure 6.2a shows the mass spectrum of benzene, in which metastable ions are observed at *m/z* 76.0, 74.1, 34.7, and 19.5.

**SAQ 6.2a**

All of the metastable ions observed in the mass spectrum of benzene arise from the molecular ion (*m/z* 78). Using the formula, $m^* = m_2^2/m_1$, assign the fragment ions ($M_2^+$). As an example, the fragmentation:

$$C_6H_6^{+\bullet} \longrightarrow C_6H_5^+ + H^\bullet$$

*m/z* 78 *m/z* 77

shows a metastable ion at an *m/z* value of $77^2/78$, i.e. 76.

From the observed metastable ions and the calculations performed in SAQ 6.2a, a fragmentation pattern for the molecular ion of benzene can be established. This is shown in Figure 6.2b.

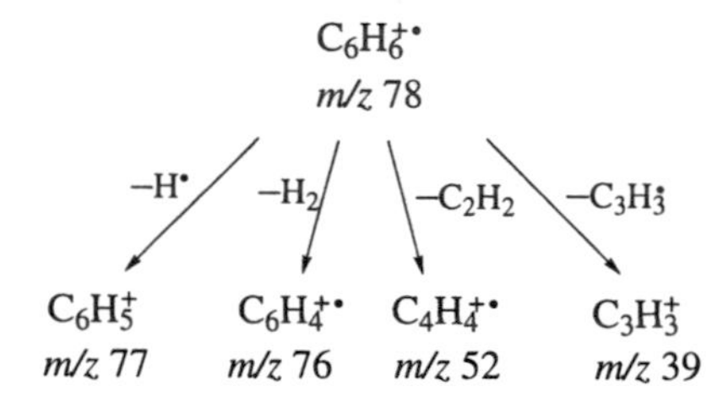


**Figure 6.2b** Partial fragmentation pattern for benzene

**SAQ 6.2b**

Other metastable ions are observed in the benzene spectrum:

| $M^*$ | $M_1^+$ | $\longrightarrow$ | $M_2^+$ |
|---|---|---|---|
| 50.0 | 52 | $\longrightarrow$ | 51 |
| 48.1 | 52 | $\longrightarrow$ | 50 |
| 33.8 | 77 | $\longrightarrow$ | 51 |
| 32.9 | 76 | $\longrightarrow$ | 50 |

Use these metastable ions, and those used in SAQ 6.2a for the $C_6H_6^{+\bullet}$ ion, to draw up a complete fragmentation pattern for benzene.

In the double-focusing instrument, there are two main field-free regions.

Π Where are these regions located?

Between the ion source and electric sector, and between the electric sector and magnetic sector. Metastable ions formed in the first region are not normally detected.

Π Why aren't they detected?

This is because they are deflected into the plates by the electric sector.

It is important to note that evidence of metastable ions provides positive identification of a particular fragmentation route. However, lack of metastables does not necessarily mean that a particular fragmentation pathway does not occur, merely that the machine isn't sensitive enough to pick up the peaks or that the internal energy of the ions isn't appropriate for metastables to be able to form.

**SAQ 6.2c**

Figure 6.1a is actually the mass spectrum of aniline, $C_6H_5NH_2$. The two metastable ions shown at *m/z* 46.8 and 45.9 correspond to fragmentations of the molecular ion ($M^{+\bullet}$) and the fragment ion $(M-H)^+$, respectively. What are the daughter ions formed in these two fragmentation processes?

Can you suggest a formula for the neutral species that is lost?

## 6.3 MODES OF FRAGMENTATION

### 6.3.1 Energy Factors

You may recall that in electron-impact ionisation the molecular ions that are formed have up to *ca.* 60 eV extra (internal) energy, for a typical 70 eV electron beam. Covalent bond energies lie typically in the range 10–12 eV, so it is possible for the $M^{+\bullet}$ species to fragment several times in succession until the remaining positive ion no longer has sufficient excess energy to break a bond. We will now consider how such *cascade* processes (as shown in Figure 6.3a) can occur.

Another condition that must be met if the ion is to fragment concerns the excess vibrational energy given to the ion by the ionising electron. This energy must be moved *rapidly* about the molecule and concentrated in the bond, which breaks. Therefore, the vibrating atoms must move apart beyond a bonding distance, with the vibrations having a frequency of $10^{10}$–$10^{12}$ Hz at 0 to 200°C.

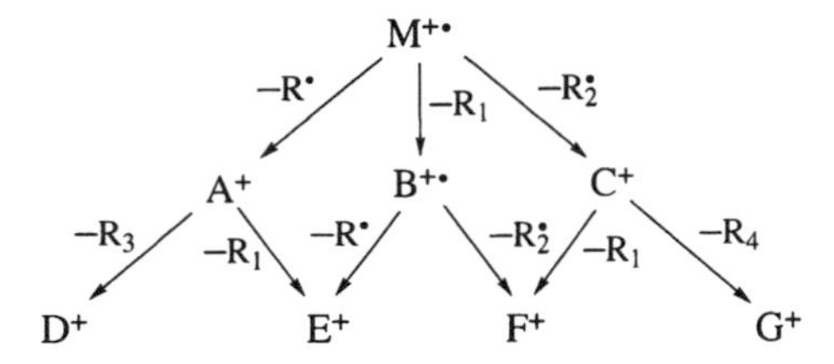


**Figure 6.3a** Illustration of a typical cascade process

Π If ions have a lifetime of, say, $10^{-5}$ s in the mass spectrometer source, and vibrations occur at $10^{11}$ Hz, what is the maximum number of vibrations that can take place in the molecule before the molecular ion fragments?

The answer is around one million vibrations.

The initial excess energy is distributed rapidly around the various bonds of the molecular ion, and in theory any of the bonds might break. The theory behind the behaviour of molecules after ionisation was developed in the 1950s by Rosenstock and coworkers, and is called the *Quasi-Equilibrium Theory* (QET).

Π Do all bonds in a molecular-ion fragment in practice?

From an observation of some of the spectra given previously, you are undoubtedly aware that the answer to this question is 'no'. Some bonds break easily to produce fragment ions, while others do not.

We will now consider the reasons behind this phenomenon.

You might assume that there simply is not enough energy in the electron beam for some molecular ions to break up. However, only 10–20 eV is needed to break most organic bonds, so 70 eV should be plenty, with the QET predicting that the necessary energy can be rapidly moved about the molecule through the vibrations. Therefore, this is obviously not the answer.

It turns out that if we increase the electron beam energy beyond 70 eV, there isn't a great change in the spectra, apart from the fact that we can form $M^{2+}$ ions. This further backs up the QET predictions.

A second explanation might be that as soon as the redistribution of energy causes the bond energy of a weak bond in the molecule to be exceeded, then that bond will break. The QET predicts that the redistribution of smaller amounts of energy will occur more rapidly, and therefore when the weaker bonds break, the overall energy of the system is reduced and there may be insufficient vibrational energy left for any stronger bonds to be broken.

This appears to be the case. Atoms such as Cl, Br, or I are easily lost in mass spectra, and their C—X bonds (where X = Cl, Br, or I) are weaker than C—C, C—H and C—N bonds, and much weaker than C=N and C=O bonds.

It can be generally said that:

*fragments (and daughter fragments) resulting from the breaking of weak bonds are prominent in mass spectra.*

### 6.3.2 Stabilisation Factors

In some cases, unexpected fragmentations can arise, e.g. in the spectrum of aniline (see Figure 6.1a), the weakest bond is $C—NH_2$, yet the $(M–NH_2)^+$ ion, *m/z* 77, formed by the loss of $H_2N^•$, is very weak. Instead, the $M^{+•}$ fragments by loss of HCN to give *m/z* 66, as is confirmed by a metastable ion at *m/z* 46.8. It seems that other factors need to be considered.

Π What do you think these other factors might be?

The answer lies in the formation of a more stable positive ion, or a stable neutral species, or indeed both. Stable neutral species are usually small molecules with multiple bonds such as HC≡N, HC≡CH, $H_2C=CH_2$, C=O, O=C=O, S=O, O=S=O, and $CH_2=C=O$, or less commonly, small singly bonded molecules such as $H_2O$, $H_2S$, $CH_3OH$, HCl, HBr, and HI. The multiple bonds in particular can remove large amounts of energy from the parent ions if they are formed in a high-vibrational state.

Positive ions can be stabilised in two ways:

(a) *Inductive Effect*

First, if the neighbouring groups are electron-donating, they can try and counteract the positive charge by partially neutralising it. This is termed the *inductive effect.* Atoms or groups which donate electron density are called **+I** groups for short (**−I** groups withdraw electron density). The most common **+I** group is the alkyl group. This helps to explain the order of stability of carbocations ($C^+$ species) which, in order of decreasing stability, is as follows:

$$(CH_3)_3C^+ > (CH_3)_2CH^+ > CH_3CH_2^+ \gg CH_3^+$$

The arrows on the bonds show the direction of the **I**-effect. Clearly, the ion with three alkyl groups, will be more stable than the ion which has none. The ions, *m/z* 57 ($(CH_3)_3C^+$) and *m/z* 43 ($(CH_3)_2CH^+$), are usually intense in the spectra of compounds containing these end-groups, while those at *m/z* 29 ($CH_3CH_2^+$) and *m/z* 15 ($CH_3^+$) are weak.

There are two useful rules to remember about **I**-effects.

First, atoms which lie in groups to the left of carbon in the Periodic Table (or compounds consisting largely of such atoms), are **+I**, while those that lie to the right of carbon are **−I**. The **+I** groups *stabilise* positive ions, whereas the **−I** groups *destabilise* positive ions.

Secondly, **±I** effects only operate at the most over two bonds. The effect falls off rapidly with distance along a carbon chain.

(b) *Mesomeric Effect*

The second way that a neighbouring group can help to stabilise a positive centre is by conjugation with the latter through multiple bonds. Conjugation exists when a compound possesses alternate single and multiple bonds. Stabilisation is achieved by a spreading (also called *delocalisation*) of the positive charge about the ion. Similar molecular species that have their charge spread on different atoms are called *resonance hybrids*. This delocalisation via conjugation and resonance is known as the *mesomeric* effect (±**M**) and is far more powerful than the inductive effect. A **+M** group will stabilise a positively charged system by spreading the positive charge about that ion.

Tables 6.3a and 6.3b show a range of organic groups classified according to their **I** and **M** behaviour. It is not vital that you remember all of these groups — they are given for reference when needed. Note that a particular organic group does not have to be both **+I** and **+M** (or −**I** and −**M**).

Π Take the case of a positively charged ion which contains an organic group that is both **+M** and −**I**. How stable would this ion be?

Well fortunately, in the vast majority of cases, the **+M** effect is stronger than the −**I** effect, so the positive ion would be stabilised. The reason behind the **+M**-effect's strength is that it operates through the more weakly held π-electrons in the double (or triple) bonds of the ion. These are more readily distorted and the effect is transmitted over any number of conjugated bonds, unlike the **I**-effect which is short range.

**Table 6.3a** Inductive effects of various organic groups

| −**I** groups[a] | | | +**I** groups[b] |
|---|---|---|---|
| $-NO_2$ | $-CHO$ | $-RC{=}CR_2$ | $-CH_3$ |
| $-C{\equiv}N$ | $-COR$ | $-C{\equiv}CH$ | $-CH_2R$ |
| $-COOH$ | $-F$ | $-C{\equiv}CR$ | $-CHR_2$ |
| $-COOR$ | $-Cl$ | $-SO_2OH$ | $-CR_3$ |
| $-OH$ | $-Br$ | $-SH$ | |
| $-OR$ | $-I$ | $-SR$ | |
| $C_6H_5$ | $-CH{=}CH_2$ | $-NH_2$ | |

[a]−**I** groups are better electron attractors than H—
[b]+**I** groups are poorer electron attractors than H—

**Table 6.3b** Mesomeric effects of various organic groups

| **+M, −I** groups | | **−M, −I** groups[a] | | **+M, +I** groups[b] |
|---|---|---|---|---|
| −F | −SH | $-NO_2$ | $-CONH_2$ | $CH_3$ |
| −Cl | −SR | −C≡N | $-SO_2R$ | $-CH_2R$ |
| −Br | $-NH_2$ | −CHO | $-CF_3$ | $-CHR_2$ |
| −I | −NHR | −COR | $-CCl_3$ | $-CR_3$ |
| −OH | $-NR_2$ | −COOH | | |
| −OR | −NHCOR | −COOR | | |
| −OCOR | $-C_6H_5$ | | | |
| $-CH{=}CH_2$ | $-CH{=}CR_2$ | | | |

[a]**−M** and **−I** groups withdraw electron density from conjugated systems
[b]**+M** and **+I** groups supply electron density to conjugated systems

The following section presents a number of examples of these stabilisation effects 'in action'.

### *(i) Benzene and its molecular ion, $C_6H_6^{+\cdot}$*

Benzene has two Kekulé resonance forms, namely:

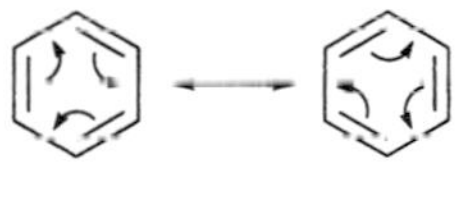

The *double-headed curly arrows* indicate movement of two π-electrons in the direction of the arrows. Both forms are, of course, identical to each other. The molecular ion is formed by loss of a π-electron to leave one carbon atom with a positive charge and one carbon with an unpaired electron. This gives the radical cation, compound A:

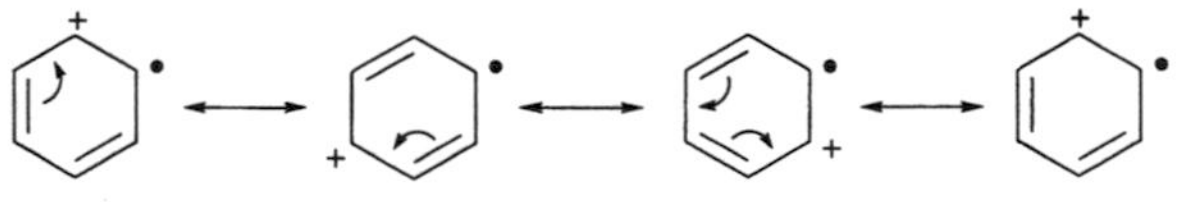

A B C A

The resonance hybrids, i.e. compounds B and C, can be interconverted with the cation A by movement of the electrons.

However, we could have placed the positive charge on the other carbon atom of the ionised double bond, thus giving compound D:

D E F D

This has two resonance hybrids, i.e. compounds E and F.

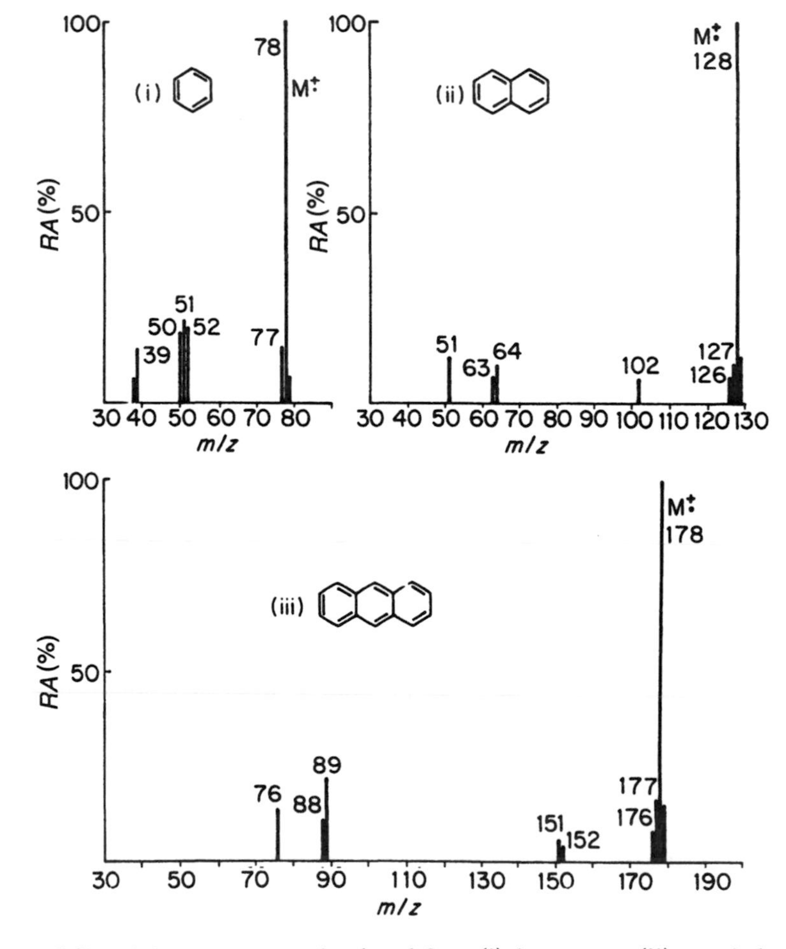


**Figure 6.3b** Mass spectra obtained for: (i) benzene; (ii) naphthalene; (iii) anthracene

Therefore, there are many possible resonance hybrids, with each containing:

$$\overset{|}{\mathrm{HC}} - \overset{|}{\mathrm{CH}}\ \text{units}$$

When these ions fragment, they tend to lose HC≡CH (*m*/*z* 26) units (see Figure 6.3b).

**SAQ 6.3a** Figure 6.3b shows the mass spectra of (i) benzene, (ii) naphthalene, and (iii) anthracene. In each case, which ions are formed from the loss of HC≡CH?

Incidentally, there is an alternative method for displaying the interconversion of the (benzene)$^{+\bullet}$ resonance forms, using single-electron shifts. These are represented by *singly headed curly arrows*, sometimes called the 'fish-hook' notation, in which the direction of the arrow refers to the direction of electron movement. This type of electron rearrangement is used regularly in mass spectrometry.

A G H

This shows that the positive charge (+) and the free electron (.) can be positioned on the benzene ring in any combination.

**SAQ 6.3b** Compare the resonance hybrids, A to H, given above. Which hybrids are equivalent?

*(ii) Methoxybenzene (anisole), $C_6H_5OCH_3$*

This could be ionised on the benzene ring, or on the oxygen, but as we shall see, this is not important. We start with an electron being lost from the π-system, giving:

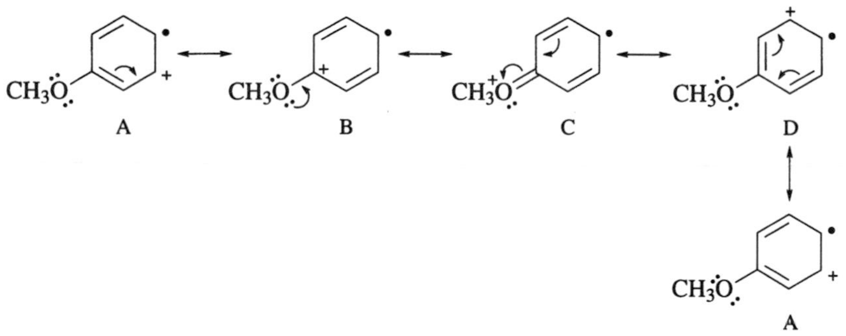

CH₃O A ⟷ CH₃O B ⟷ CH₃O C ⟷ CH₃O D ⟷ CH₃O A

Ions A and B are similar to those formed from benzene in the last example, but C is a key ion. It is formed by the use of one of the lone pair of p-electrons on the oxygen atom. This is how the oxygen atom exerts its **+M** effect to help stabilise the methoxybenzene molecular ion.

If the molecular ion is formed by the loss of one of the relatively easily ionised oxygen p-electrons, then the following structures are possible:

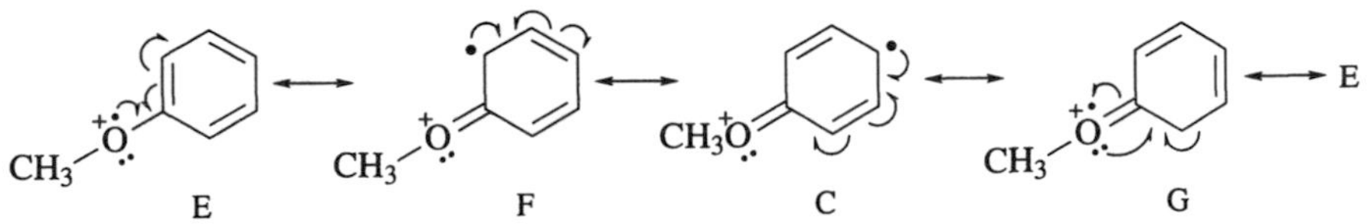

Ion C is formed in this sequence too! Therefore, it doesn't matter where the initial loss of the electron occurs. All of the possible forms of $M^{+\bullet}$ are interconverted by resonance. This result is quite general for the **+M** groups shown above in Table 6.3b.

**SAQ 6.3c**

Show how the $M^{+\bullet}$ ion of phenyl ethanoate:

$$C_6H_5O\overset{\overset{\displaystyle O}{\|}}{C}CH_3$$

might be stabilised.

**SAQ 6.3d**

Would you expect the $M^{+\bullet}$ ion of benzyl ethanoate:

$$C_6H_5CH_2OC(=O)CH_3$$

to be as stable as the $M^{+\bullet}$ ion of phenyl ethanoate?

*(iii) 1-Phenylethanone (acetophenone), $C_6H_5COCH_3$*

This compound could be ionised on the benzene ring, or on the carbonyl group. Let us start with the benzene ring:

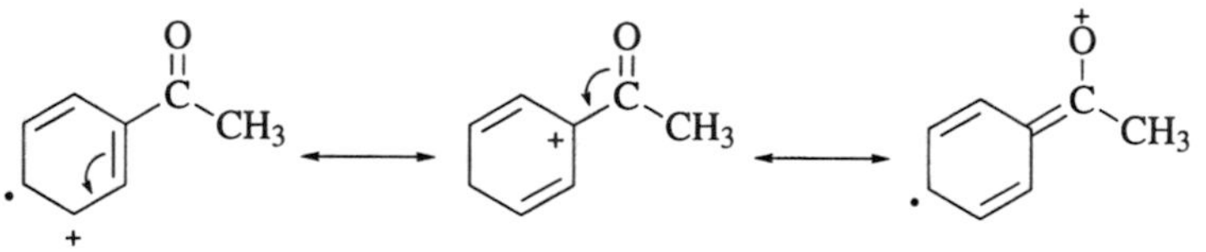

A ⟷ B ⟷ C

It would appear that we can form C by using the π-electrons of the carbonyl group, which *are conjugated* with the benzene ring. Therefore, A would be more stabilised. However, isn't oxygen more electron-attracting than carbon? It would be more likely to attract the electrons of the benzene ring itself, as follows:

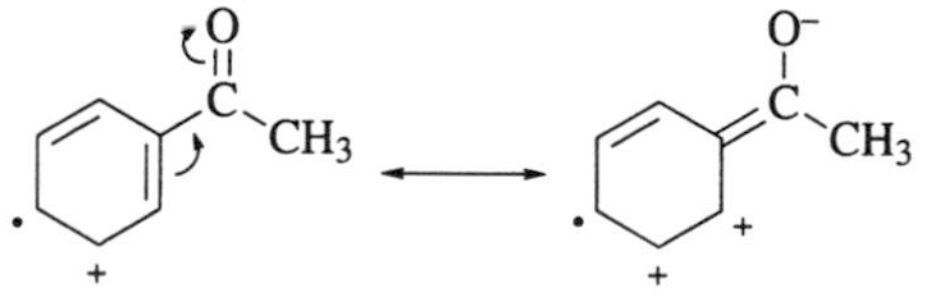

A ⟷ D

In form D, you can see that two positive charges are formed on adjacent carbon atoms, which is *a highly unstable arrangement*. We can conclude that **−M** groups, such as those shown in Table 6.3b, will destabilise molecular ions such as A.

Ionisation in 1-phenylethanone could also take place on the carbonyl-oxygen atom to give form E:

E can be stabilised by the **+M**-effect of the phenyl group as shown, so it is far more likely that the $M^{+\bullet}$ of this type of compound will exist as E-type ions, with the positive charge occurring on the **−M** group.

**SAQ 6.3e**

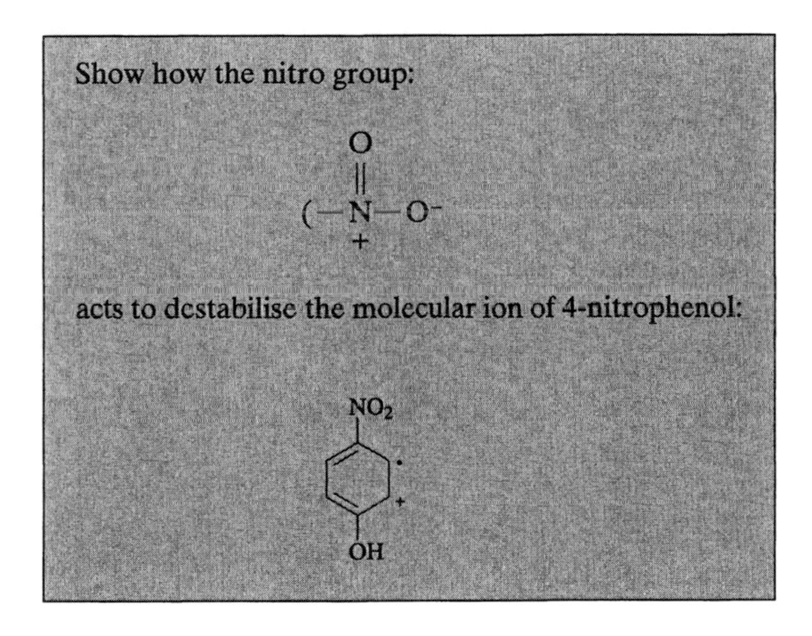

*(iv) The acyl ion,* $R-\overset{+}{C}=O$

The acyl ion itself is $CH_3\overset{+}{C}=O$, *m/z* 43. Such ions are frequently found as intense fragments in the mass spectra of such compounds as aldehydes, ketones and esters. This occurs as a result of the stability of the molecular ions, which are resonance stabilised by the lone pair of electrons on the oxygen atom:

$$R-\overset{+}{C}=\ddot{O}: \longleftrightarrow R-C\equiv O^{+}$$

If R is a **+I** group, this will also help to stabilise the acyl ion.

*(v) The phenyl ion,* $C_6H_5^{+}$ *(m/z 77)*

The phenyl ion is frequently found (as you might expect) in the mass spectra of phenyl compounds and is usefully diagnostic, since few other compounds produce ions of *m/z* 77.

Π Can you write resonance structures for the phenyl cation?

You probably wrote:

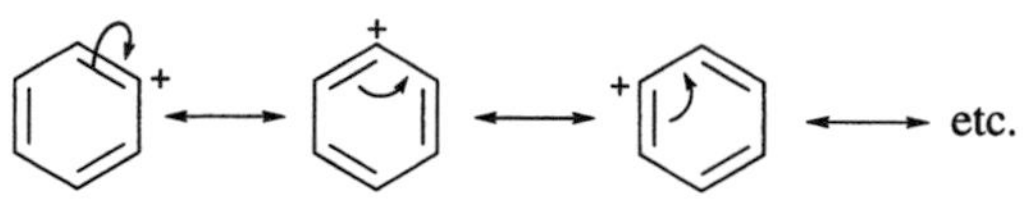

etc.

which shows that the positive charge is delocalised around the benzene ring. However, this is not possible — you cannot delocalise the positive charge in $C_6H_5^{+}$ because the charge *is in the plane of the ring*! Figure 6.3c attempts to illustrate this fact by using phenylethanone as the source of the $C_6H_5^{+}$ ion.

When the bond between the benzene ring and the acyl group breaks, the positive charge occupies the position left by the departing group, as shown in Form B (Figure 6.3c). This is an empty $sp^2$ orbital, shown by the dashed line in Form C. Electrons in the $\pi$-orbitals of the

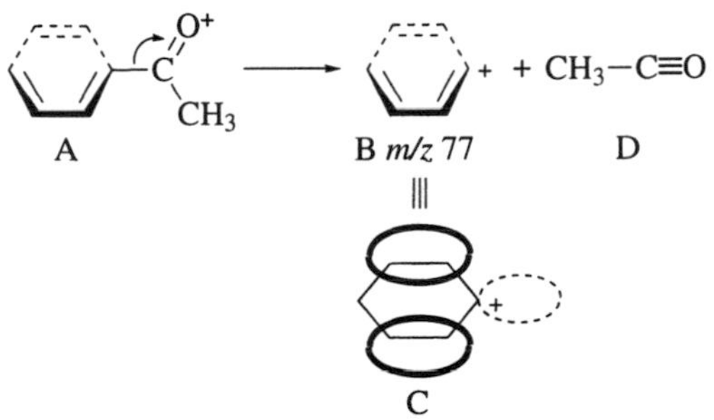

A → B $m/z$ 77 + $CH_3{-}C{\equiv}O$ (D)

B ≡ C

**Figure 6.3c** Illustration of the formation and structure of $C_6H_5^+$

benzene ring (shown by the 'shaded' regions) cannot interact with this orbital, so the positive ion is *not* significantly stabilised. Phenyl cations are still aromatic, and so are reasonably stable when compared to alternative, non-aromatic ions which might be formed in the fragmentation of the $M^{+\bullet}$ ion.

Π Which do you think would be the most intense fragment-ion line in the spectrum of phenylethanone — $C_6H_5^+$ or $CH_3CO^+$?

The spectrum of phenylethanone is shown in Figure 6.3d, and clearly shows that $C_6H_5^+$ is more stable than $CH_3CO^+$, since the relative-abundance values are in the ratio 70 : 30 (approximately). You might have chosen $CH_3CO^+$ since it has two resonance forms to help stabilise it (see above), as opposed to $C_6H_5^+$, which we have seen does not have such stabilising forms. However, the benzene ring is itself highly stabilised.

Figure 6.3d shows that the base peak occurs at $m/z$ 105, which is due to the loss of a $CH_3$ group (15 amu). There must be something special about this ion for it to be so intense, which compared to the alternative $CH_3CO^+$ ion formed by loss of $C_6H_5$.

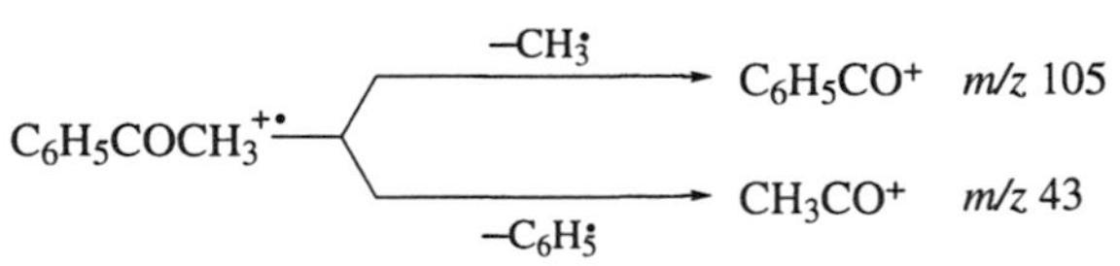

$$C_6H_5COCH_3^{+\bullet} \xrightarrow{-CH_3^{\bullet}} C_6H_5CO^+ \quad m/z\ 105$$

$$C_6H_5COCH_3^{+\bullet} \xrightarrow{-C_6H_5^{\bullet}} CH_3CO^+ \quad m/z\ 43$$

In the next example, we will consider why $CH_3CO^+$ is so preferred in the fragmentation of phenylethanone.

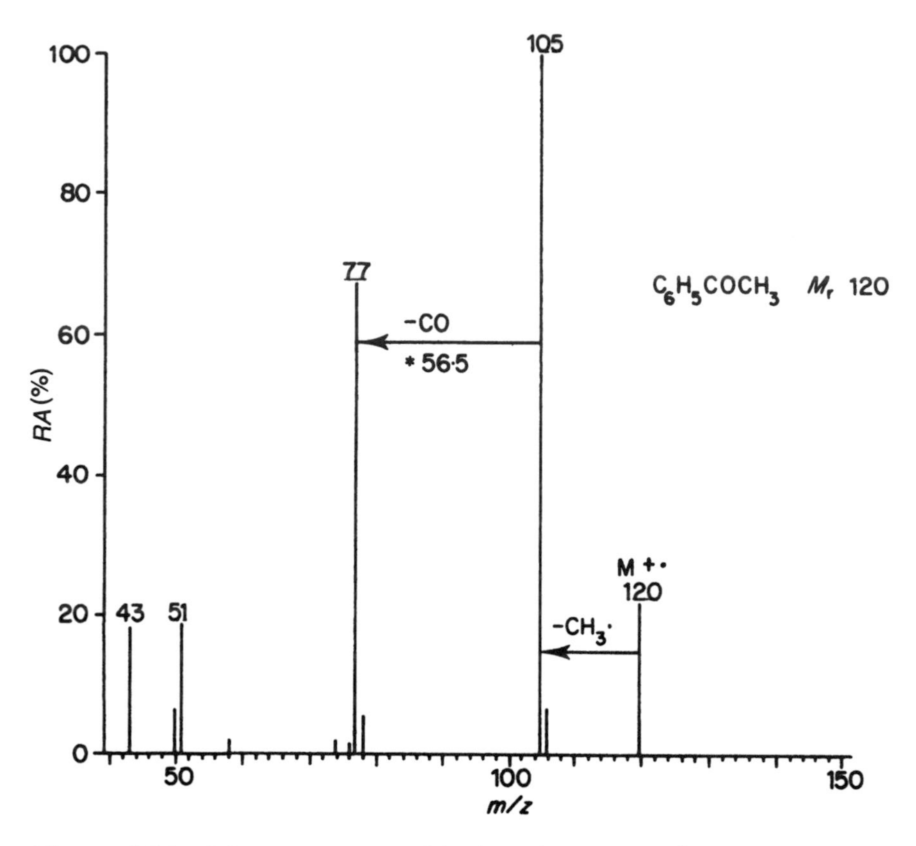


**Figure 6.3d** Mass spectrum of 1-phenylethanone (acetophenone)

*(vi) The benzoyl ion,* $C_6H_5\overset{+}{C}O$ (m/z *105)*

Π Can you draw the resonance hybrids of the benzoyl ion.

This ion, like any carbonyl ion, could have the two structures A and A′, shown in Figure 6.3e, which are then delocalised by using the π-electrons of the benzene ring to give forms B, C, D, and E.

Therefore, $C_6H_5\overset{+}{C}O$ has six resonance forms to stabilise it. Here, the phenyl group is acting as a **+M** group to stabilise the benzoyl ion, and this explains why this is the base peak in the $C_6H_5COCH_3$ mass spectrum. An interesting feature of the spectrum is a metastable (m*) ion at *m/z* 56.5 (not shown in Figure 6.3d as it is too small), which indicates that the benzoyl ion decomposes by loss of CO to give $C_6H_5^{+}$.

$$C_6H_5-\overset{+}{C}O \longrightarrow C_6H_5^+ + CO, \ m^* = \frac{77^2}{105} = 56.5$$



**Figure 6.3e** Resonance structures of the $C_6H_5^+$ ion

This route to *m/z* 77 is not perhaps what you would have expected from Figure 6.3c, which implies direct cleavage of $M^{+\bullet}$ to give $C_6H_5^+$ and $CH_3CO^\bullet$. Remember, however, from Section 6.2, that we do not necessarily observe a metastable ion for every fragmentation process and some *m/z* 77 species could have formed directly from $M^{+\bullet}$ — there is no evidence that it did not!

This is a good example of how the existence of stabilised ions influences the fragmentation products we see in our mass spectra.

**SAQ 6.3f**

In compounds of the type $RC_6H_4COX$, would the following substituents (i.e. R) stabilise, destabilise, or have little effect on the benzoyl ion that is produced:

(a) 3-methoxy;
(b) 4-methoxy;
(c) 3-cyano;
(d) 4-nitro?

Try to explain your answers by means of resonance forms of the ions concerned.

### 6.3.3 Basic Rules Concerning Resonance Structures

The basic rules for writing down the various resonance structures and qualitatively assessing their relative importance can be summarised as follows:

(1) Resonance structures are interconvertible by one or a series of short electron shifts; electrons do not *jump* across space.

(2) Resonating structures must have the same number of electrons.

(3) Resonance structures involving even numbers of electrons are more stable than those involving odd numbers of electrons.

(4) The more bonds that are involved in the resonance, then the greater is the stability of the resulting structure, e.g. benzene and naphthalene, which have resonance energies of 144 and 244 kJ $mol^{-1}$, respectively.

(5) The closer the stabilities of the resonance structures, then the

greater is the degree of resonance and the lower the energy of the system, e.g. the Kekulé forms of benzene, which are equivalent.

(6) Resonance can only occur between structures that correspond to very nearly the same relative positions of *all* of the atoms involved. Bond distances and angles should remain the same and also be compatible with the overlap of the orbitals being used, e.g. resonance cannot occur between different structural *isomers*.

**SAQ 6.3g**

(This following optional question is intended for those who feel they need practice in predicting ion stabilisation.)

Show how the following ions are stabilised:

(a) $C_6H_5\overset{+}{C}H_2$; (b) $CH_3CH{=}CH{-}CH{-}\overset{+}{C}O$; (c) $CH_3\ddot{S}{-}CH{=}CH{-}\overset{+}{C}H_2$;

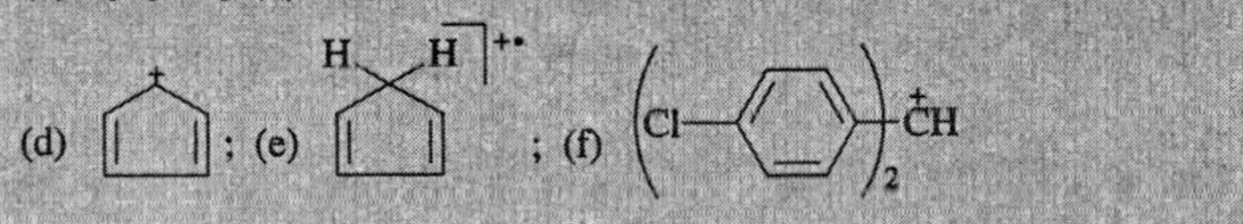

(d) [cyclohexadienyl cation]; (e) [cyclohexadiene radical cation, H H, +•]; (f) $(Cl{-}C_6H_4)_2\overset{+}{C}H$

### 6.3.4 Mechanisms of Bond Cleavages in Mass Spectrometry

(a) *Heterolytic α-Cleavage*

In Section 6.3.2(v), we showed how the phenyl cation might have been formed from $C_6H_5COCH_3^{+\bullet}$ (Figure 6.3c). This involved breaking the phenyl to carbonyl bond by placing both the electrons on the carbonyl group:

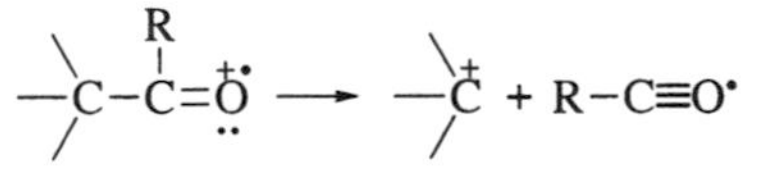

$$\geq C{-}\underset{}{\overset{R}{C}}{=}\overset{+\bullet}{\ddot{O}} \longrightarrow \geq\overset{+}{C} + R{-}C{\equiv}O^{\bullet} \qquad (6.9)$$

You may have recognised this as an example of a *heterolytic cleavage* of a covalent bond:

$$X{-}Y \longrightarrow \overset{+}{X} + Y^{-}$$

where Y is more electron-loving (electronegative) than X. In such a cleavage, one of the fragments (normally the more electronegative) gains *both* of the bonding electrons. The opposite process:

$$\text{Y}\!-\!\text{X} \longrightarrow \overset{+}{\text{Y}} + \text{X}^-$$

is theoretically possible but is much less likely if X is less electronegative than Y. In the example shown in equation (6.9), the result would be:

$$\text{-}\overset{|}{\underset{|}{\text{C}}}\text{-}\underset{\text{H}}{\text{C}}\text{=}\overset{+\bullet}{\ddot{\text{O}}} \longrightarrow \text{-}\overset{|}{\underset{|}{\text{C}}}^{-\bullet} + \text{-}\overset{+}{\text{C}}\text{=}\overset{+\bullet}{\ddot{\text{O}}} \qquad (6.10)$$

The doubly charged carbonyl species would be highly unstable. In mass spectrometry, only heterolytic cleavages corresponding to equation (6.9) are observed as these avoid creating adjacent positive charges.

Equation (6.9) shows the decomposition of a molecular ion, i.e. an odd-electron species. Daughter ions of even-electron composition may also fragment in this way. For example, the benzoyl ion in the spectrum of 1-phenylethanone (Figure 6.3d) loses CO as already described. The mechanism is as follows:

$$C_6H_5\!-\!\overset{+}{C}\!=\!O \longrightarrow C_6H_5^+ + CO$$

This produces two even-electron fragments. Such heterolytic cleavages producing fragments in which *all* of the electrons are paired are energetically favoured and are correspondingly common in mass spectra.

The heterolytic cleavage (Equation (6.9)) can be generalised to the situation —C—Z, where Z is a −**I** group, a −**I** and −**M** group, or even a −**I** and **+M** group (see Section 6.3.2), and where ionisation has occurred on Z:

$$\text{-}\overset{|}{\underset{|}{\text{C}}}\text{-}\text{Z}^{+\bullet} \longrightarrow \text{-}\overset{|}{\underset{|}{\overset{+}{\text{C}}}} + \text{Z}^\bullet$$

Figure 6.3d gave one example of the vast range of compounds whose $M^{+\bullet}$ might fragment in this way. Cleavage of the bond next to the atom or group carrying the positive charge is often called α-cleavage.

(b) *Heterolytic β-Cleavage*

Π Can you explain what is meant by the term β-cleavage and draw the resulting fragments?

This is where the next bond along the carbon chain from the α-bond breaks:

$$R-\overset{|}{\underset{|}{C}}-\overset{+\bullet}{Z} \longrightarrow \overset{+}{R} + >C=Z^{\bullet}$$

β-cleavage is usually less prevalent than α-cleavage, but nevertheless does occur to a minor extent. It is promoted if $R^+$ or $C=Z^{\bullet}$ are particularly stable, e.g. if $R^+$ is $C_6H_5CH_2^+$, which is stabilised by resonance, as shown by the following example:

$$C_6H_5-CH_2-CH_2-Br^{+\bullet} \xrightarrow{\beta} C_6H_5-\overset{+}{C}H_2 + H_2C{=}Br^{\bullet}$$

*m*/*z* 91 (100%)

$$C_6H_5-CH_2-CH_2-Br^{+\bullet} \xrightarrow{\beta} C_6H_5-CH_2\overset{+}{C}H_2 + Br^{\bullet}$$

*m*/*z* 105 (50%)

In this case, the *m*/*z* 105 ion which is formed by α-cleavage is *not* resonance stabilised and is only half as intense as the *m*/*z* 91 ion.

(c) *Homolytic α-Cleavage*

In this mechanism, when the bond stretches and breaks, one of the two electrons forming the bond is returned to X, thus forming an $X^{\bullet}$ radical, and the other to Y to give a $Y^{\bullet}$ radical:

$$X \div Y \longrightarrow X^{\bullet} + Y^{\bullet}$$

Applying the idea to the molecular ion, we have:

$$X \div Y^{+\bullet} \longrightarrow X^{\bullet} + :\overset{+}{Y}$$

In this case, X is released as a radical, but Y carries off the positive charge with the two electrons paired up, i.e. Y is an *even-electron species.*

Of course, in some cases $:X^+$ may be stable, so we could observe the opposite result, as follows:

$$Y \div X^{+\bullet} \longrightarrow Y^{\bullet} + :\overset{+}{X}$$

When this happens, the mass spectrum will contain both $:X^+$ and $:Y^+$ ions. These are said to be *complementary* ion pairs and are very useful diagnostically. One ion is very likely to be more stable than the other and thus their relative abundances may vary enormously. Returning to the example of 1-phenylethanone again (Figure 6.3d), *m/z* 43 and 77 are complementary ion pairs formed by homolytic α-cleavage:

$$\overset{+\bullet}{C_6H_5}-\overset{O}{\overset{||}{C}}-CH_3 \longrightarrow \underset{m/z\ 77}{C_6H_5^+} + CH_3\dot{C}{=}O$$

$$\underset{m/z\ 120}{C_6H_5-\overset{\overset{+\bullet}{O}}{\overset{||}{C}}-CH_3} \longrightarrow \underset{m/z\ 43}{CH_3C{\equiv}\overset{+}{O}} + C_6H_5^{\bullet}$$

We have already noted that $C_6H_5^+$ can also be formed from *m/z* 105 ($C_6H_5CO^+$) by heterolytic α-cleavage, but, in the case of 1-phenylethanone, $C_6H_5CO^+$ can be formed from $M^{+\bullet}$ by *homolytic* α-cleavage:

$$\underset{m/z\ 120}{C_6H_5-\overset{\overset{+\bullet}{O}}{\overset{|}{C}}-CH_3} \longrightarrow C_6H_5-C{\equiv}O^+ \quad CH_3^{\bullet}$$

$$\updownarrow$$

$$\underset{m/z\ 105}{C_6H_5\overset{+}{C}{=}O} \longleftrightarrow \text{etc.}$$

The formation of acyl ($RCO^+$) and aroyl ($ArCO^+$) ions from molecular ions by homolytic α-cleavage is a very important mechanism of fragmentation of carbonyl compounds.

(d) *Homolytic β-Cleavage*

Homolytic β-cleavage is also possible, although it is much less common:

$$R{-}CH_2{-}\underset{\substack{|\\R'}}{C}{=}\ddot{O}^{+\bullet} \longrightarrow R^\bullet + CH_2{=}\underset{\substack{|\\R'}}{C}{-}\ddot{O}^+$$

The $(M{-}R)^+$ species produced here is not very stable. However, the same process occurring in an acylium ion will produce ketene ions, i.e. $CH_2{=}C{=}C{:}^{+\bullet}$, as follows:

$$R{-}CH_2{-}C{\equiv}O^+ \longrightarrow R^\bullet + \underset{m/z\ 42}{CH_2{=}C{=}\ddot{O}^{+\bullet}}$$

These are more stable and are occasionally observed in mass spectra.

### 6.3.5 The Even-electron Rule

The mass spectra of carbon-, hydrogen-, and oxygen-containing compounds tend to be dominated by fragment ions of *odd* mass. This is because of the *Even-electron Rule*, which follows from the bond cleavages we have just described.

This rule states that *odd-electron* ions ($M^{+\bullet}$ is such an ion) decompose by loss of radicals or *even-electron* molecules, but *even-electron* ions may fragment only by loss of neutral molecules, and not of radicals, as follows:

$$A^{+\bullet} \longrightarrow C^+ + N_1^\bullet \qquad (6.11a)$$

$$A^{+\bullet} \longrightarrow D^{+\bullet} + N_2 \qquad (6.11b)$$

$$B^+ \longrightarrow E^+ + N_3 \quad (6.11c)$$

$$B^+ \not\longrightarrow F^{+\bullet} + N_4^\bullet \quad (6.11d)$$

In this scheme, $A^{+\bullet}$ may be a molecular ion or the daughter ion formed from it by the loss of a neutral molecule, $N_1^\bullet$ and $N_4^\bullet$ are neutral radicals, and $N_2$ and $N_3$ are neutral molecules.

Processes such as that shown in equation (6.11d) do sometimes occur in practice, but they are not energetically favoured processes because an even-electron ion is more stable than an odd-electron one, and so there is no overall reduction in energy when such a process occurs.

Radicals lost from carbon-, hydrogen-, and oxygen-containing compounds are always of odd masses, so since the relative molecular mass of such compounds is even, the positive ion produced must have an odd mass too. This applies for both heterolytic and homolytic cleavages and explains why the mass spectra of such compounds have so many odd-mass ions.

For example, in the case of 1-phenylethanone, we find the following occurs:

$$\underset{m/z\ 120}{C_6H_5COCH_3^{+\bullet}} \xrightarrow[-15\ \text{amu}]{-CH_3^\bullet} \underset{m/z\ 105}{C_6H_5C\overset{+}{O}} \xrightarrow[-28\ \text{amu}]{-CO} \underset{m/z\ 77}{C_6H_5^+} \xrightarrow[-26\ \text{amu}]{-C_2H_2} \underset{m/z\ 51}{C_4H_3^+}$$

Once the even-electron *m/z* 105 has been formed it fragments by successive losses of even-mass molecules of CO and HC≡CH, so that the positive ions are all odd-mass species apart from the $M^{+\bullet}$ itself.

Π Would you expect the typical positive ions in the mass spectrum of a C-, H-, O-, and N-containing compound to be of odd or even mass when the number of N atoms is (a) odd, and (b) even?

When there are an *odd* number of N atoms in the molecule, the relative molecular mass will be odd (the nitrogen rule). Hence, the loss of an odd-mass radical from $M^{+\bullet}$ would lead to an *even*-mass fragment by the even-electron rule, and subsequent losses of *even-*

mass molecules from this would give further *even*-mass fragments, i.e. $M^{+\bullet}$ odd-mass fragments even-mass.

When there are an *even* number of N atoms in the molecule, the relative molecular mass will be even (the nitrogen rule), so the even-electron rule predicts fragments of *odd* masses as in C-, H-, and O-containing compounds.

**SAQ 6.3h**

How would you expect the molecular ion of 1-bromo-1-phenylethane, $C_6H_5CHBrCH_3$, to fragment? Which of the ions formed by heterolytic and homolytic cleavages (α- and β-) would be stabilised by **+M**- and/or **+I**-effects?

**SAQ 6.3i**

How would you expect the molecular ions of propan-2-ol, $CH_3CHOHCH_3$, to fragment? Which of the ions formed by heterolytic and homolytic cleavages (α- and β-) would be stabilised by **+M**- and/or **+I**-effects?

## Summary

You should now be aware of the origin of metastable ions and understand how they are formed. You have also seen how such ions can be used in elucidating fragmentation pathways by using a simple equation. Molecules fragment when the vibrational energy, concentrated in a particular bond, exceeds the bond energy. The Quasi-Equilibrium Theory assumes that the transfer of vibrational energy through the ion is rapid, relative to the rates of the subsequent fragmentations, and therefore any bond may break. In practice, the weaker bonds break first and the process of fragmentation continues until the last ion no longer possesses sufficient energy to break a bond. The mesomeric (**M**) and inductive (**I**) effects, which have a bearing on the stability of the positive ions that are formed, have been discussed. The relative abundances of some aliphatic and aromatic ions have also been outlined, along with the reason why the $C_6H_5^{+}$ ion is not as stable as might at first be expected.

The four main ways of cleaving covalent bonds have been described, namely α- and β-heterolytic cleavages, and α- and β-homolytic cleavages. The α-cleavage mechanism is the process most commonly observed in mass spectrometry.

The even-electron rule was introduced as a useful generalisation which follows from the modes of cleavage.

## Objectives

When you have finished this chapter, you should be able to:

- explain the origin of metastable ions;
- explain where these ions are formed;
- derive the equation, $m^* = m_2^2/m_1$;
- use this equation to assign metastable ions to parent and daughter ions;
- use metastable-ion data to give information about fragmentation patterns;
- appreciate that molecular ions are unstable, higher-energy species, which fragment by breaking bonds until the last daughter ion no longer has sufficient excess energy to cleave again;
- explain qualitatively what happens to the energy given to the ionised molecules before fragmentation occurs;

- predict which bonds in a molecular ion, based on their relative bond strengths, are most likely to break, i.e. the weakest break most readily;

- recognise and give examples of molecules which show that bond strength is not the only factor governing fragmentation, and that ions are often formed due to their enhanced stability and/or because of the expulsion of small, stable multiply bonded molecules;

- define what is meant by a **±I** and a **±M** group and give examples of each;

- appreciate that **±I**, **±M**, and **+M** and **−I** groups stabilise positive ions by donating electron density to them, and thereby enhance the stability of such ions in mass spectrometry;

- recognise and give examples of positive ions stabilised and destabilised by some of the common functional groups found in organic molecules;

- recognise the correct uses of $\curvearrowright$ and $\curvearrowright$ symbols to show the movement of electrons through the bonds of an ion, both to (de)stabilise or to cleave a bond:

- (optional) demonstrate how **±M** effects work by the use of $\curvearrowright$ and $\curvearrowright$ in order to show the movement of electrons through the double bonds of an ion;

- predict which ions from a simple compound would be expected to give intense peaks in the mass spectrum on the basis of their stabilisation by **±M** and **−I** effects, as well as ease of bond cleavage based on energetic grounds;

- recognise situations where no clear-cut predictions are possible because all of the expected fragmentations lead to ions of similar stabilities;

- describe the four main modes of bond cleavage which occur in mass spectrometry, namely:

(a) heterolytic α-cleavage;
(b) heterolytic β-cleavage;
(c) homolytic α-cleavage;
(d) homolytic β-cleavage;

and appreciate that (a) and (c) are the most common;

- recognise the results of the four types of bond cleavage in the mass spectra of simple compounds;
- state the even-electron rule, and use it to decide the most likely sequence of ions in a cascade process from both even- and odd-mass molecular ions.

# 7. Typical Fragmentation Patterns of Common Functional Groups

In this chapter, we shall consider the mass-spectral fragmentations of some selected functional groups commonly met with in organic chemistry. These are compounds such as alcohols, ethers, phenols, carbonyl compounds, amines, hydrocarbons, halo-compounds, nitro-compounds, heterocycles and sulphur derivatives. Finally, we shall look briefly at some molecules found in biology.

It is advised that you to study Section 7.1 carefully, as this presents a general approach to identifying unknown compounds from their mass spectra by using correlation tables. Next, go to Section 7.2, as this not only covers the fragmentation of alcohols in some detail, but also introduces you to a number of basic concepts which are common to the fragmentations of many other organic compounds. Therefore, it is important to study this section even if you are interested in other functional groups. After this, you can then proceed to your own particular functional group(s) of interest.

Section 7.5.1 describes a common rearrangement process which occurs particularly in carbonyl compounds; this is well worth studying because the ions resulting from such a process are diagnostically very useful in interpreting mass spectra.

## 7.1 INTERPRETATION PROCEDURES

Before we deal with the features associated with common functional groups, we will present a procedure for interpreting a mass spectrum, and introduce two types of correlation tables which you can use. When you become experienced at interpretation you may be able to skip some of the steps and be able to identify the class of compound straight away.

You should bear in mind that correlation tables cannot possibly include all of the types of functional groups and structures which have been investigated. Those structures that you are interested in may not be quite so common. In this case, we suggest that you do a literature search for journal articles and review papers in your particular field and then construct a table for yourself.

### 7.1.1 General Procedure for the Interpretation of a Mass Spectrum

(a) Identify the molecular ion, $M^{+\bullet}$.

(b) Identify the typical background peaks which are present in all spectra, e.g. $H_2O$ (18), $N_2$ (28), $O_2$ (32), and $CO_2$ (44), and count the spectrum from low to high mass.

(c) Identify the most intense fragment, i.e. the base peak.

The next three steps, (d), (e), and (f), need only be carried out if your mass spectrum cannot be manipulated by a computer.

(d) Measure the intensities of the ion peaks. Set the most abundant ion as 100 arbitary units, and then measure all other ions $>2\%$ relative to this ion. Tabulate this 'normalised' spectrum.

(e) Identify and tabulate any metastable peaks, and 'half-mass' peaks.

(f) Plot a bar graph of relative abundance against $m/z$ on square graph paper.

(g) Mark abundant odd-electron ions, i.e. those at even mass numbers if the molecule contains even numbers of N atoms, and check that the molecular ion is one of these.

(h) Examine the general appearance of the spectrum for molecular stability, i.e. the number of peaks, and thus labile bonds.

(i) Identify neutral fragments accompanying high-mass formation from a table of (M–X) values (see Table 7.1a) — correlate these with metastable transitions if observed.

(j) Use a mass-composition table (see Table 7.1b) to assign likely formulae to the major fragment ions.

Coupled with information about the origin of the sample compound, this should lead to structures for the most abundant ions and the respective molecular ions. Remember to allow for hydrogen transfers and rearrangements which lead to ions one mass unit greater than expected.

If the results are still not conclusive, you should then check the following:

(k) The counting of the original mass spectrum.

(l) Is $M^{+\bullet}$ the highest mass ion observed? Determine the mass again accurately and assign the molecular formula, and/or run the sample through again, using a chemical-ionisation source to reveal the $MH^+$ ion.

(m) The presence of any peaks which are 14 amu greater or smaller than the supposed molecular ion; these may be due to the presence of minor amounts of homologues of the compound.

Table 7.1a shows correlations between compounds and neutral fragments produced in the following process:

$$M^{+\bullet} \longrightarrow (M\text{–}X)^+ \qquad (7.1)$$

in which X is the neutral fragment.

**Table 7.1a** Table of correlations between compounds and neutral fragments produced in the process, $M^{+\bullet} \longrightarrow (M-X)^+$, where X is the neutral fragment

| X | Compound |
|---|---|
| 1 | Aldehydes, acetals, compounds with aryl $-CH_3$ groups, compounds with $N-CH_3$ groups, compounds with $-CH_2CN$ groups, alkynes |
| 2 | Fused-ring aromatic compounds |
| 15 | Acetals, methyl derivatives, *t*-butyl and *i*-propyl compounds, compounds with aryl $-C_2H_5$ groups, $(CH_3)_3SiO$ derivatives |
| 16 | Aromatic nitro compounds, *N*-oxides, *S*-oxides, aromatic amides |
| 17 | Carboxylic acids, aromatic compounds with a functional group containing oxygen *ortho* to one containing hydrogen, e.g. *o*-nitrotoluene |
| 18 | Straight-chain aldehydes ($\geqslant C_6$), primary straight-chain and aromatic alcohols, alcohol derivatives of cyclic alkanes, steroid alcohols and ketones, aliphatic ethers with one group having eight or more carbons, carboxylic acids |
| 20 | Aliphatic alcohols (thermal products), *n*-fluoroalkanes |
| 26 | Aromatic hydrocarbons |
| 27 | Aromatic amines, nitrogen heterocycles, aromatic nitriles |
| 28 | Diaryl ethers, phenols and naphthols, aldehydes, keto-derivatives of cycloalkanes, quinones and polycyclic ketones, aliphatic nitriles, ethyl esters |
| 29 | Aromatic aldehydes, keto-derivatives of cycloalkanes, phenols and naphthols, poly(hydroxybenzene)s, propionals, diaryl ethers, aliphatic nitriles, ethyl derivatives |
| 30 | Aromatic nitro compounds, aromatic methyl ethers |
| 31 | Methoxy derivatives, methyl esters |
| 32 | *o*-Methylbenzoates |
| 33 | Short-chain unbranched primary alcohols, alcohol derivatives of cycloalkanes, steroid alcohols, secondary and tertiary thiols, thio-derivatives of cycloalkanes, RNCS compounds |

| X | Compound |
|---|---|
| 34 | Thiols |
| 35 | Secondary and tertiary chloroalkanes, chloroaryls |
| 36 | *n*-Chloroalkanes (loss of HCl) |
| 40 | Aliphatic nitriles and dinitriles |
| 41 | Propyl esters |
| 42 | Acetates, *N*-acetyl compounds (loss of $CH_2{=}C{=}O$), butyl ketones |
| 43 | Propyl derivatives, aliphatic nitriles, *t*-amides, methyl ketones, acetyl derivatives |
| 44 | Aliphatic aldehydes containing $\gamma$-hydrogens, esters, anhydrides, acids |
| 45 | Carboxylic acids, ethoxy derivatives, ethyl esters |
| 46 | Aromatic esters with the ethyl group *ortho* to COOH, nitro compounds, long-chain unbranched alcohols, ethyl esters |
| 47 | Alkyl nitro compounds, phosphoryl compounds |
| 48 | Sulphoxides |
| 55 | Butyl esters |
| 56 | Pentyl ketones, ArOBu compounds |
| 57 | Ethyl and butyl ketones, butyl-X compounds |
| 58 | Straight-chain primary mercaptans, RCNO and RNCO compounds, methyl ketones containing $\gamma$-hydrogens |
| 59 | *n*-Propyl esters, methyl esters of 2-hydroxycarboxylic acids, acetyl derivatives |
| 60 | Methyl esters of short chain dibasic carboxylic acids, *o*-methyltoluates, acetates |
| 63 | Methyl esters of short-chain dibasic carboxylic acids |
| 64 | Methyl esters of all aliphatic dibasic carboxylic acids (excluding dimethyl adipate), sulphones, sulphonyl derivatives |
| 73 | *n*-Butyl esters, methyl esters of all aliphatic dibasic acids |
| 77 | Phenyl derivatives |
| 79 (and 81) | Bromo compounds |
| 87 | Pentyl esters |
| 91 | Benzyl and tolyl compounds |
| 93 | Phenoxy derivatives |
| 107 | Benzyloxy derivatives |
| 127 | Iodo compounds |

The above is essentially an arithmetic approach to solving a mass spectrum. It assumes no knowledge of the mechanisms of the fragmentation processes. The extent to which X is lost will depend, of course, on the factors mentioned previously in Chapter 6, and there must also be a reasonable mechanism to account for the bonding changes that have occurred. However, you can often identify a compound of simple structure by using the (M–X) table alone without any of these considerations, and therefore the arithmetic approach is as good as any in compound identification.

Π Correlation tables are often used in compound identification for a number of analytical techniques. Can you name any of these techniques and list some reasons why these tables often fail to find the correct identity?

(a) The table has no entry for that class of compound, as it is an uncommon one. This is particularly the case when mass spectrometry is used as a research tool.

(b) Two or more classes of compounds may show the same X-loss.

(c) The same numerical value of X may arise from more than one molecular formulae, e.g. CO and $C_2H_4$, which are both 28 amu, and $CH_3CO$ and $(CH_3)_2CH$, which are both 43 amu.

(d) The structure concerned might exist in a number of isomeric forms, with each showing the same X-loss, e.g. 2-, 3-, and 4-disubstituted benzene derivatives.

(e) Where a compound has two (or more) functional groups, the expected characteristic X-loss from one group may not be observed, as another more favourable process, i.e. loss of X from another functional group, may have occurred. Thus, signals for intense ions from only one of the substituents are observed.

Therefore, we can conclude that the (M–X) table may give us details about the class of compound we are dealing with, but not necessarily its exact structure.

A mass composition table is the second table that we can use to help us in this task (see Table 7.1b). This lists the *m/z* values of the ions that are commonly observed in mass spectra, along with possible groups associated with that particular mass and some possible interferences. As with the (M–X) table, this table is not completely infallible for structural identification.

**Table 7.1b** Mass-composition table of common fragment ions

| *m/z* | Possible formula | Possible compound/group type |
|---|---|---|
| 26 | $C_2H_2$ | Hydrocarbons (particularly unsaturated) |
| 27 | $C_2H_3$ | Hydrocarbons (perhaps unsaturated) |
| 28 | $CO, C_2H_4, N_2$ | Carbonyl, ethyl, and azo compounds |
| 29 | $CHO, C_2H_5$ | Aldehydes, ethyl compounds |
| 30 | $CH_2{=}NH_2$, NO | Primary amines, nitro compounds |
| 31 | $CH_2{=}OH$ | Primary alcohols, methoxy compounds |
| 35/37 (3/1) | $^{35}Cl, ^{37}Cl$ | Chloro compounds |
| 36/38 (3/1) | $^{35}ClH, ^{37}ClH$ | Chloro compounds |
| 39 | $C_3H_3$ | Hydrocarbons (particularly aromatic) |
| 40 | Ar, $C_3H_4$ | Background from air, hydrocarbons |
| 41 | $C_3H_5$ | Hydrocarbons (particularly unsaturated) |
| 42 | $CH_2{=}C{=}O, C_3H_4$ | Acetates and acetyl compounds, hydrocarbons |
| 43 | $CH_3CO$ | $CH_3COX$ |
| 43 | $C_3H_7$ | $C_3H_7X$ (particularly iso-species) |
| 44 | $CO_2$ | Background from air, carbonates, anhydrides |
| 44 | $CH_3CH{=}NH_2$ | Aliphatic amines |
| 44 | $O{=}C{=}NH_2$ | Primary amides |
| 44 | $CH_2{=}CH(OH)$ | Aldehydes containing γ-hydrogens |
| 45 | $CH_3CH{=}OH$ | Secondary alcohols |

*(continued overleaf)*

**Table 7.1b** *(continued)*

| *m/z* | Possible formula | Possible compound/group type |
|---|---|---|
| 45 | $CH_2{=}OCH_3$ | Certain others |
| 45 | $CO_2H$, $OCH_2CH_3$ | Acids, ethoxy compounds |
| 46 | $NO_2$ | Nitro compounds |
| 47 | $P{=}O$, $CH_2{=}SH$ | Phosphoryl compounds, primary thiols |
| 49/51 (3/1) | $CH_2Cl$ | Chloromethyl compounds |
| 50 | $C_4H_2$ | Aromatic compounds |
| 51 | $C_4H_3$ | $C_6H_5X$ |
| 55 | $C_4H_7$, $C_3H_3O$ | Unsaturated hydrocarbons, $C_5$ and $C_6$ cyclic ketones |
| 56 | $C_4H_8$ | Hydrocarbons |
| 57 | $C_4H_9$, $CH_3CH_2CO$ | $C_4H_9X$, $CH_3CH_2COX$ |
| 58 | $CH_2{=}C(OH)CH_3$ | Methyl ketones containing γ-hydrogens, certain dialkyl ketones |
| 58 | $(CH_3)_2N{=}CH_2$ | Aliphatic amines |
| 59 | $(CH_3)_2COH$, $COOCH_3$ | $(CH_3)C(OH)X$, methyl esters |
| 59 | $CH_2{=}C(OH)NH_2$, $C_2H_5CH{=}OH$ | Primary amides, $C_2H_5CHOHX$ |
| 60 | $CH_2{=}C(OH)OH$ | Aliphatic acids containing γ-hydrogens, acetate esters |
| 61 | $CH_3CO(OH_2)$ | Acetate esters, i.e. $CH_3COOC_nH_{2n+1} (n > 1)$ |
| 65 | $C_5H_5$ | Benzyl and tolyl compounds, phenols, anilines |
| 66 | $C_5H_6$ | Aromatic compounds |
| 68 | $C_4H_4N$ | Pyrroles (monosubstituted) |
| 69 | $C_5H_9$ | Hydrocarbons (particularly unsaturated) |
| 70 | $C_5H_{10}$ | Hydrocarbons (perhaps unsaturated) |
| 71 | $C_5H_{11}$, $C_3H_7CO$ | $C_5H_{11}X$, propyl ketone, butanoate ester |
| 72 | $CH_2{=}C(OH)C_2H_5$ | Ethyl ketones containing γ-hydrogens |

| *m/z* | Possible formula | Possible compound/group type |
|---|---|---|
| 72 | $C_3H_7CH{=}NH_2$ | Amines |
| 73 | $C_3H_7CH{=}OH$, $C_3H_7OCH_2$ | Alcohols, ethers |
| 73 | $CO_2C_2H_5$ | Ethyl esters |
| 73 | $CH_2{=}CHC(OH){=}OH$ | Aliphatic acids |
| 73 | $(CH_3)_3Si$ | Trimethylsilyl derivatives |
| 74 | $CH_2{=}C(OH)OCH_3$ | Methyl esters containing γ-hydrogens |
| 75 | $(CH_3)_2Si{=}OH$ | $(CH_3)_3SiOX$ |
| 75 | $C_2H_5CO(OH_2)$ | $C_2H_5COOC_nH_{2n+1}$ $(n > 1)$ |
| 76 | $C_6H_4$ | Benzene derivatives (mono- or disubstituted) |
| 77 | $C_6H_5$ | $C_6H_5X$ |
| 78 | $C_6H_6$, $C_5H_4N$ | $C_6H_5X$, with X containing β- or γ-hydrogens, pyridines (monosubstituted) |
| 79 | $C_6H_7$ | $C_6H_5X$ |
| 79/81 (1/1) | $^{79}Br$, $^{81}Br$ | Bromo compounds |
| 80/82 (1/1) | $^{79}BrH$, $^{81}BrH$ | Bromo compounds |
| 82 | $C_5H_6N$ | Methylpyrroles, monoalkylpyrroles |
| 83 | $C_4H_3S$ | Thiophenes (monosubstituted) |
| 85 | $C_6H_{13}$, $C_4H_9CO$ | $C_6H_{13}X$, $C_4H_9COX$ |
| 86 | $CH_2{=}C(OH)C_3H_7$ | Propyl ketones containing γ-hydrogens |
| 86 | $C_4H_9CH{=}NH_2$ | Amines |
| 87 | $CH_2{=}CHC(OH)OCH_3$ | $XCH_2CH_2COOCH_3$ |
| 88 | $CH_3CH_2CH_2COOH$ | $C_3H_7COOC_nH_{2n+1}$ $(n > 1)$ |
| 89 | $C_7H_5$ | N- and O-containing heterocyclics |
| 90 | $C_7H_6$ | N- and O-containing heterocyclics |
| 91 | $C_7H_7$ | $C_6H_5CH_2X$, $CH_3C_6H_4$-X |
| 92 | $C_7H_8$, $C_6H_6N$ | $C_6H_5CH_2R$, monoalkylpyridines |
| 93/95 (1/1) | $^{79}BrCH_2$, $^{80}BrCH_2$ | $BrCH_2X$ |
| 93 | $C_6H_7N$ | $C_6H_5NHX$, with X containing hydrogen |

*(continued overleaf)*

**Table 7.1b** *(continued)*

| *m/z* | Possible formula | Possible compound/group type |
|---|---|---|
| 93 | $C_6H_5O$ | Phenols, nitrobenzenes |
| 93 | $C_7H_9$ | Mono- and sesquiterpenes |
| 94 | $C_6H_6O$ | $C_6H_5OR$ (excluding $C_6H_5OCH_3$) |
| 95 | $C_7H_{11}$ | Mono- and sesquiterpenes |
| 96 | $C_5H_4NO$ | 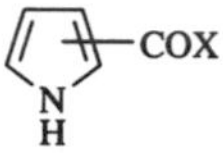 COX, N, H |
| 97 | $C_5H_5S$ | Methylthiophenes, monoalkylthiophenes |
| 99 | $C_7H_{15}$ | $C_7H_{15}X$ |
| 103 | $C_6H_5CH{=}CH$ | $C_6H_5CH{=}CHX$ |
| 105 | $C_6H_5CO$, $C_8H_9$ | $C_6H_5COX$, $CH_3C_6H_4CH_2X$ |
| 107 | $C_7H_7O$ | $C_6H_5CH_2OX$, $HOC_6H_4CH_2X$ |
| 107/109 (1/1) | $C_2H_4Br$ | $BrCH_2CH_2{-}X$ |
| 111 | $C_5H_3OS$ | 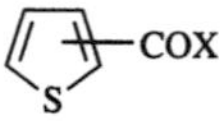 COX, S |
| 121 | $C_8H_9O$, $C_6H_5CO_2$ | $CH_3OC_6H_4CH_2X$, $C_6H_5CO_2X$ |
| 122 | $C_6H_5COOH$ | Alkyl benzoates |
| 123 | $C_6H_5COOH_2$ | Alkyl benzoates |
| 127 | $C_{10}H_7$ | Naphthyl derivatives |
| 127 | I | Iodo compounds |
| 128 | HI | Iodo compounds |
| 131 | $C_6H_5CH{=}CHCO$ | $C_6H_5CH{=}CHCOX$ |
| 141 | $ICH_2$ | $ICH_2X$ |
| 147 | $(CH_3)_2Si{=}OSi(CH_3)_3$ | $((CH_3)_3SiO)_n$ derivatives ($n > 1$) |
| 149 | 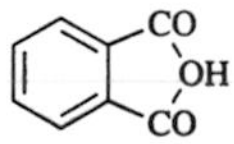 CO, OH, CO | Dialkyl phthalates (plasticisers) |

Use the space below to add other ions as you come across them.

There are a number of points to remember when using these two types of correlation tables.

First, can you remember the nitrogen rule?

Π What does this rule state? (Refer back to Chapter 1 if you are uncertain about this.)

This rule states that all compounds containing only carbon, hydrogen and oxygen atoms, or these atoms plus an even number of nitrogen atoms, must have an even $M_r$ value. If an odd number of nitrogen atoms are present, then the $M_r$ value must be odd.

Secondly, mass losses of 3–14 from *any* ion are not observed. This means that losses of C, CH or $CH_2$ species are unlikely.

It is not always the best approach to analysing a spectrum to subtract each *m*/*z* from that of the immediately higher *m*/*z* to obtain X values, since the cascade process shown previously in Figure 6.3a outlines a variety of diverging pathways to the daughter ions, and a lower *m*/*z* ion may therefore not be formed from the higher *m*/*z* ion in question, but from the molecular ion (or other daughters) itself. The correct procedure is to begin the analysis process by subtracting all of the observed higher *m/z* values from that of $M^{+\bullet}$, to see if significant X values result, before moving down the spectrum to ions of lower masses. This is the order in which they were formed.

Thirdly, small mass losses are likely to be more specific. The loss of 15 amu is almost certainly a $CH_3$ group, whereas the loss of 57 amu could be $C_4H_9$, $C_3H_5O$, $C_3H_7N$, or various other groups. Only the first two structures are given in our tables because they are the commonest. Sometimes, a particular higher mass loss is very indicative of the compound present, e.g. a loss of 77 amu is very probably a phenyl ($C_6H_5$) group.

Fourthly, inferences drawn from the tables are more likely to be significant if the peak in question has a high relative abundance (*RA*). We saw in Section 6.3 that aromatic molecular ions are, in general, more stable than aliphatic ones. Therefore, there will be a small number of high *m/z* peaks with a high *RA*. Aliphatic molecular ions are often weak or absent, and break down to give many daughters with low *m/z* values. In either case, work out the structures of the intense ions first, then confirm likely breakdowns with any metastable ions present, and finally try to gain further information from the minor fragments.

Let us now apply the processes we have just discussed in order to try to elucidate an unknown spectrum, as shown in Figure 7.1a.

The highest mass observed in the spectrum of *Unknown 1* is 107. This is almost certainly based upon the $^{13}C$ isotope of a molecular ion of *m/z* 106. The base peak is *m/z* 105, i.e. a loss of 1 amu, clearly H from the (M–X) table. The latter lists aldehydes, acetals, aryl$-CH_3$ compounds, $-CH_2CN$, alkynes, and $N-CH_3$ compounds as possibilities. We can discard $-CH_2CN$ and $N-CH_3$, as under the nitrogen rule, $M_r$ would be odd. This leaves four possibilities, so let us now turn to the 77 ion—how has this been formed and what is it?

If *m/z* 77 is formed directly from $M^{+\bullet}$, then X is 29, indicating an aldehyde or an ethyl compound (Table 7.1a). If it is formed from 105, X is 28, again indicating a CHO or a $CH_2-CH_2$ compound.

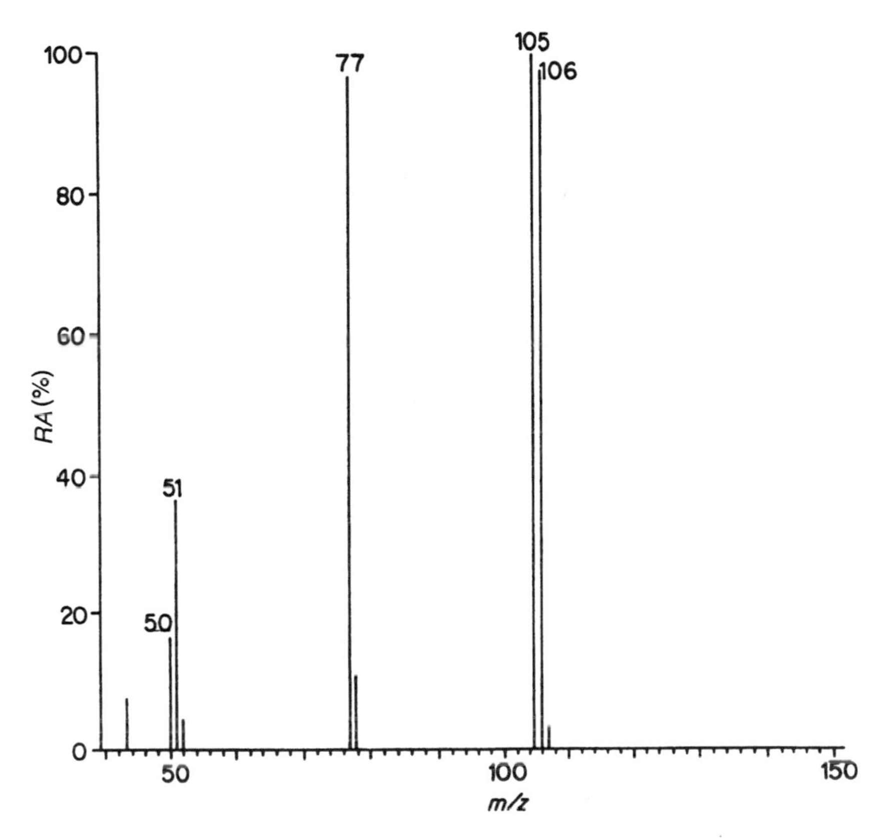


**Figure 7.1a** Mass spectrum of *Unknown 1*

Therefore, losses of 28 and 29 amu **and** loss of H is consistent only with the compound being an aldehyde. The base peak (*m/z* 105) could be $C_6H_5CO^+$ or $CH_3C_6H_4CH_2^+$ (from Tabel 7.1b). Clearly, only $C_6H_5CO^+$ fits in with our other deductions, and therefore *Unknown 1* is benzaldehyde, $C_6H_5CHO$.

These deductions can be conveniently presented in a table (see Table 7.1c). Note that in this table, the molecular ion appears at the head of each postulated fragmentation pathway, twice in this present case. The loss in the X column is then derived from the entry immediately above it.

Π Metastable ions are observed in the spectrum of *Unknown 1* at *m/z* 104, 56.5, and 33.8. What do these tell us about the fragmentation routes of benzaldehyde?

**Table 7.1c** A method of presenting the analysis of a mass spectrum

| *m/z* | Possible structure | Associated X-loss | Inference |
|---|---|---|---|
| 106 | | – | $M^{+\bullet}$ |
| 77 | $C_6H_5$ | CHO, $CH_3CH_2$ | $C_6H_5CHO$, $C_6H_5CH_2CH_3$ |
| 106 | | – | $M^{+\bullet}$ |
| 105 | $C_6H_5CO$, $CH_3C_6H_4CH_2$ | H | $C_6H_5CHO$, $CH_3C_6H_4CH_2$ |
| 77 | $C_6H_5$ | CO, $CH_2{=}CH_2$ | CO or $CH_3CH_2$ present |
| 51 | $C_4H_3$ | $CH{=}CH$ | $C_6H_5$ present |
| | | | Deduction: $C_6H_5{-}CHO$ |

Metastable ($m^*$) 104 corresponds to 106 $\longrightarrow$ 105 (($105^2/106$) = 104.01)).

$m^*$ 56.5 could correspond to 106 $\longrightarrow$ 77, or 105 $\longrightarrow$ 77, so both must be considered:

$77^2/106 = 55.93$, while $77^2/105 = 56.47$, so the latter is correct.

Hence, the pathway which has the metastable ion support is:

$$C_6H_5CHO^{+\bullet} \xrightarrow{-H} C_6H_5CO^{+} \xrightarrow{-CO} C_6H_5^{+}$$

rather than:

$$C_6H_5{-}CHO^{+\bullet} \xrightarrow{-CHO} C_6H_5^{+\bullet}$$

$m^*$ 33.8 should correspond to:

$$106 \longrightarrow 51, \text{ i.e. } m^* = 51^2/77 = 33.78.$$

This is the answer, thus indicating that the following takes place:

$$C_6H_5^{+\bullet} \xrightarrow{-CH\equiv CH} C_4H_3^+$$

Hence, the likely fragmentation route of the benzaldehyde molecular ion is as follows:

$$C_6H_5CHO^{+\bullet} \xrightarrow{-H} C_6H_5CO^+ \xrightarrow{-CO} C_6H_5^+ \xrightarrow{-CH\equiv CH} C_4H_3^+$$

All of the above has been deduced without the use of any mechanisms. Here are the actual mechanisms to help you visualise the processes involved:

(a) $C_6H_5{-}C(H){=}\ddot{O}^{+\bullet} \longrightarrow C_6H_5{-}C{\equiv}O^+ + H^\bullet$

(b) $C_6H_5{-}C{\equiv}O^+ \longrightarrow C_6H_5^+ + :C{=}O$

(c) $C_6H_5^+ \longrightarrow H{-}C{\equiv}C{-}H + C_4H_3^+$, $m/z$ 51

in which (a) and (c) are homolytic cleavage reactions, with (b) being a heterolytic cleavage reaction.

Since the double bonds in the $C_6H_5^+$ ion are delocalised, any consecutive pair of carbon atoms can be eliminated as HC≡CH.

**SAQ 7.1a**

Look below at the spectrum of *Unknown 2* (shown in Figure 7.1b) and try to analyse this in the same way as *Unknown 1*.

→

**SAQ 7.1a**
(contd)

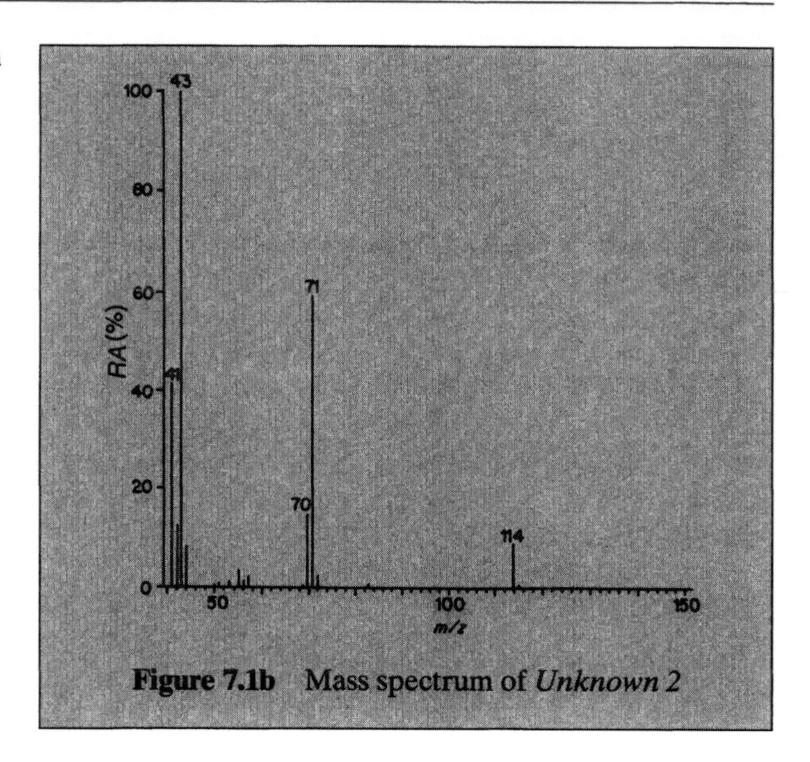


**Figure 7.1b** Mass spectrum of *Unknown 2*

**SAQ 7.1a**

**SAQ 7.1a**

**SAQ 7.1b**

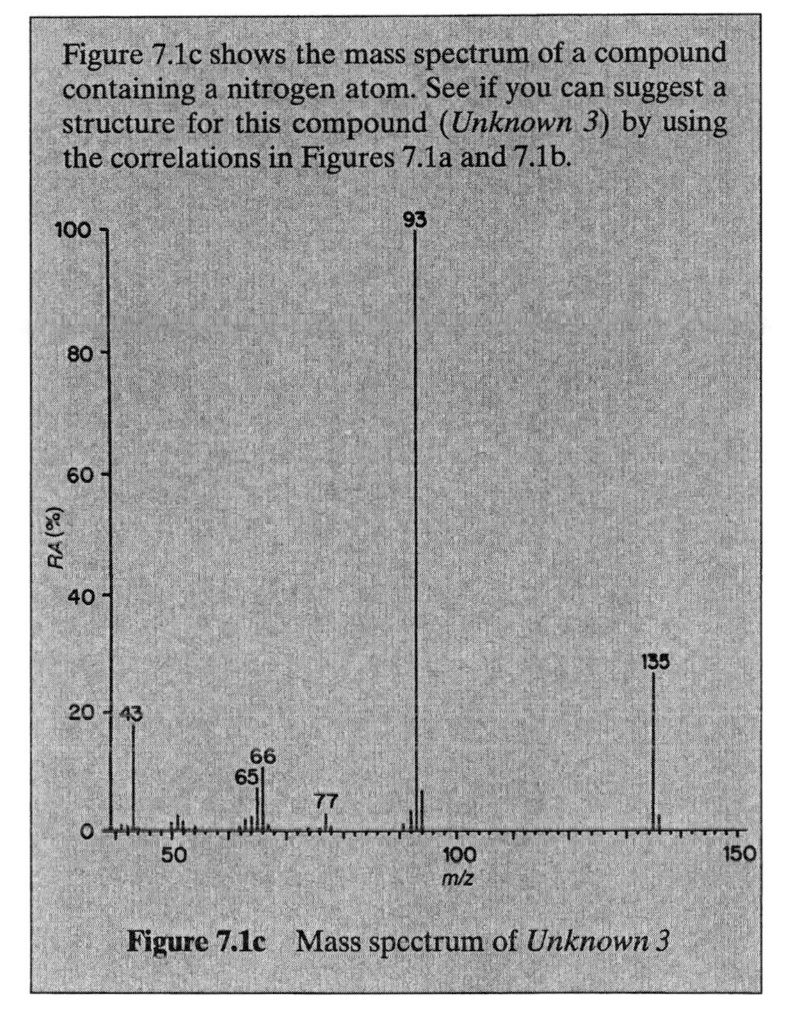

Figure 7.1c shows the mass spectrum of a compound containing a nitrogen atom. See if you can suggest a structure for this compound (*Unknown 3*) by using the correlations in Figures 7.1a and 7.1b.



**Figure 7.1c** Mass spectrum of *Unknown 3*

**SAQ 7.1b**

**SAQ 7.1b**

You may not have been entirely convinced from these examples that mass spectrometry leads to unambiguous structure determination. It is indeed quite correct to state that often no one spectroscopic method can solve the problem. You will often need to use infrared or nuclear magnetic resonance spectroscopies as well.

**Summary**

You have now used two tables of data to interpret mass spectra, i.e. the (M–X) table (Table 7.1a) and the mass-composition table (Table 7.1b). These are a basic starting point for interpretation of an unknown, although do not always expect to find your compound in these lists!

When analysing a mass spectrum, the most intense ion should be assigned first and any ambiguities of structure recognised, e.g. $CH_3CO^+/CH_3CH_2CH_2^+$. Compounds giving intense $M^{+\bullet}$ ions and a few intense higher-mass ions are usually aromatic derivatives.

Metastable ions are often helpful in distinguishing between different fragmentation routes.

We will now go on to look at some simple fragmentation patterns of a number of common functional groups.

## 7.2 FRAGMENTATIONS OF ALCOHOLS

The mass spectra of methanol and ethanol were described in Chapter 1. You should recall that the base peak in both of these alcohols is *m/z* 31, $CH_2{=}\overset{+}{O}H$. In methanol, this is due to the loss of $H^\bullet$:

$$H_2C-\overset{+\bullet}{O}H \longrightarrow H_2C=\overset{+}{O}H + H^{\bullet}$$

It is conventional to shorten this mechanism to:

$$H-H_2C-\overset{+\bullet}{O}H \longrightarrow H_2C=\overset{+}{O}H + H^{\bullet}$$

which shows the movement of only one of the electrons involved, in order to simplify the picture. In ethanol (see Figure 7.2a), the same process can occur:

$$CH_3CH(H)-\overset{+\bullet}{O}H \longrightarrow CH_3CH=\overset{+}{O}H + H^{\bullet}$$

$$m/z\ 45\ (20\%)$$

but the base peak is formed by the alternative loss of $CH_3^{\bullet}$:

$$H-C(CH_3)(H)-\overset{+\bullet}{O}H \longrightarrow H_2C=\overset{+}{O}H + CH_3^{\bullet}$$

$$m/z\ 31\ (100\%)$$

These processes of homolytic α-cleavage are competitive. In longer-chain primary alcohols e.g. *n*-butanol (Figure 7.2b(i)), the amount of $(M-H)^+$ is much smaller, *ca.* 1% and $CH=\overset{+}{O}H$ is still the base peak.

Π Why should the loss of the alkyl fragment be preferred over the loss of $H^{\bullet}$ in the α-cleavage of primary-alcohol molecular ions?

The radical $CH_3CH_2^{\bullet}$ would be more stable than $CH_3^{\bullet}$, just as $CH_3CH_2^+$ is more stable than $CH_3^+$.

The question was answered by Stevenson in the early 1950s. He was the first to notice this tendency for the largest alkyl fragment to be preferentially lost as a radical. The molecular ion is a high-energy species trying to get rid of as much energy as quickly as possible. An alkyl radical has several bonds which can be in higher vibrational states, thus absorbing the excess energy of the molecular ion and then leaving with it. The hydrogen atom can only depart with kinetic energy. The more bonds the radical species has, the better. Therefore, we have a useful empirical rule named after Stevenson which states the following:

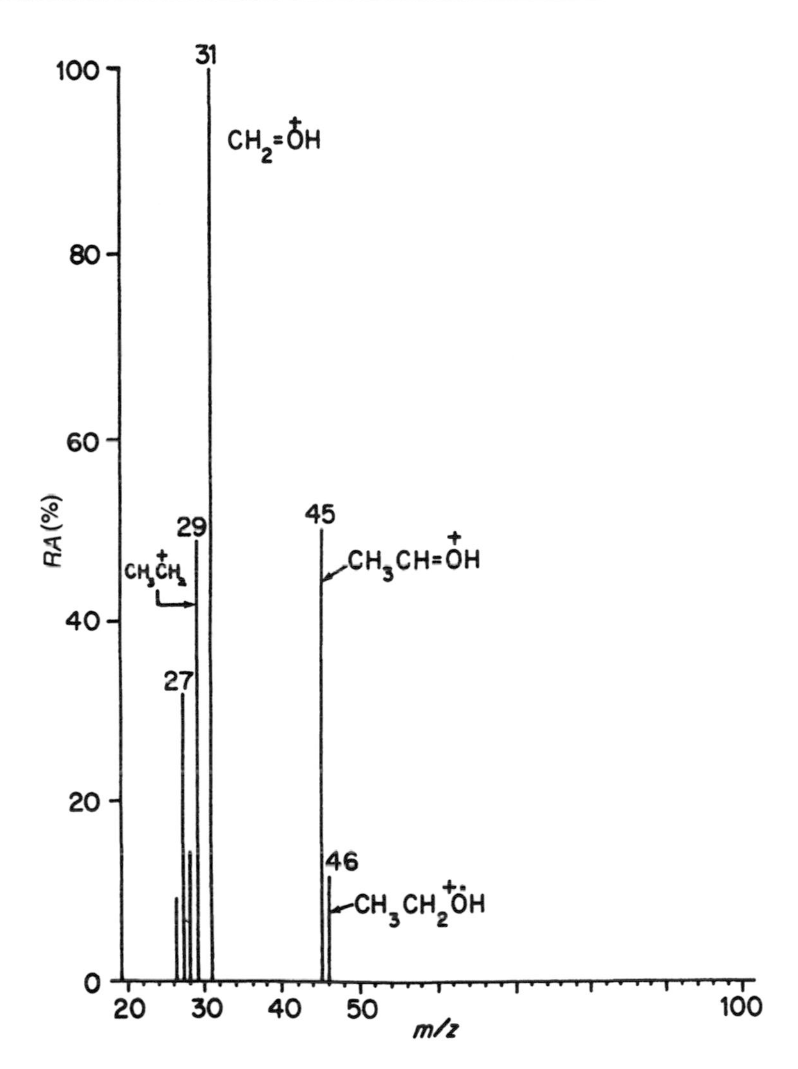


**Figure 7.2a** Mass spectrum of ethanol

*In a fragmentation, the largest radical is lost preferentially.*

In Figure 7.2b, the bond that has been cleaved is shown broken by a wavy line, and the ion formed and its intensity are shown above or below the horizontal straight line. This is a very useful way of summarising the main features of a mass spectrum.

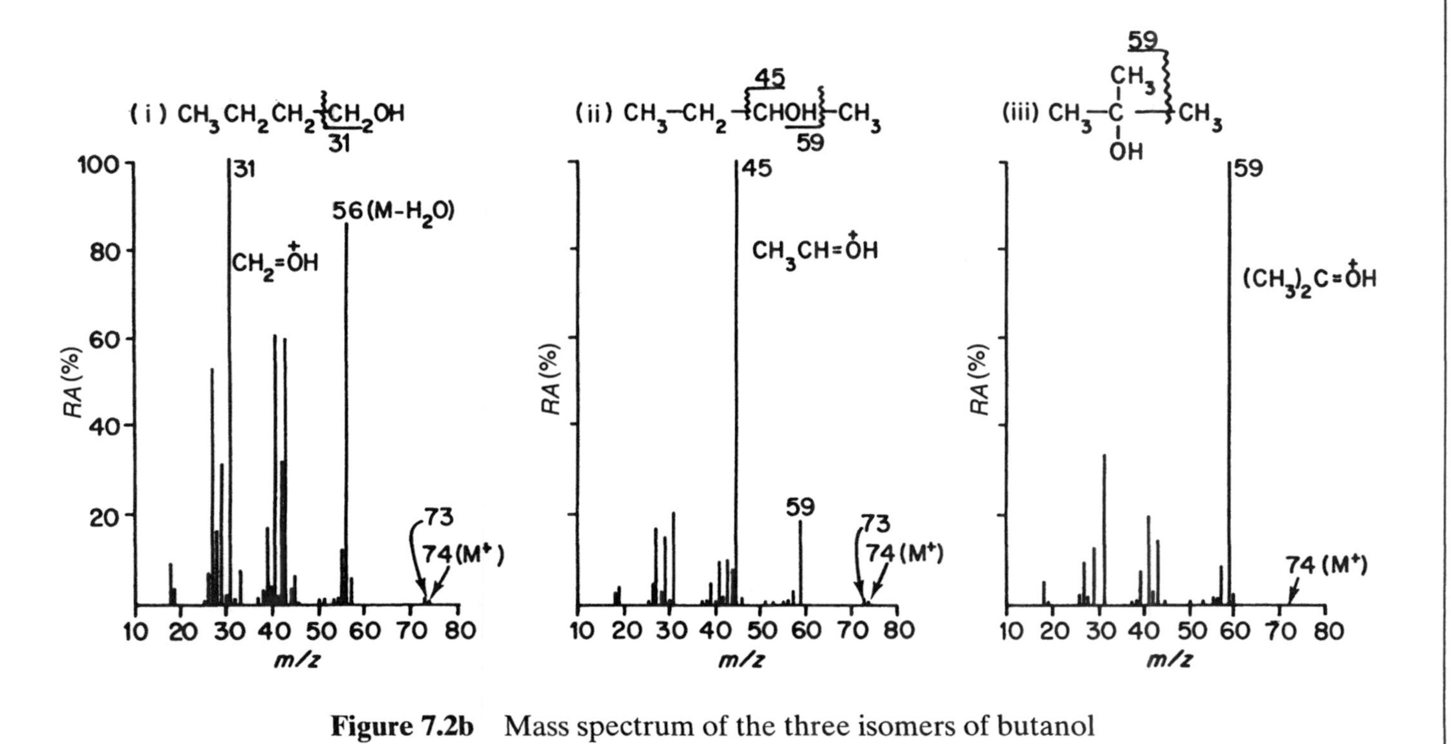


**Figure 7.2b** Mass spectrum of the three isomers of butanol

**SAQ 7.2a**

Examine the mass spectra of the three isomers of butanol shown in Figure 7.2b and see if Stevenson's rule applies.

Therefore, in aliphatic alcohols the base peak is usually formed by the loss of the largest radical attached to the α-carbon atom:

$$R_1-\underset{R_2}{\overset{R_3}{\underset{|}{\overset{|}{C}}}}-\overset{+\bullet}{O}H \longrightarrow \underset{R_2}{\overset{R_1}{\diagdown}}C=\overset{+}{O}H + R_3^{\bullet}$$

where $R_3 > R_2 \geqslant R_1$

The α- and β-bonds in alcohols can also break heterolytically to give hydrocarbon ions, which can also be seen in the spectra shown in Figures 7.2a and 7.2b:

$$CH_3-CH_2-\overset{+\bullet}{O}H \longrightarrow CH_3CH_2^+ + \dot{O}H$$
$$m/z\ 29$$

$$CH_3-CH_2-\overset{+\bullet}{O}H \longrightarrow CH_3^+ + CH_2=\overset{+}{O}H$$

$$CH_3CH_2CH_2-CH_2-\overset{+\bullet}{O}H \longrightarrow CH_3(CH_2)_3^+ + \dot{O}H$$
$$m/z\ 57$$

$$CH_3CH_2CH_2-CH_2-\overset{+\bullet}{O}H \longrightarrow CH_3CH_2CH_2^+ + CH_2=\overset{+}{O}H$$
$$m/z\ 43$$

and generally:

$$R-CH_2-\overset{+\bullet}{O}H \longrightarrow R^+ + CH_2=\overset{+}{O}H$$

These hydrocarbon ions, and the daughter ions derived from them, are usually of less relative abundance than the $R_1R_2C=\overset{+}{O}H$ ions, but increase in importance in long-chain, particularly branched alcohols, which come to resemble the corresponding hydrocarbons (see Section 7.7).

In secondary or tertiary alcohols, where there might be two or three different $R^+$ ions, the highest order is most intense, as this follows from the order of stability of carbocations. Revise Section 6.2, if you are unsure about this. In some alcohol mass spectra, the $R^+$ ions are as intense or more intense than the $R_1R_2C=\overset{+}{O}H$ ions if R is secondary or tertiary.

You may have noticed that there is another prominent peak in the spectrum in Figure 7.2b(i); this also appears to a lesser extent in the other spectra (ii and iii)

Π What is this other peak?

It is *m/z* 56 in butan-1-ol and is formed from the loss of water.

The order of ease of dehydration of alcohols is tertiary>secondary>primary. Thus, 2-methylpropan-2-ol would be expected to have the most intense *m/z* 56 peak. However, this is not always the case in practice.

The mechanism for the loss of water involves hydrogen transfer from the third or fourth carbon atom down the chain to the $-\overset{+\bullet}{O}H$ group.

The transfer from the fourth carbon is favoured because it involves a six-membered-ring intermediate, but it can also occur to a limited extent from the third carbon. Butan-2-ol has such a carbon atom, from which a hydrogen *could* be transferred, but shows little loss of $H_2O$ in practice. The relevant mechanisms are as follows:

1,4 elimination

$$\text{H–CH(–CH}_2\text{–CH}_2\text{–CH}_2\text{–}\overset{+\bullet}{\text{O}}\text{H)} \longrightarrow \text{H}_2\dot{\text{C}}\text{–CH}_2\text{–CH}_2\text{–}\overset{+}{\text{C}}\text{H}_2 + \text{H}_2\text{O}$$

$m/z$ 56

and 1,3 elimination:

$$\text{H}_2\text{C(H)–CH}_2\text{–C(CH}_3\text{)(H)–}\overset{+\bullet}{\text{O}}\text{H} \longrightarrow \text{H}_2\dot{\text{C}}\text{–CH}_2\text{–}\overset{+}{\text{C}}\text{HCH}_3 + \text{H}_2\text{O}$$

$m/z$ 56

Therefore, since ethanol and 2-methylpropan-2-ol have neither three not four carbons, let alone hydrogens, they show little loss of $H_2O$.

Alcohols with five carbons or more can lose both $CH_2{=}CH_2$ and water in a single concerted step, involving a metastable ion:

$$\text{H}_2\text{C}(\overset{+\bullet}{\text{O}}\text{H})\text{–H}_2\text{C–CH}_2\text{–CHR–H} \longrightarrow \text{H}_2\text{C{=}CHR}^{+\bullet} + \text{H}_2\text{O} + \text{H}_2\text{C{=}CH}_2$$

$(M-46)^{+\bullet}$

Both this and the 1,4-$H_2O$ elimination are typical examples of hydrogen-transfer processes induced by electron-impact sources, which involve a six-centred-ring transition state.

In cyclopentanols and hexanols, a similar six-centred-ring process involves the loss of ethene and $CH_3^{\bullet}$. The fragmentation starts by α-cleavage next to the $-OH$ group, followed by further bond reorganisation and cleavages. Figures 7.2c and 7.2d show, respectively, the mass spectrum of cyclohexanol and its proposed fragmentation pattern. Water can also be lost from the molecular ion, to give a $m/z$

100
80
60
40
20
0
RA (%)
57
OH
$M_r$ 100
+OH
82
(M-H2O)
44
41
67
-CH3
27
39
54
M+•
85
100
50
100
150
m/z

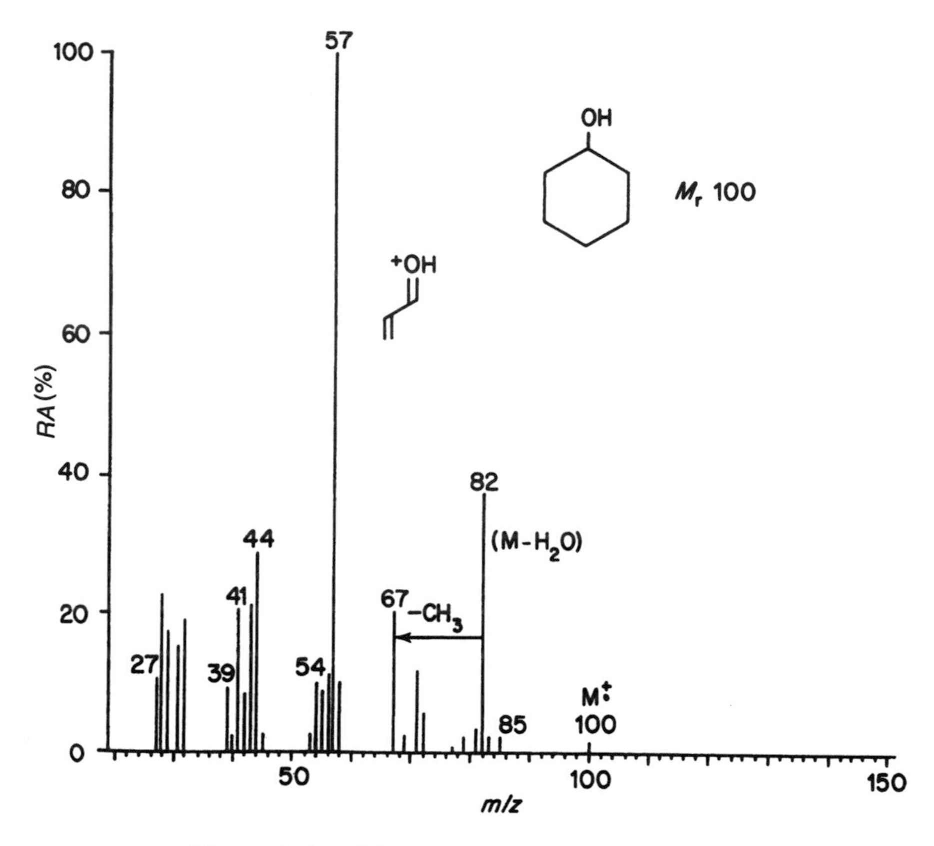

**Figure 7.2c** Mass spectrum of cyclohexanol

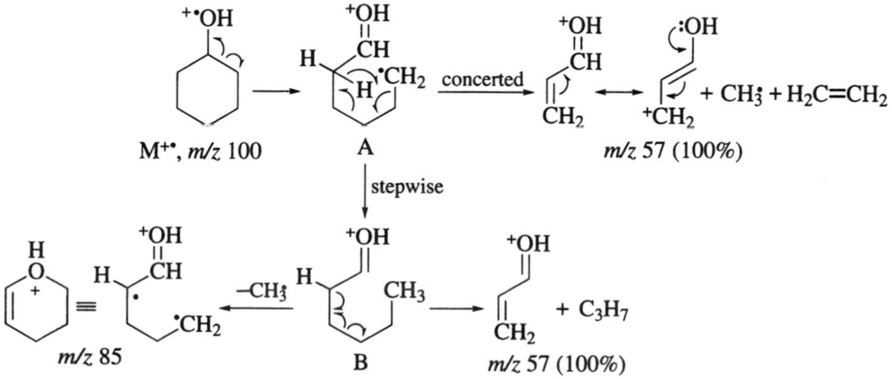

+•OH
M+•, m/z 100
+OH
CH
H
•CH2
H
A
concerted
+OH
CH
CH2
:OH
+CH2
+ CH3• + H2C=CH2
m/z 57 (100%)
stepwise
+OH
H
CH3
B
+OH
CH2
+ C3H7
m/z 57 (100%)
−CH3•
+OH
CH
H
•CH2
H
O
+
m/z 85

**Figure 7.2d** Proposed fragmentation pattern for cyclohexanol

82 peak. Deuterium labelling has shown that the second hydrogen (the first one being from OH) comes from positions 3 or 4 in the ring, as a direct result of free rotation in the acyclic intermediate (B in Figure 7.2d).

**SAQ 7.2b**

Why do you think the *m/z* 57 peak is so abundant in most cyclanol mass spectra?

**SAQ 7.2c**

Show how the $M^{+\bullet}$ of 3-methylcyclohexanol can give rise to both *m/z* 57 and *m/z* 71 peaks in its mass spectrum. Why is *m/z* 71 more intense than *m/z* 57 in this spectrum?

**SAQ 7.2c**

Aromatic alcohols show much stronger molecular ions, $M^{+\bullet}$, in their spectra than aliphatic alcohols, and they do not follow Stevenson's rule. Look at Figure 7.2e, which shows the mass spectrum of benzyl alcohol.

Π What fragment would you expect to be lost from the molecular ion of $C_6H_5CH_2OH$ (benzyl alcohol) based on the predictions of Stevenson?

Stevenson's rule would predict the loss of $C_6H_5^{\bullet}$ from benzyl alcohol by α-cleavage, since this is the largest radical.

Instead, β-cleavage gives $C_6H_5^{+}$. Loss of $H^{\bullet}$ gives *m/z* 107, i.e. $C_7H_7O^{+}$. This hydrogen is lost randomly from all of the carbon atoms, rather than specifically from the $-CH_2-$ group. This indicates that the *m/z* 107 ion has a somewhat different structure than we at first thought. It is in fact one of a family of ions called *tropylium ions* which have the form $RC_7H_6^{+}$, where R is $-OH$ in our case, i.e. the hydroxytropylium

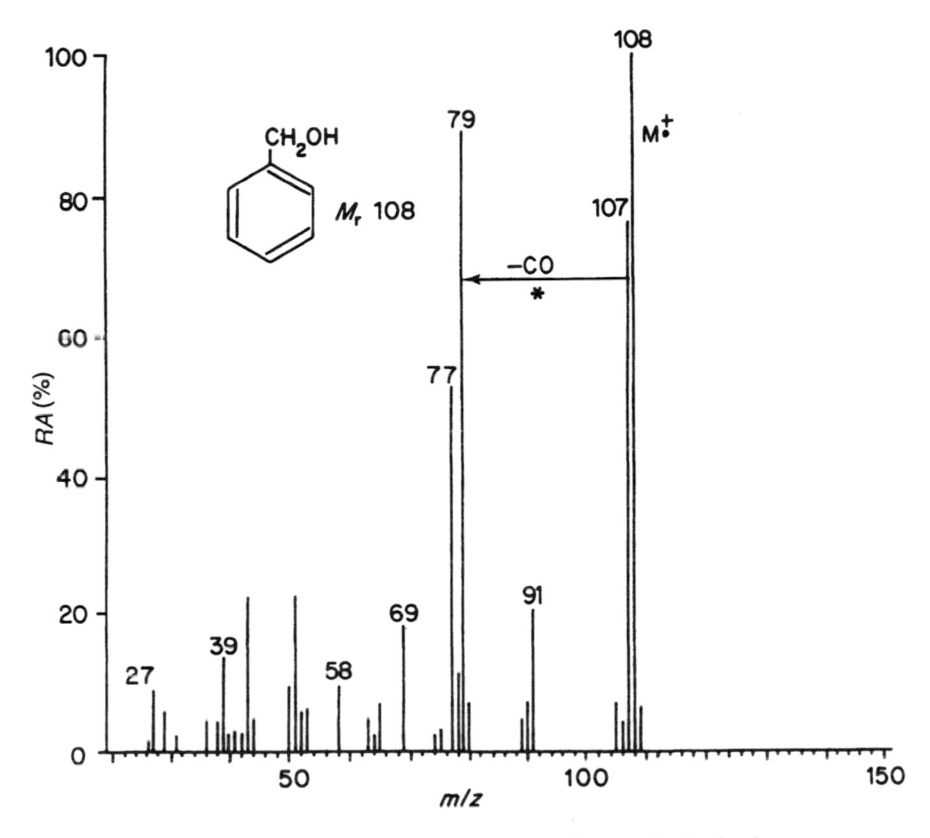


**Figure 7.2e** Mass spectrum of benzyl alcohol

ion (A). Tropylium ions are very stable species, especially where R is a **+M** group. Ion A (the *enol* form) is, of course in equilibrium with its *keto* form (B), which can lose :C=O to give the benzenium ion, *m/z* 79, as shown in Figure 7.2f.

$CH_2OH^{+\bullet}$ $\xrightarrow{-H^{\bullet}}$ $^{+}OH$ A (enol) $\rightleftharpoons$ B (keto) $\xrightarrow[*58.3]{-CO}$ $\xrightarrow[*75.1]{-H_2}$

*m/z* 108 — A (enol) *m/z* 107 — B (keto) — *m/z* 79 — *m/z* 77

**Figure 7.2f** Fragmentation pattern of benzyl alcohol (*represents a metastable ion)

**SAQ 7.2d**

Tropylium ions have a great number of resonance forms. Can you draw resonance forms for the hydroxytropylium ion (A in Figure 7.2f)?

**SAQ 7.2e**

Figure 7.2g shows the mass spectrum of an alcohol (*Unknown 4*). Interpret this spectrum and identify the alcohol. Note that *m/z* 31 (not shown in the figure) is actually 45%.

**SAQ 7.2e**
(Contd)

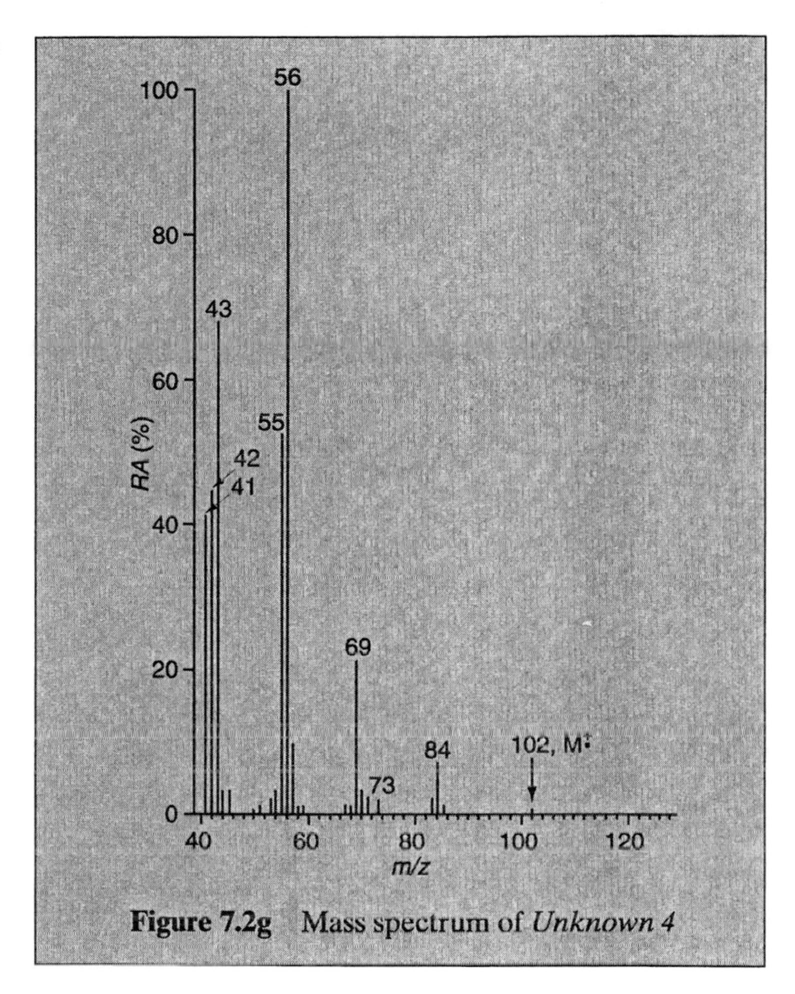


**Figure 7.2g** Mass spectrum of *Unknown 4*

**SAQ 7.2e**

**SAQ 7.2f**

Figure 7.2h shows the mass spectrum of an alcohol (*Unknown 5*). Interpret this spectrum and identify the alcohol.

**SAQ 7.2f**
(Contd)

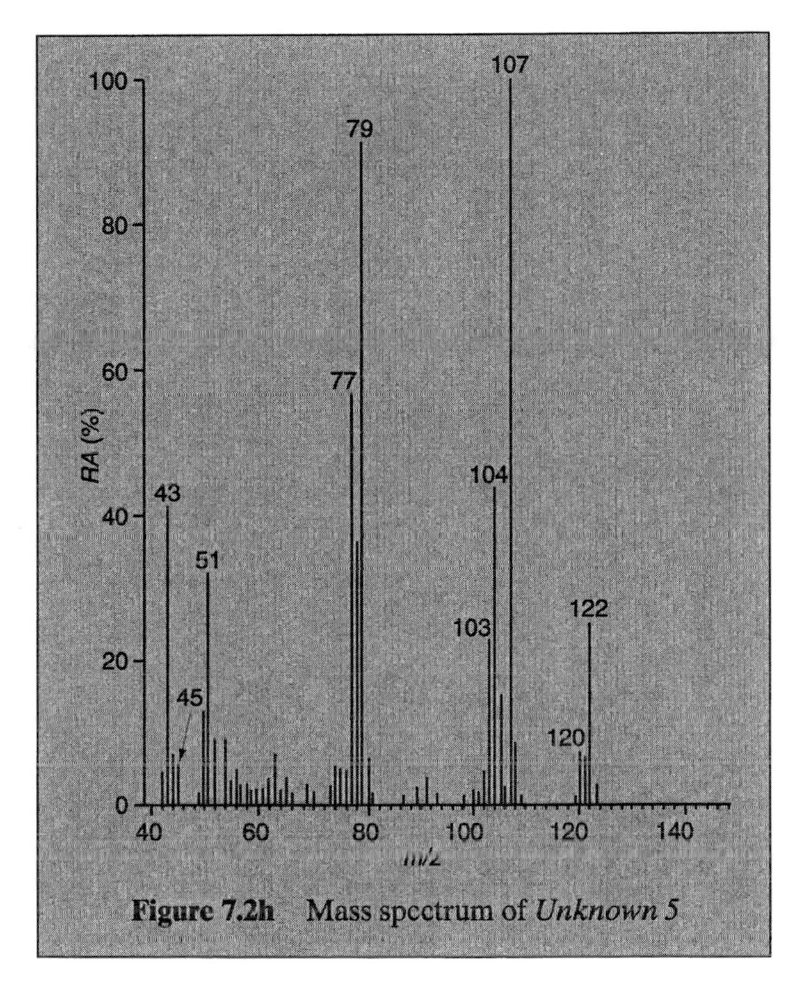


**Figure 7.2h** Mass spectrum of *Unknown 5*

**SAQ 7.2f**

## 7.3 FRAGMENTATIONS OF ETHERS

Ethers and alcohols are isomeric compounds, e.g. diethyl ether, $CH_3CH_2OCH_2CH_3$ and butanol, $CH_3CH_2CH_2CH_2OH$, and benzyl alcohol, $C_6H_5CH_2OH$, and methoxybenzene, $C_6H_5OCH_3$. However, do they have similar mass spectra?

First, ethers tend to show more intense $M^{+\bullet}$ than the isomeric alcohols, although these are still rather weak. They fragment by primary cleavages similar to those occurring in alcohols:

$$R-CH_2-\overset{+\bullet}{O}-CH_2-R^1 \xrightarrow{-R} CH_2{=}\overset{+}{O}-CH_2R^1 \quad (A)$$

$$R-CH_2-\overset{+\bullet}{O}-CH_2-R^1 \xrightarrow[-R^1]{} R-CH_2-\overset{+}{O}{=}CH_2 \quad (B)$$

to give oxonium ions A or B of *m/z* 45, 59, 73, 87, etc. which are isomeric with the $R_1R_2C{=}\overset{+}{O}H$ ions found in the spectra of alcohols.

However, ethers show a further easy fragmentation which is not found in most isomeric alcohols. This is the loss of a neutral alkene, which is derived from the remaining alkyl substituent in A or B:

$$\overset{+}{\mathrm{C}H_2=O}-CH_2 \quad H-O-R_2 \quad H \;(\mathrm{A}) \longrightarrow CH_2=\overset{+}{O}H + CH_2=CHR_2$$

*m/z* 31

In this process, transfer of a β-hydrogen atom occurs; hence for this to be observed, $R^1$ (or R) must have at least one carbon atom with a hydrogen on the β-carbon. The final products of this cascade will still be members of the series, *m/z* 31, 45, 59, 73, 87, etc.

To illustrate these processes, consider the spectrum of ethyl 1-methylpropyl ether, $CH_3CH_2OCH(CH_3)CH_2CH_3$ (Figure 7.3a).

Note that this compound is isomeric with hexanol (*Unknown 4*, Figure 7.2g). In accordance with Stevenson's rule, there will be α-cleavage of the largest radical ($CH_3CH_2$) to give *m/z* 73 (51%), i.e. pathway (i) in

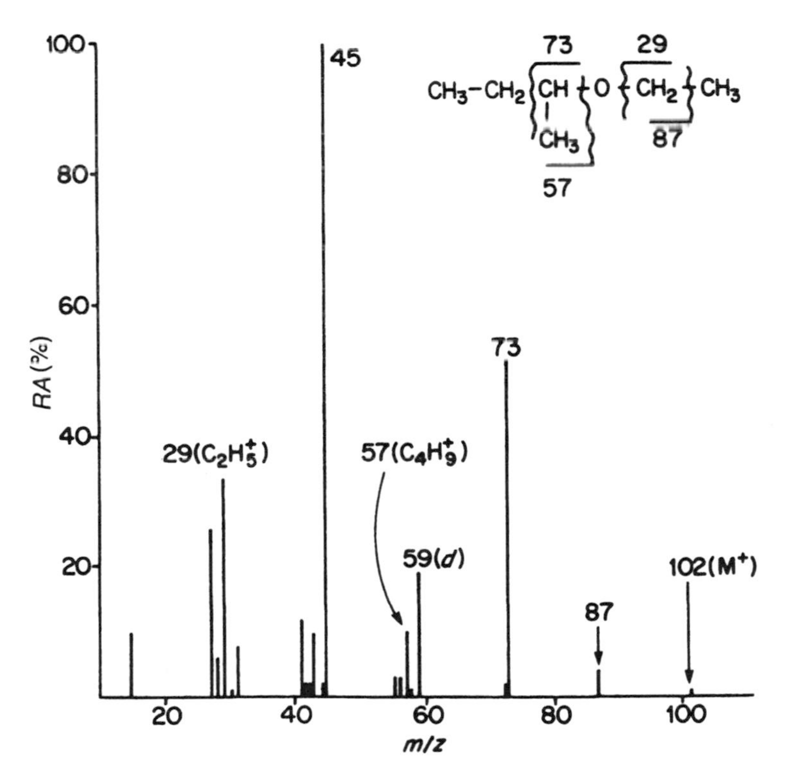


**Figure 7.3a** Mass spectrum of ethyl 1-methylpropyl ether

the fragmentation pattern shown in Figure 7.3b. Alternatively, loss of $CH_3^{\bullet}$, (pathway (ii)) gives *m/z* 87 to a minor extent (4%).

Loss of $CH_2{=}CH_2$ from *m/z* 73 gives the base peak *m/z* 45, while *m/z* 59 results from a similar loss from *m/z* 87. Fragmentation of the C—O bonds results in the hydrocarbon ions, $CH_3CH_2^+$ and $CH_3CH_2\overset{+}{C}HCH_3$ at *m/z* 29 and 57, respectively. Can you see the differences in the spectra of hexanol (Figure 7.2g) and ethyl 1-methylpropyl ether (Figure 7.3a)?

The behaviour of aromatic ethers is quite distinctive. The charge tends to be retained by the aromatic fragments, leading to a loss of alkyl and alkene fragments, as shown in Figure 7.3c for methoxybenzene. In this figure, the symbol * indicates that a metastable ion of that mass is observed for the fragmentation shown, e.g. 56.3 arises from $78^2/108$. Fragmentations tend to start from $M^{+\bullet}$, where the positive charge is on the aromatic ring.

Note that the mechanism for the formation of *m/z* 78 occurs via the following route. Other mechanisms (incorrectly) lead to benzene having no charge:

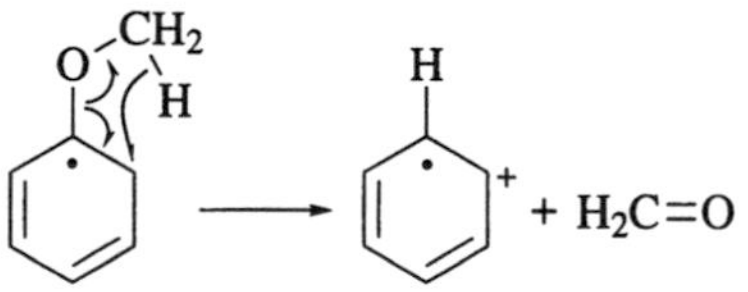

$CH_2$, O, H, H, + $H_2C{=}O$

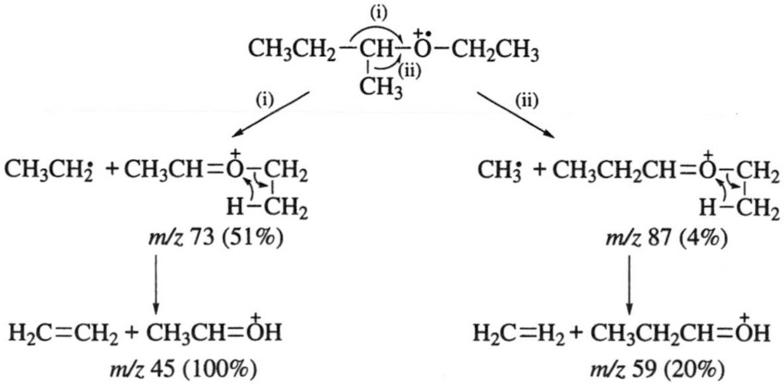

(i)

$CH_3CH_2{-}CH{-}\overset{+\bullet}{O}{-}CH_2CH_3$ (ii) $CH_3$

(i) (ii)

$CH_3CH_2^{\bullet}$ + $CH_3CH{=}\overset{+}{O}{-}CH_2$ H—$CH_2$
*m/z* 73 (51%)

$CH_3^{\bullet}$ + $CH_3CH_2CH{=}\overset{+}{O}{-}CH_2$ H—$CH_2$
*m/z* 87 (4%)

$H_2C{=}CH_2$ + $CH_3CH{=}\overset{+}{O}H$
*m/z* 45 (100%)

$H_2C{=}H_2$ + $CH_3CH_2CH{=}\overset{+}{O}H$
*m/z* 59 (20%)

**Figure 7.3b** Fragmentation pattern of ethyl 1-methylproply ether

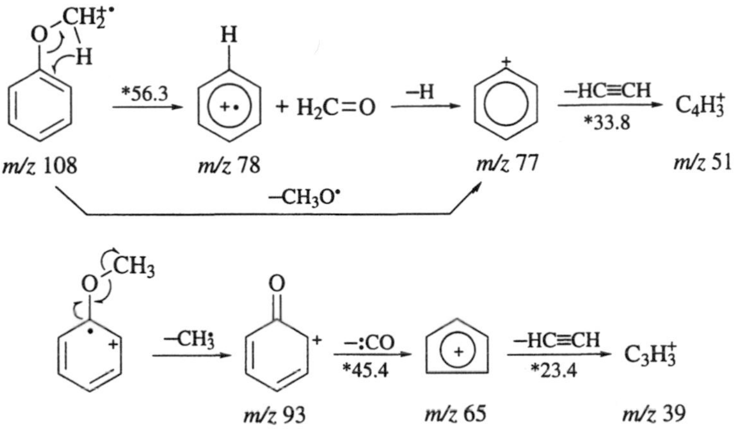

*56.3 —H —HC≡CH *33.8
$C_4H_3^+$
+ $H_2C=O$
m/z 108 m/z 78 m/z 77 m/z 51
$-CH_3O^•$

$-CH_3^•$ —:CO *45.4 —HC≡CH *23.4 $C_3H_3^+$
m/z 93 m/z 65 m/z 39

**Figure 7.3c** Fragmentation pattern of methoxybenzene

Longer-chain ArOR compounds (R = $CH_3CH_2$, etc.) can also eliminate an alkene from the aliphatic group with the corresponding formation of the molecular ion of a phenol:

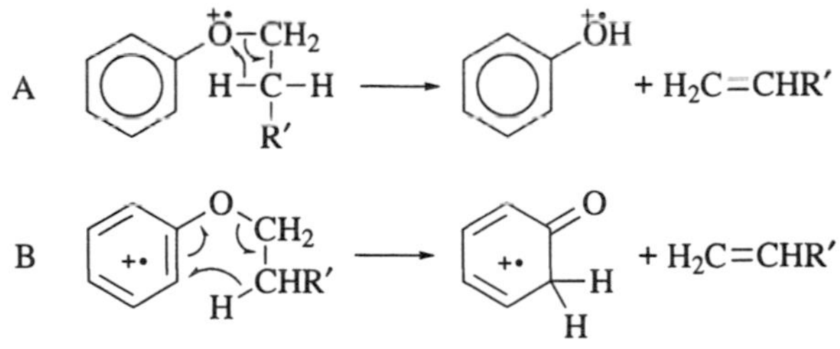

A  + $H_2C=CHR'$

B  + $H_2C=CHR'$

**SAQ 7.3**

Figure 7.3d shows the mass spectrum of an aliphatic ether (*Unknown 6*). Interpret this spectrum and suggest a structure for the ether.

→

**SAQ 7.3**
(Contd)

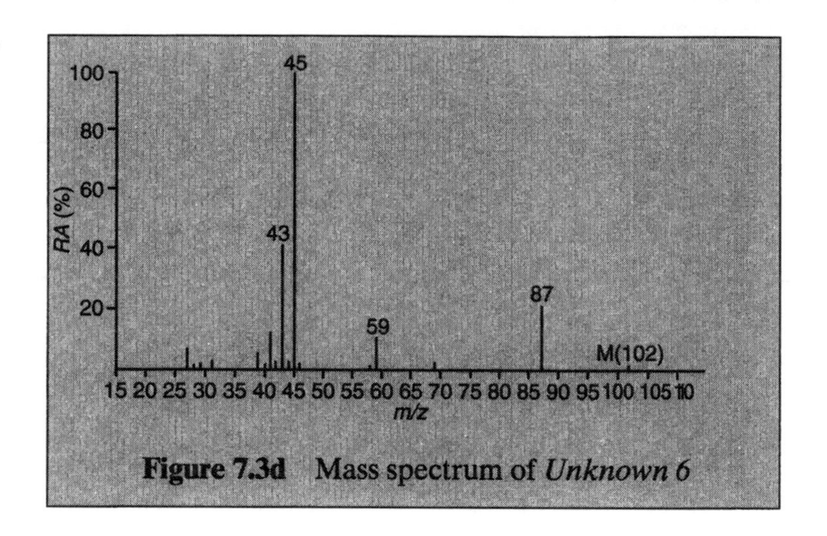


**Figure 7.3d** Mass spectrum of *Unknown 6*

**SAQ 7.3**

## 7.4 FRAGMENTATIONS OF HYDROXYBENZENES (PHENOLS)

Hydroxybenzenes have intense molecular ions and do not usually fragment to any great degree by the loss of either —OH hydrogen or the —OH group itself (see Figure 7.4a).

In the case of phenol, there is only a very weak $C_6H_5^+$ ion and there are additional ions present at *m/z* 66 and 65. These are due to elimination of neutral CO and the $CHO^•$ radical:

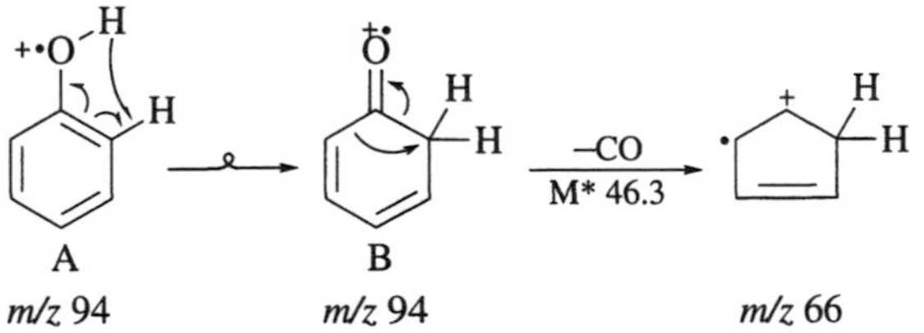

A *m/z* 94 → B *m/z* 94 —(–CO, M* 46.3)→ *m/z* 66

Ions A and B are keto–enol tautomers

Most substituted hydroxybenzenes show similar patterns of loss.

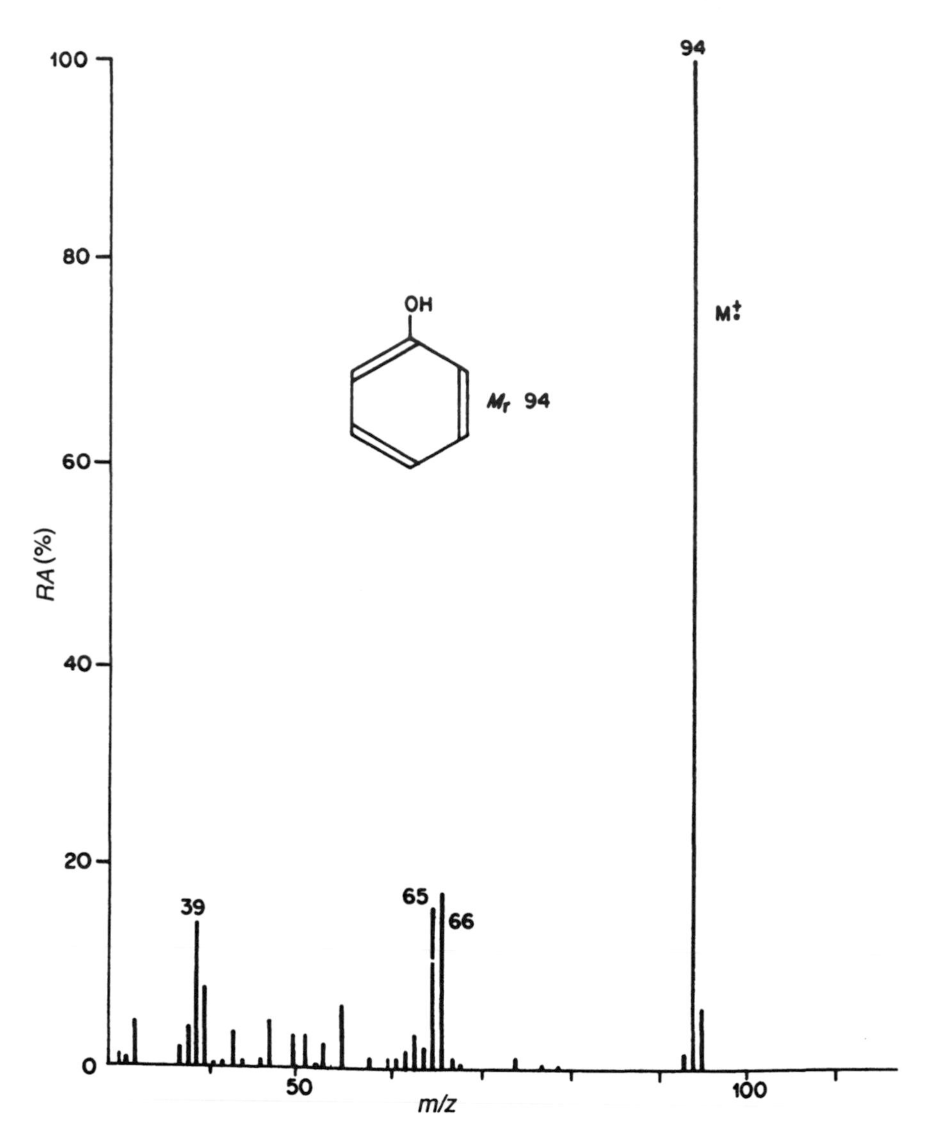


**Figure 7.4a** Mass spectrum of hydroxybenzene (phenol)

Alkyl-substituted compounds also show intense $(M-H)^+$ ions (see Figures 7.4b–7.4d, which show, respectively, the mass spectra of 2-, 3-, and 4-methoxybenzenes (the cresols)). This is again achieved via a resonance-stabilised hydroxytropylium ion (ions D and E in the following):

OH OH OH O H H
$-H^{\bullet}$
$CH_3$ $^{+}CH_2$ + +
C D E
*m/z* 108 *m/z* 107

Loss of CO, as before, then gives *m/z* 79.

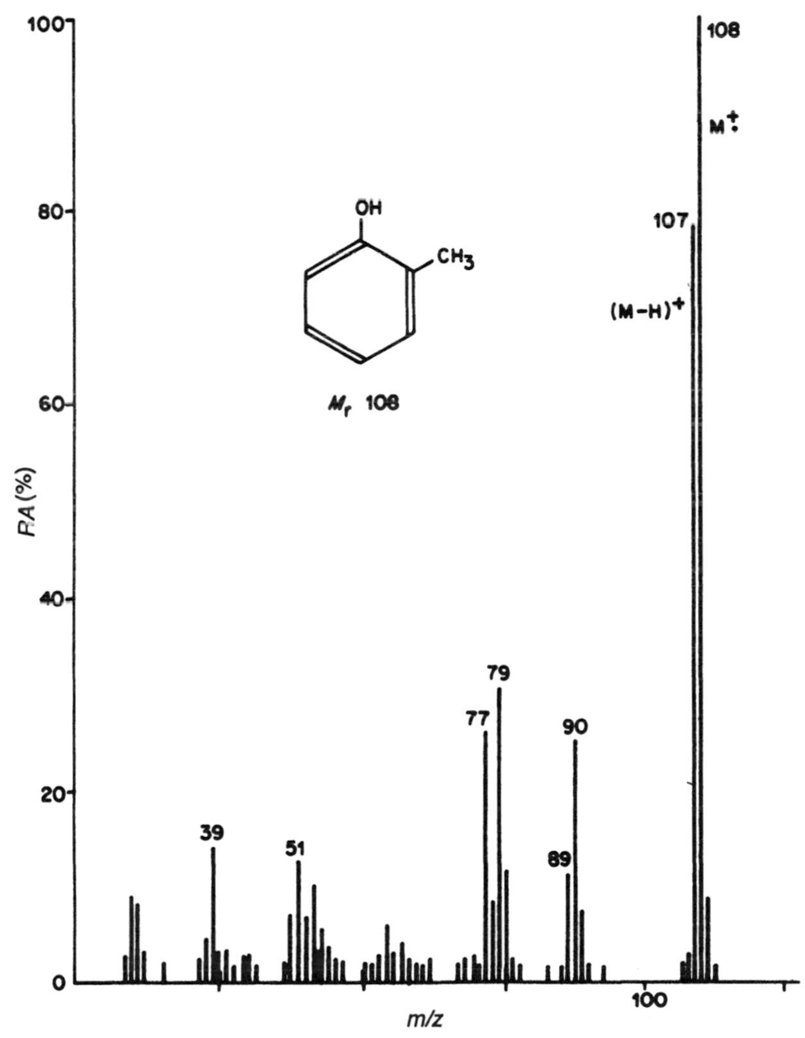


**Figure 7.4b** Mass spectrum of 2-methylhydroxybenzene

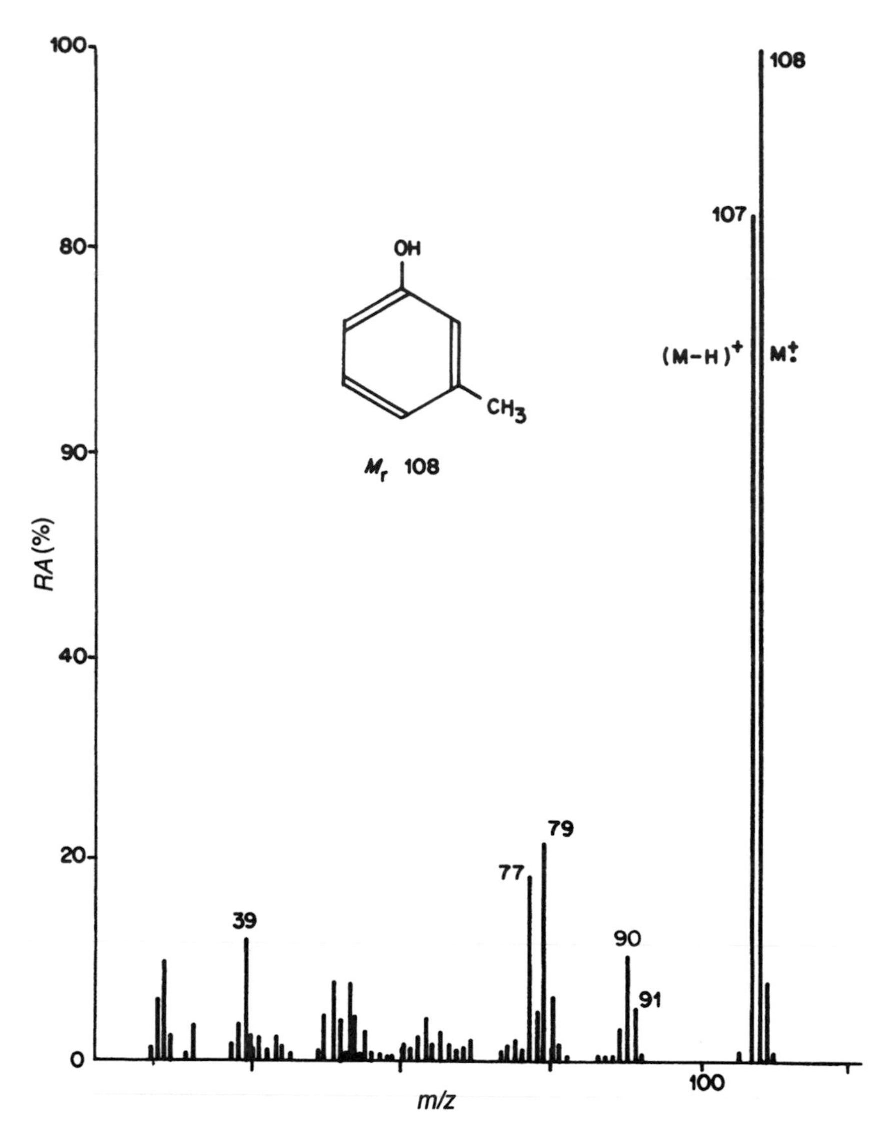


**Figure 7.4c** Mass spectrum of 3-methylhydroxybenzene

Π What other compound have we come across where a tropylium ion occurs in its fragmentation mechanisms?

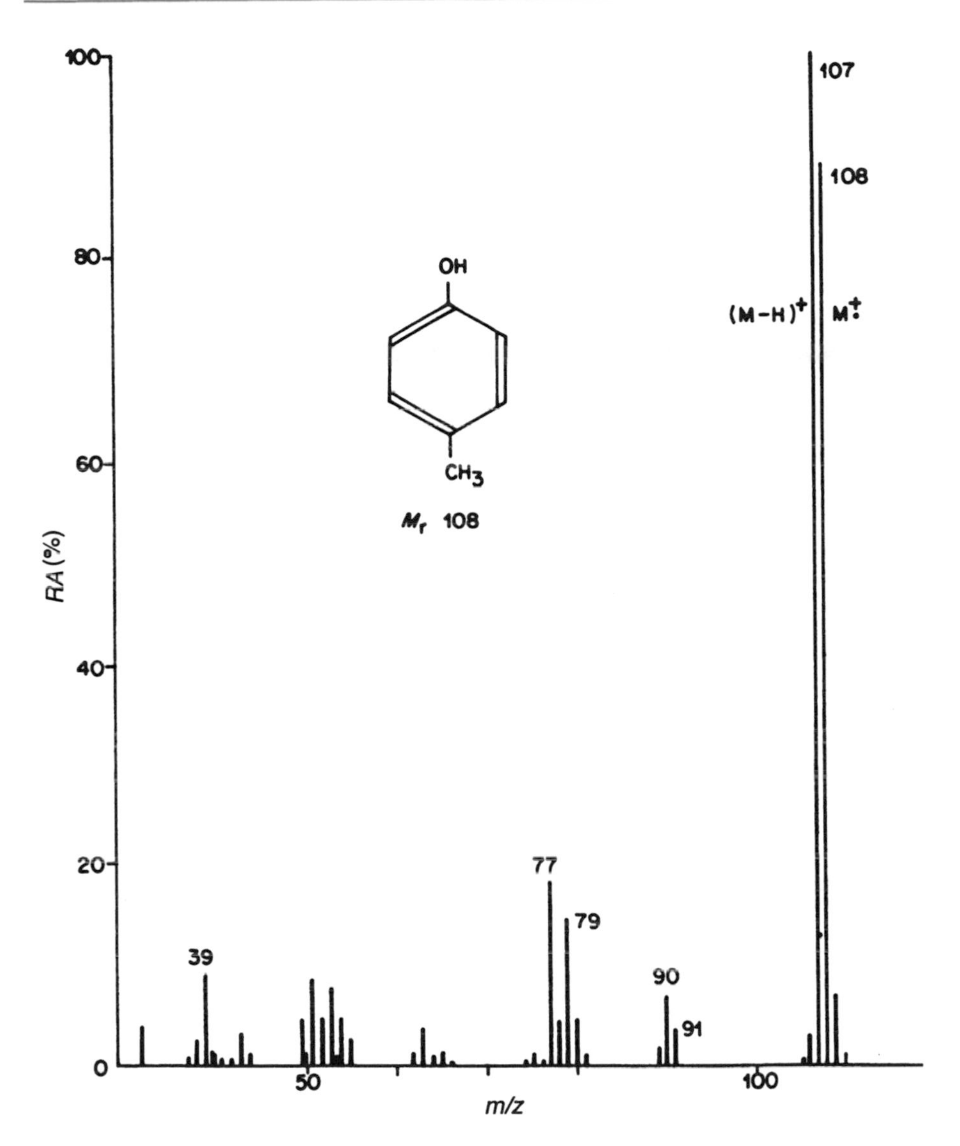


**Figure 7.4d** Mass spectrum of 4-methylhydroxybenzene

The answer is benzyl alcohol (see Figure 7.2e).

There are some similarities in the spectra of benzyl alcohol and the methoxybenzenes, but they are not completely identical because the *m/z* 107 ions formed from the four isomers will have differing amounts of excess energy and will fragment to different degrees.

2-Methyl hydroxybenzene shows an *ortho*-effect which causes the elimination of $H_2O$. By this means, one can distinguish this compound from the other two isomers:

OH H CH₂ → CH₂

*m/z* 108 *m/z* 90 (28% in Fig 7.4b)

Hydroxybenzenes with long saturated carbon chains may undergo both of the following:

(a) *benzylic cleavage*, i.e. cleavage of the α–β bond in the alkyl chain with loss of R• instead of H• (Stevenson's rule):

HO–C₆H₄–CH₂-R → HO–C₆H₄–$CH_2^+$ ≡ HO–C₇H₆⁺ + R•

*m/z* 107

(b) a *six-centred γ-hydrogen transfer process*, i.e. elimination of an alkene and hydrogen transfer to the benzene ring:

HO–C₆H₄–CH₂–CH₂–CH(H)–R¹ → HO–C₆H₄(=CH₂)(H)(H) + $R^1CH{=}CH_2$

*m/z* 108

This means that both *m/z* 107 and *m/z* 108 may be present in these spectra.

**SAQ 7.4**

Figure 7.4e shows the mass spectrum of a phenol (*Unknown 7*). Interpret this spectrum and suggest a structure for the compound.

**SAQ 7.4**
(Contd)

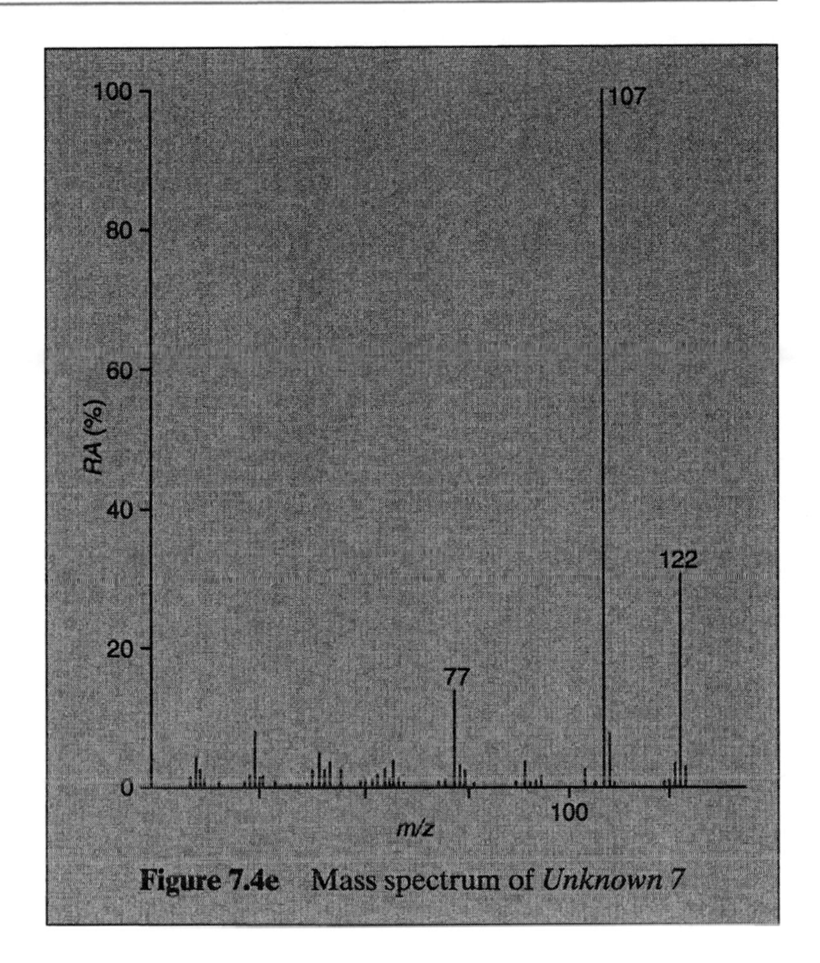


**Figure 7.4e** Mass spectrum of *Unknown 7*

**SAQ 7.4**

## 7.5 FRAGMENTATIONS OF CARBONYL COMPOUNDS

### 7.5.1 The McLafferty Rearrangement

The principle cleavage of carbonyl compounds is $\alpha$ to the C=O group, which in most $M^{+\bullet}$ will carry the positive charge:

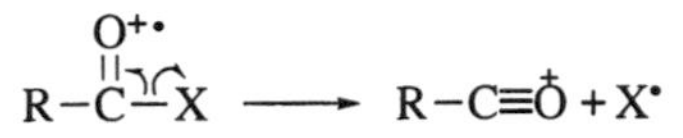

$$R{-}C({=}O^{+\bullet}){-}X \longrightarrow R{-}C{\equiv}\overset{+}{O} + X^{\bullet}$$

The formation of acylium ions by loss of $H^{\bullet}$, $R^{1\bullet}$, $R^1O^{\bullet}$, $HO^{\bullet}$, and $H_2N^{\bullet}$ are important diagnostic features of the mass spectra of aldehydes, ketones, esters, acids, and amides, respectively. There is another important rearrangement that is useful in identifying carbonyls which involves the concerted loss of an alkene from the R group, with hydrogen-transfer to the carbonyl oxygen. This rearrangement is named after its 'discoverer', Fred McLafferty, and is therefore termed the *McLafferty Rearrangement*, sometimes shortened to 'McL'.

FIgure 7.5a shows the general rearrangement procedure.

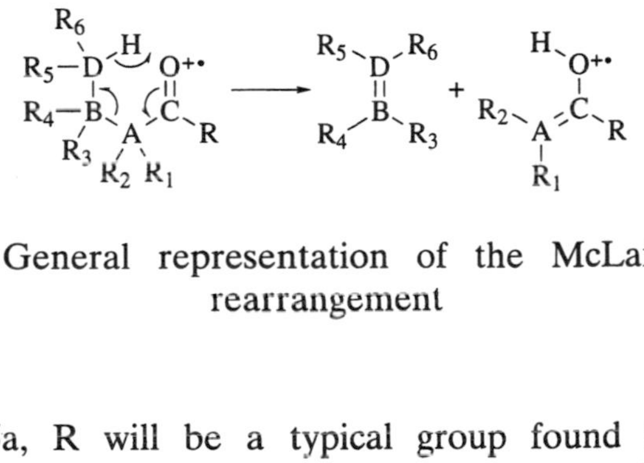

**Figure 7.5a** General representation of the McLafferty ('McL') rearrangement

In Figure 7.5a, R will be a typical group found in a carbonyl compound, e.g. $R_1$–$R_6$ can be a variety of alkyl, aryl, halo, alkoxy, aryloxy and other groups, while A, B and D could be various combinations of carbon, oxygen or sulphur.

Since the eliminated group is a neutral molecule (often an alkene), 'McL' rearrangements give rise to *even*-mass ions from C, H, O containing compounds, i.e. to molecular ions of a smaller molecule. Even-mass ions are unusual in spectra of these compounds, so their presence is usually distinctive, even if they are weak. In the case of *odd*-mass compounds, e.g. amides, the 'McL'-rearrangement ions will stand out because they will have *odd m/z* values. Table 7.5a summarises the common 'McL' peaks found in the spectra of various carbonyl compounds.

**Table 7.5a** Assignment of 'McL' peaks commonly observed in the spectra of various carbonyl compounds

| Compound type | Substituent (R) | 'McL' peak ($m/z$) | Structure[a] |
|---|---|---|---|
| Aldehyde | H | 44 | $CH_2{=}C(\overset{+\bullet}{O}H)H$ |
| Methyl ketone | $CH_3$ | 48 | $CH_2{=}C(\overset{+\bullet}{O}H)CH_3$ |
| Amide | $H_2N$ | 59 | $CH_2{=}C(\overset{+\bullet}{O}H)NH_2$ |
| Acid | HO | 60 | $CH_2{=}C(\overset{+\bullet}{O}H)OH$ |
| Ethyl ketone | $CH_3CH_2$ | 72 | $CH_2{=}C(\overset{+\bullet}{O}H)CH_2CH_3$ |
| Methyl ester | $CH_3O$ | 74 | $CH_2{=}C(\overset{+\bullet}{O}H)OCH_3$ |
| Propyl ketone | $CH_3CH_2CH_2$ | 86 | $CH_2{=}C(\overset{+\bullet}{O}H)CH_2CH_2CH_3$ |
| | | 58[b] | $CH_2{=}C(\overset{+\bullet}{O}H)CH_2$ |
| Ethyl ester | $CH_3CH_2O$ | 88 | $CH_2{=}C(\overset{+\bullet}{O}H)OCH_2CH_3$ |
| | | $(M-CH_2{=}CH_2)$[c] | $R-C(\overset{+\bullet}{O}H){=}O$ |
| Phenyl ketone | $C_6H_5$ | 120 | $CH_2{=}C(\overset{+\bullet}{O}H)C_6H_5$ |
| Phenyl ester | $C_6H_5O$ | 136 | $CH_2{=}C(\overset{+\bullet}{O}H)OC_6H_5$ |

[a]The $HC_2{=}$ group in any of the ions listed in this column may be substituted either singly or doubly, e.g. $CH_3CH{=}$ ions would all be 14 amu higher
[b]This ion arises because the propyl group itself contains a γ-hydrogen; the 'McL' rearrangement can thus occur *twice* consecutively.
[c]This ion arises because the 'McL' rearrangement can utilise γ-hydrogens from either the RCO or $OCH_2CH_3$ groups, thus giving 'McL' ions of different structures and/or masses

**SAQ 7.5a**

In which of the following compounds would you expect to see a 'McI' peak? Give its *m/z* value and structure.

(a) $CH_3COCH_2CH_3$;

(b) $CH_3COCH_2CH_2CH_3$;

(c) $CH_3COCH(CH_3)_2$;

(d) $(CH_3)_2CHCH_2CHO$;

(e) $CH_3CH_2COOCH_2CH_3$;

(f) $CH_3CH_2CH_2CONHCH_3$;

(g) $CH_3CH(CH_3)CH_2CH(CH_3)COOH$;

(h) $CH_3CH_2CH(CH_3)COC_6H_5$;

(i) $CH_3CH{=}CHCH_2COCH_2CH_3$;

(j) $(C_6H_5)_2\overset{\overset{\large O}{\|}}{P}SCH_2CH_3$

**SAQ 7.5a**

Sometimes, we can have a double 'McL' situation, as is the case with propyl ketones (simplest form, propyl ketone — see footnote (b) in Table 7.5a) or esters containing γ-hydrogens in both of the acid and the alcohol portions of the molecule (simplest form, ethyl butanoate — see footnote (c) in Table 7.5a).

Π Can you draw the resonance forms of the first 'McL' ion (*m/z* 86) of dipropyl ketone (4-heptanone) and show how it might fragment to give second 'McL' isomeric ions of *m/z* 58?

The second 'McL' rearrangement can be depicted as follows:

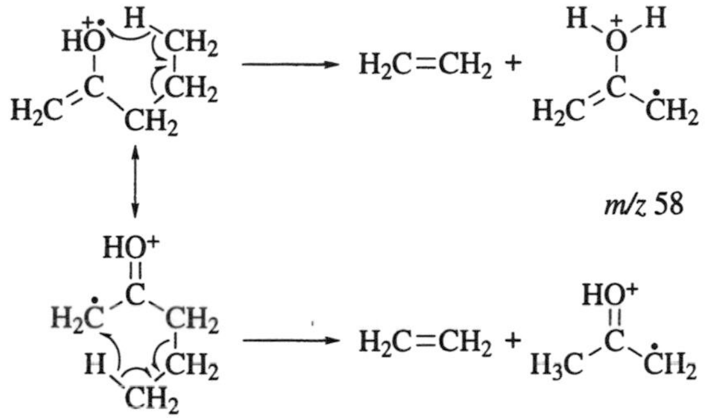

$\longrightarrow H_2C{=}CH_2 +$

*m/z* 58

$\longrightarrow H_2C{=}CH_2 +$

Note that in the formation of the second 'McL' ions (*m/z* 60) in ethyl butanoate (mass spectrum shown in Figure 7.5b), we do not always require an $X{=}Y^{+\bullet}$ terminus; $C{-}\overset{+\bullet}{O}{-}H$ will do, for example.

The mechanisms for the formation of the two structures with *m/z* 88 and *m/z* 60 are shown below:

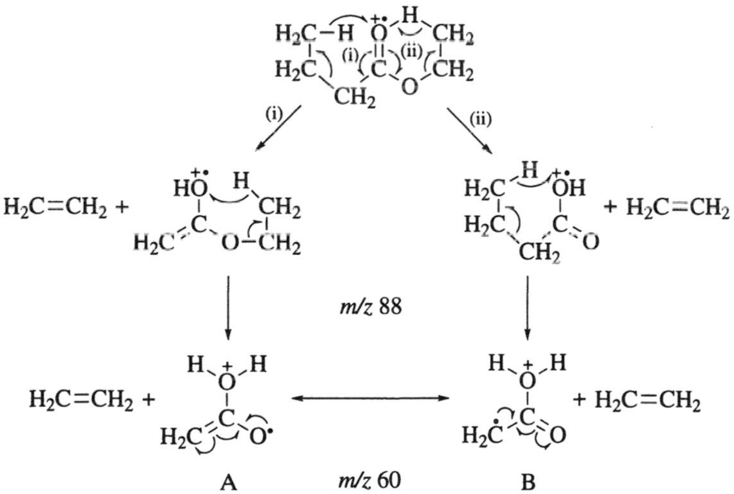

(i) (ii)

$H_2C{=}CH_2 +$ $+ H_2C{=}CH_2$

*m/z* 88

$H_2C{=}CH_2 +$ $+ H_2C{=}CH_2$

A *m/z* 60 B

## 7.5.2 Fragmentations of Aldehydes

(a) α-*Cleavage*

The molecular ions of aliphatic aldehydes are frequently weak, while aromatic aldehydes show quite intense molecular ions. The

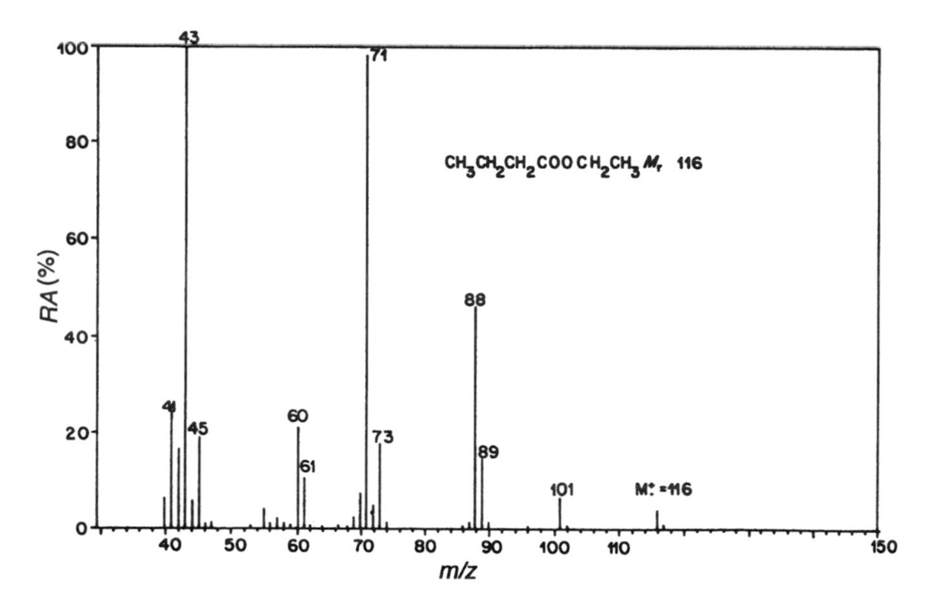


**Figure 7.5b** Mass spectrum of ethyl butanoate

characteristic feature of either class of aldehydes is the loss of α-hydrogens:

$$\text{R}-\underset{\text{H}}{\overset{|}{\text{C}}}=\overset{+\bullet}{\text{O}} \longrightarrow \text{R}-\text{C}\equiv\overset{+}{\text{O}} + \text{H}^{\bullet}$$

(M–1)

(R = alkyl or aryl)

Benzoyl ions (R = Ar) are particularly stabilised by resonance. Loss of $R^{\bullet}$ or $Ar^{\bullet}$ also occurs, giving rise to $H-C=O^{+}$, *m/z* 29. The presence of both $(M-1)^{+}$ and *m/z* 29 ions are therefore often useful indicators of the presence of aldehydes. (Note that acetals, $RCH(OR^{1})_2$, and alcohols can also give $(M-1)^{+}$ peaks occasionally.)

(b) β-*Cleavage*

Aliphatic aldehydes often undergo β-cleavage. For aldehydes with a $CH_2-CHO$ end group, this gives rise to a characteristic (M–43) peak:

$$R{-}CH_2{-}C{=}\overset{+\bullet}{O} \longrightarrow \underset{(M{-}43)}{R^+} + CH_2{=}CHO^{\bullet}$$

or via homolytic β-cleavage to give $m/z$ 43 and $R^{\bullet}$:

$$R{-}CH_2{-}\underset{H}{C}{=}\overset{+\bullet}{O} \longrightarrow R^{\bullet} + \underset{m/z\ 43}{CH_2{=}\underset{H}{C}{-}\overset{+}{O} \longleftrightarrow \overset{+}{C}H_2{-}\underset{H}{C}{=}O}$$

α-Substituents will of course cause shifts in the mass loss from $M^{+\bullet}$, and the $m/z$ 43 peak; higher aliphatic aldehydes may have resultant $R^+$ groups that fragment in ways which are typical of hydrocarbons.

### (c) *'McL' Rearrangement*

Aliphatic aldehydes can also undergo 'McL' rearrangements, e.g. to give a characteristic $m/z$ 44 structure in the case of $CH_2{=}CH\overset{+\bullet}{O}H$:

$$\text{R(H)CH–CH}_2\text{–CH}_2\text{–CH}{=}\overset{+\bullet}{O} \longrightarrow \text{RCH}{=}\text{CH}_2 + \underset{m/z\ 44}{H_2C{=}CH{-}HO^{+\bullet}}$$

### (d) *Other Mechanisms*

In addition, many higher aldehydes without branching show ions having the formula $C_nH_{2n}^{+\bullet}$, which occur by γ-hydrogen transfer processes:

$$\text{R(H)CH–CH}_2\text{–CH}_2\text{–CH}{=}\overset{+\bullet}{O} \longrightarrow \text{RCH}^{\bullet}\text{–CH}_2\text{–CH}_2\text{–CH}{=}\text{HO}^+ \longrightarrow \underset{\substack{(M{-}44) \\ C_nH_{2n}^{+\bullet}}}{R\dot{C}H{-}\overset{+}{C}H_2} + H_2C{=}CH{-}HO$$

These features are illustrated in the mass spectrum of butanal (Figure 7.5c). The spectrum of benzaldehyde was discussed in Section 7.1 (as *Unknown 1*).

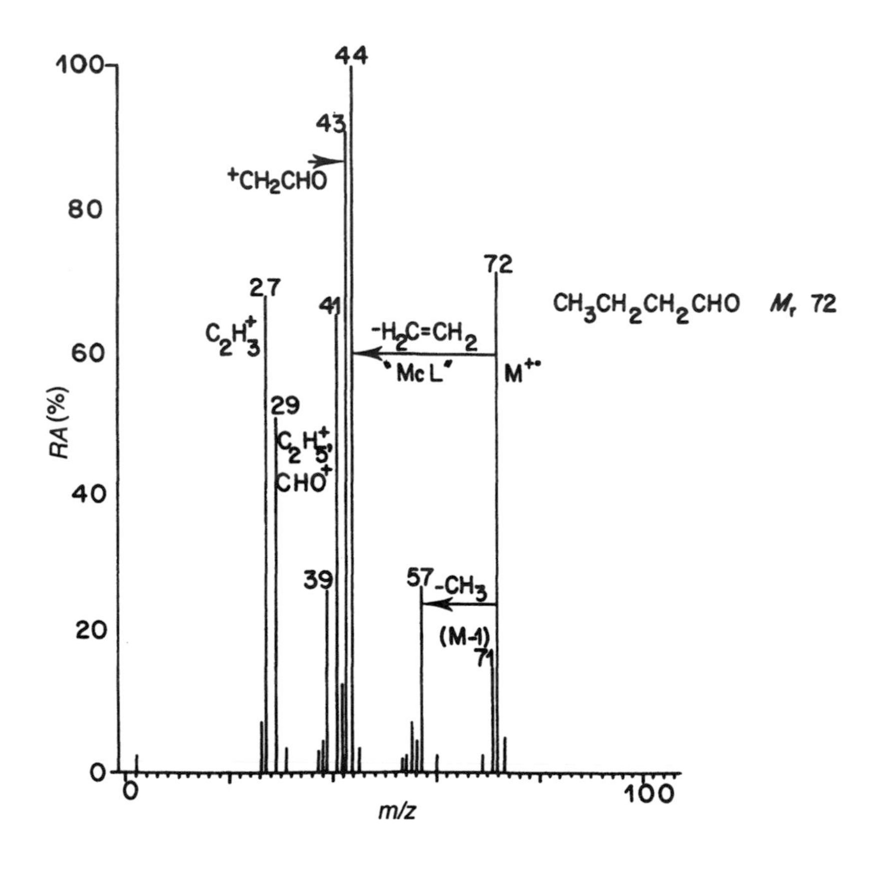


**Figure 7.5c** Mass spectrum of butanal

**SAQ 7.5b**

> Figure 7.5d shows the mass spectrum of an aldehyde (*Unknown 8*). Interpret the spectrum and identify the aldehyde.

**SAQ 7.5b**
(Contd)

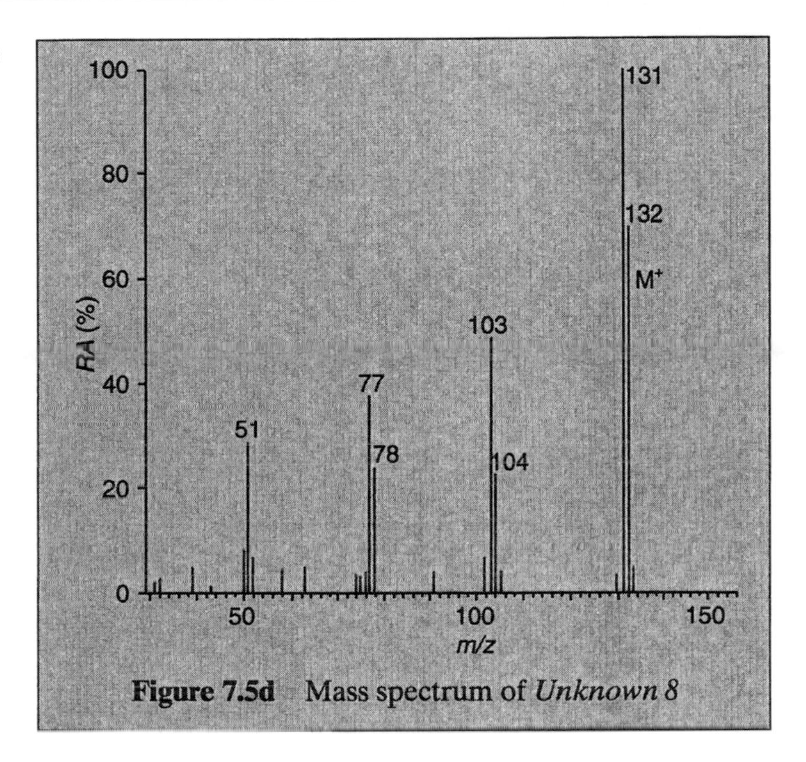


**Figure 7.5d** Mass spectrum of *Unknown 8*

**SAQ 7.5b**

### 7.5.3 Fragmentations of Ketones

We have already said a great deal about the ways in which ketones fragment, e.g. in Sections 6.3.2, 7.1, and 7.5.1, so this section will only briefly recap some of the more important points.

The $M^{+\bullet}$ peaks of most ketones are quite intense, indeed more so than aldehydes. The loss of alkyl groups by α-cleavage is an important mode of fragmentation. The larger of the two alkyl groups is most readily lost (Stevenson's rule), a trend which increases if one of the R groups increases in size at the expense of the other. For example, in the mass spectrum of 2-butanone, the ratio of the two acylium ions, $m/z$ 43 and $m/z$ 57, is 10/1, while in 2-hexanone, the corresponding ratio of *m/z* 43 and *m/z* 85 is 20/1. Hydrocarbon ions, formed by heterolytic α-, β-, or γ-cleavages, are usually weak (unless there is a branch point in the chain), thus giving secondary or tertiary carbocations.

The 'McL' rearrangement of ketones has already been discussed in Section 7.5.1. If a γ-hydrogen is present, it gives rise to ions of the type $R_1R_2C{=}C(OH)R^{+\bullet}$, where R, $R_1$ and $R_2$ may be a great variety of substituents. If R is long enough to possess γ-hydrogen atoms itself, a double 'McL' process may take place.

Cyclic ketones fragment as a result of α-cleavage, to give mass spectra of the type shown in Figure 7.5e for cyclohexanone.

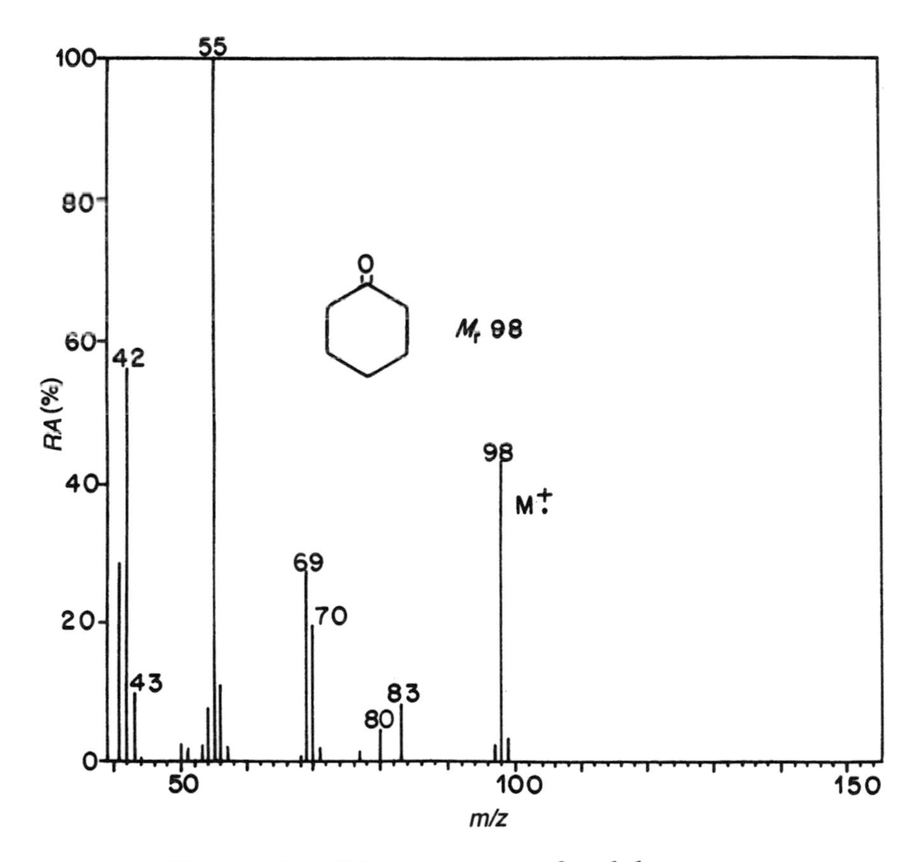


**Figure 7.5e** Mass spectrum of cyclohexanone

Aromatic ketones give benzoyl ions:

$$\mathrm{Ar\overset{+\bullet}{C}OR} \longrightarrow \underset{(100\%)}{\mathrm{ArCO^{+} + R^{\bullet}}}$$

although the corresponding acylium ion will usually be observed:

$$\mathrm{Ar\overset{+\bullet}{C}OR} \longrightarrow \underset{(\leqslant 20\%)}{\mathrm{RCO^{+} + Ar^{\bullet}}}$$

If R contains γ-hydrogens, the 'McL' ion *m/z* 120, $C_6H_5C(OH){=}CH_2^{+\bullet}$, or a substituted analogue, will appear (see section 7.5.1).

**SAQ 7.5c**

Figure 7.5f shows the mass spectrum of a ketone (*Unknown 9*). Interpret this spectrum and identify the ketone.

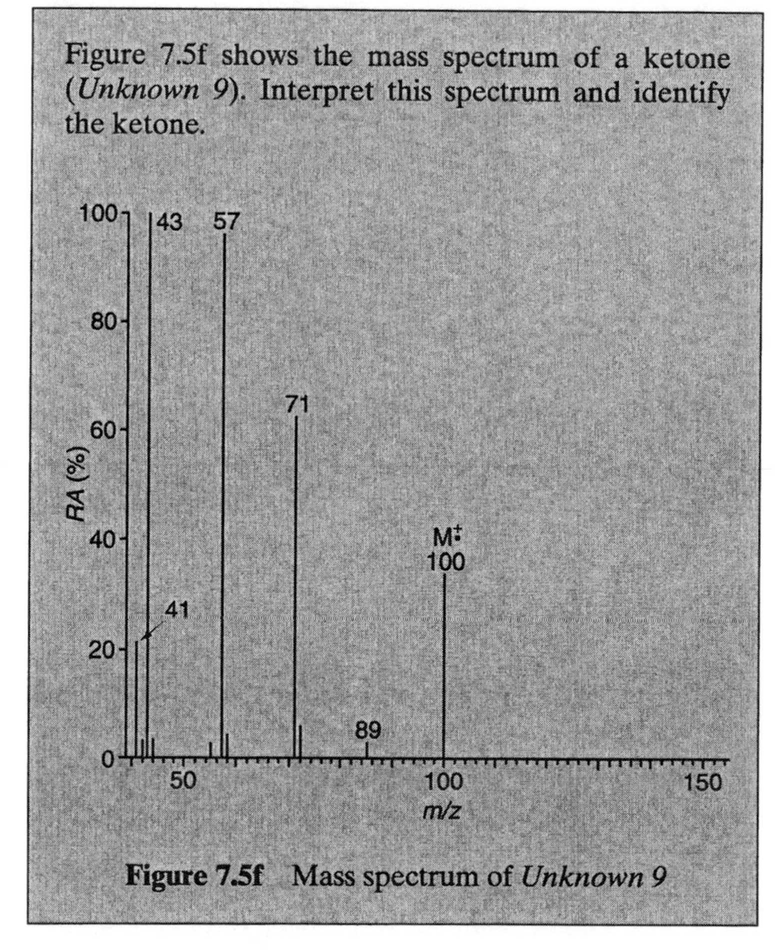


**Figure 7.5f** Mass spectrum of *Unknown 9*

**SAQ 7.5c**

**SAQ 7.5c**

### 7.5.4 Fragmentations of Esters

$\Pi$ What types of ions would you expect to be formed from the esters $RCO_2R_1$, $ArCO_2R_1$ and $RCO_2Ar$?

Laying out the molecules, so that all of the α-bonds are clear, you may have come up with the following:

| | $RCO_2R_1$ | | | $ArCO_2R_1$ | | | $RCO_2Ar$ | |
|---|---|---|---|---|---|---|---|---|
| | R | | | Ar | | | R | |
| $R_1OCO^+$ ← | | → $R^+$ | $R_1OCO^+$ ← | | → $Ar^+$ | $ArOCO^+$ ← | | → $R^+$ |
| | C=O | | | C=O | | | C=O | |
| $R_1O^+$ ← | | → $RCO^+$ | $R_1O^+$ ← | | → $ArCO^+$ | $ArO^+$ ← | | → $RCO^+$ |
| | O | | | O | | | O | |
| $RCO_2^+$ ← | | → $R_1^+$ | $R_1^+$ ← | | → $ArCO_2^+$ | $RCO_2^+$ ← | | → $Ar^+$ |
| | $R_1$ | | | $R_1$ | | | Ar | |

There will be even-mass 'McL' ions from esters whose R and $R_1$ groups have γ-hydrogens, e.g. $CH_2{=}C(OH)OR_1^{+\bullet}$ and $RCOOH^{+\bullet}$. If both R and $R_1$ have a γ-hydrogen, then a double 'McL' rearrangement is possible, leading to a further 'McL' ion, e.g. $CH_2{=}C(\overset{+}{O}H_2)O^\bullet$ at *m/z* 60.

Some α-cleavage ions are either unstable or very weak in most mass spectra. These are $R^+$, $R_1^+$, $R_1OCO^+$, $R_1O^+$, $RCO_2^+$, and $ArCO_2^+$. The spectra of esters tend to be dominated by the $RCO^+$ and $ArCO^+$ ions and the 'McL' rearrangement ion peaks. $R_1OCO^+$ ions (*m/z* 59, 73, 87, 101, etc.), although weak, can be useful in identifying the alcohol portion of an ester.

(a) γ-*Cleavage*

γ-Cleavage of the acyl chain gives ions in the series, *m/z* 87, 101, 115, etc., for methyl, ethyl, propyl esters, and so on:

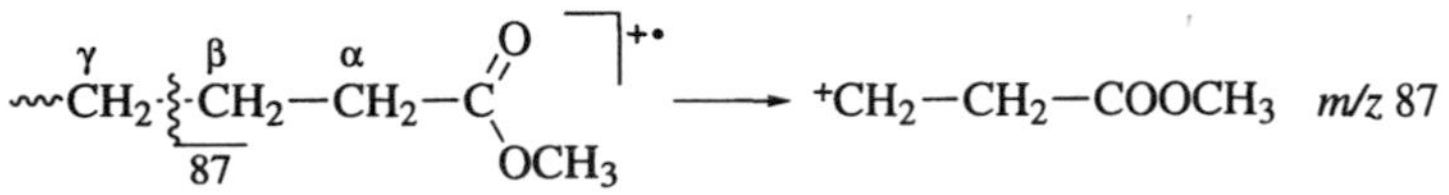

$$\sim CH_2 \cdot CH_2-CH_2-C(=O)OCH_3^{+\bullet} \longrightarrow {}^{+}CH_2-CH_2-COOCH_3 \quad m/z\ 87$$

(b) *Double H-Transfer (DHT)*

In this process, ions of the structure $R-C=\overset{+}{O}H(OH)$, at 1 amu higher than the corresponding 'McL' peaks, are produced.

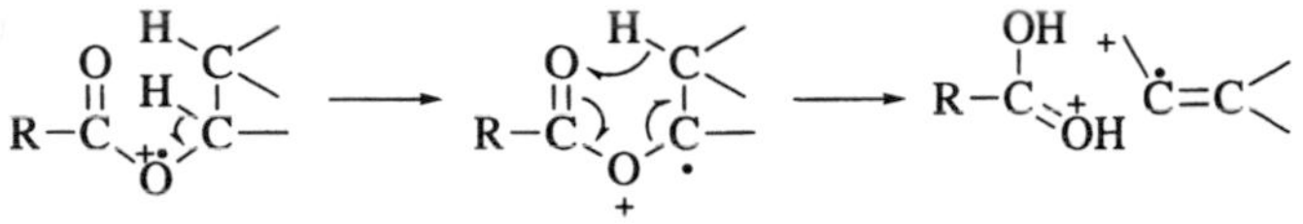

DHT becomes increasingly common as the alkyl chain of the alcohol moiety lengthens. These peaks are of *odd* mass in the spectra of C, H, O esters.

Esters of aromatic acids may also show both 'McL' and DHT peaks in their mass spectra if the alkyl group has the appropriate β- and γ-hydrogens e.g.

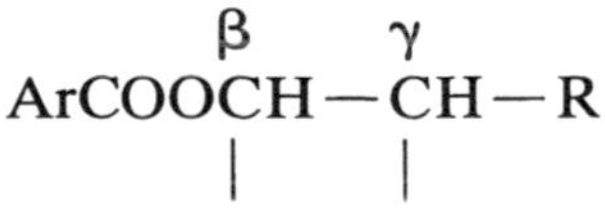

$$\overset{\beta}{ArCOOCH}-\overset{\gamma}{CH}-R$$

The resulting ions are $ArCOOH^{+}$ and $ArC(OH)_2^{+}$.

*Ortho*-substituted esters lose ROH rather than RO•. However, this is not the case with the 3- and 4-disubstituted isomers:

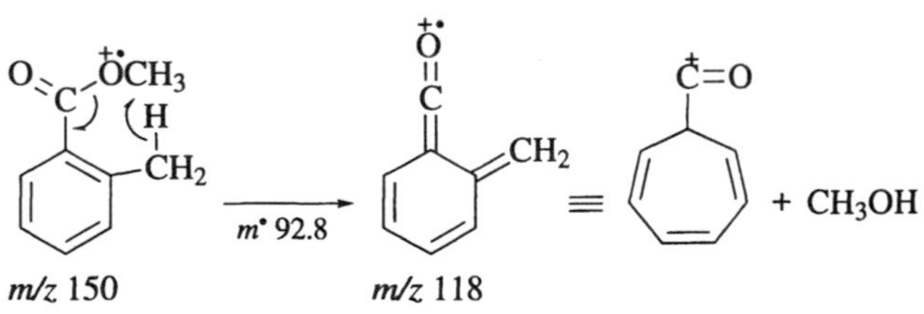

$m/z$ 150 $\xrightarrow{m^{*}\ 92.8}$ $m/z$ 118 + $CH_3OH$

Benzyl esters eliminate a neutral ketene molecule and form $C_6H_5CH_2OH^{+\bullet}$, $m/z$ 108. Further loss of H• and CO may also occur (see Section 7.2):

$$C_6H_5-CH_2-\overset{+\bullet}{O}-C(=O)-CHR-H \longrightarrow C_6H_5CH_2OH^{+\bullet} + RCH{=}C{=}O$$

Aryl esters (ArOCOR), such as phenyl ethanoate, also rearrange via a six-membered transition state to eliminate a ketene and give the keto form of the corresponding phenol. This is often the base peak:

$$\underset{m/z\ 136}{C_6H_5\overset{+\bullet}{O}C(=O)CH_2H} \xrightarrow[m^*\ 64.9]{-CH_2=C=O} \underset{m/z\ 94}{C_6H_6O^{+\bullet}} \xrightarrow[m^*\ 46.3]{-CO} \underset{m/z\ 66}{C_5H_6^{+\bullet}}$$

The fragmentation of di- and polyesters is too complex to discuss in detail here, but one such class of esters, namely the dialkylphthalates, is important. All phthalates give intense *m/z* 149 peaks, which are formed from cyclisation of the first formed acylium ion from one of the —COOR groups. Elimination of alkene, and hydrogen transference to the central oxygen then takes place:

$$C_6H_4(\overset{+\bullet}{O}{=}C{-}OR)(COOR) \xrightarrow{-RO^\bullet} C_6H_4(C{\equiv}O^+)(C(=O)\ddot{O}R) \longrightarrow C_6H_4(CO)_2\overset{+}{O}{-}R \xrightarrow{-(R-H)} \underset{m/z\ 149}{C_6H_4(CO)_2\overset{+}{O}H}$$

Π Why might you find this peak at *m/z* 149 in your mass spectra?

Dialkylphthalates are widely used as plasticisers and may be leached out by common solvents from the walls of plastic containers and tubing, etc. Therefore, it is important that you use glass containers wherever possible when submitting samples for MS analysis.

**SAQ 7.5d**

Figure 7.5g shows the mass spectrum of an ester, (*Unknown 10*). Interpret this spectrum and suggest a structure for the compound.

**SAQ 7.5d**
(Contd)

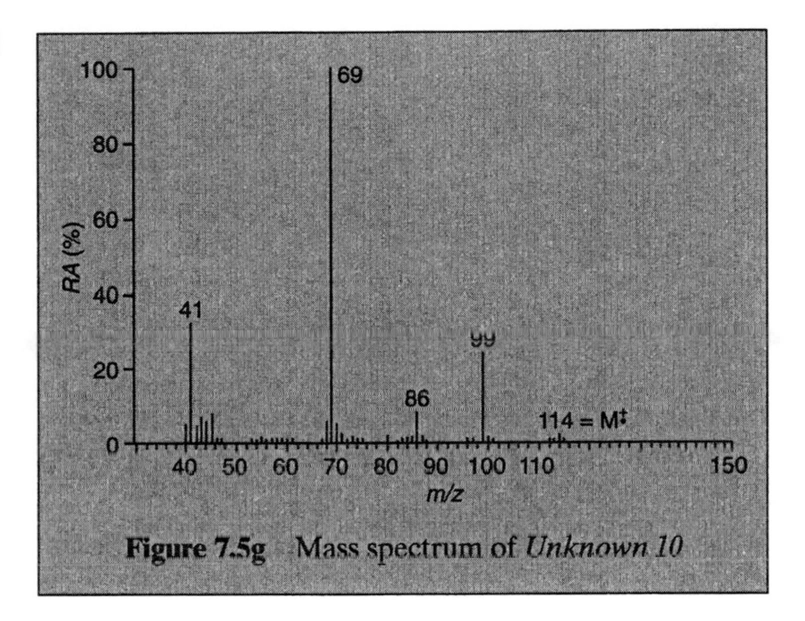


**Figure 7.5g** Mass spectrum of *Unknown 10*

**SAQ 7.5d**

### 7.5.5 Fragmentations of Acids

Only weak $M^{+\bullet}$ peaks are observed for aliphatic acids. In a similar way to methyl esters, short-chain acids lose both $^{\bullet}CO_2H$ and $^{\bullet}OH$. Loss of the OH hydrogen does not occur, but loss of the $R^{\bullet}$ group from RCOOH, leaving COOH (*m/z* 45), is quite apparent. If a $\gamma$-hydrogen is present, the 'McL' peak, ($CH_2{=}C(OH_2^{+\bullet})$) (*m/z* 60), is usually very intense (see Figure 7.5h).

Aromatic acids, however, give intense $M^{+\bullet}$ peaks and usually fragment by the loss of $OH^{\bullet}$, followed by CO from the aroyl ion to give $Ar^{+}$:

$$ArC(=O^{+\bullet})-OH \longrightarrow \dot{O}H + Ar-C{\equiv}O^{+\bullet} \longrightarrow CO + Ar^{+}$$

*Ortho*-substituted aromatic acids can eliminate $H_2O$ from $M^{+\bullet}$ if there is a suitably located hydrogen in the *ortho*-group:

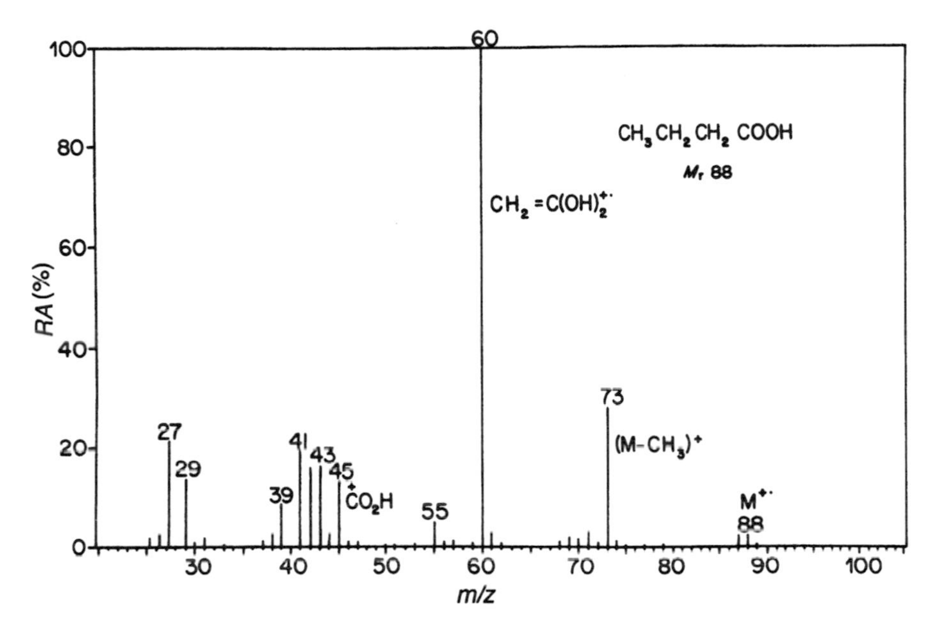


**Figure 7.5h** Mass spectrum of butanoic acid

salicyclic acid $\longrightarrow$ $H_2O$ + $\xrightarrow[m^*\ 70.5]{-CO}$

salicyclic acid

*m/z* 138 *m/z* 120 *m/z* 92

In most analytical work, acids are base-extracted, neutralised and converted to esters. For example, from the trimethylsilyl, $-OSi(CH_3)_3$, (TMS) derivative you will find ions which are 57 and 72 amu higher than expected for the acid alone.

### 9.5.6 Fragmentations of Amides

Π Do you remember what the nitrogen rule states?

Amides, as compounds containing nitrogen, must follow this rule, i.e. if they contain an *odd* number of nitrogen atoms, they will have an *odd* $M_r$ value.

There are often intense peaks in the spectra of both aliphatic and aromatic amides. The fragmentation patterns are quite similar to those of the corresponding acids and esters. α-Cleavage occurs by loss of R•, which in primary amides gives $(O{=}C{=}NH_2)^+$ in two resonance structures. *N*-substituted ions follow a series, *m/z* 44, 58, 72, 86, etc.

$$R{-}C({=}O^{+\bullet})NH_2 \longrightarrow R^{\bullet} + \overset{+}{O}{\equiv}C{-}\ddot{N}H_2$$

or $\updownarrow$ *m/z* 44

$$R{-}C({=}O)NH_2 \longrightarrow R^{\bullet} + \ddot{O}{=}C{=}\overset{+}{N}H_2$$

The acylium ion ($RCO^+$) peaks are usually weak in aliphatic-amide mass spectra, but the corresponding aroyl ($ArCO^+$) peaks are stronger for aromatics, as the π-orbital system helps to stabilise the carbonyl group's positive charge. 'McL' rearrangements can occur readily in amides; sometimes they undergo DHT rearrangements as well by using the β- and γ-hydrogen atoms from the *N*-alkyl group, in the same way as esters; however, these ions are relatively weak. The spectra also show ions typical of the $-NR_1R_2$ group, similar to those formed in amine spectra (see Section 7.6), such as *m/z* 58 $(CH_3CH_2\overset{+}{N}H{=}CH_2)$ from $-N(CH_3CH_2)_2$ amides.

Cleavage between the β- and γ-atoms in the acid portion of an aliphatic amide is usually quite important:

$$\sim\overset{\gamma}{C}H_2\text{-}\wr\text{-}\overset{\beta}{C}H_2\overset{\alpha}{C}H_2CONR_1R_2\rceil^{+\bullet} \longrightarrow \sim CH_2^{\bullet} + {}^{+}CH_2CH_2CONR_1R_2$$

In primary amides, this ion occurs at *m/z* 72 $(\overset{+}{C}H_2CH_2CONH_2)$. β-Cleavage also occurs in the N-alkyl chains:

$$CH_3CO{-}\overset{+\bullet}{\underset{H}{N}}{-}CH_2{-}R \longrightarrow R^{\bullet} + \underset{\substack{|\\H_2C-H}}{\overset{\substack{O\\\|}}{C}}{-}\overset{+}{N}H{=}CH_2 \longrightarrow CH_2{=}C{=}O + H_2\overset{+}{N}{=}CH_2$$

*m/z* 72 *m/z* 30

This is followed by the expulsion of a ketene and the formation of *m/z* 30.

Aromatic amides, such as acetanilide, $C_6H_5NHCOCH_3$, fragment almost entirely by the loss of ketene-type compounds, giving base peaks of the corresponding anilines:

$\longrightarrow$ + O=C=CHR

*m/z* 93 (100%)

The rest of the mass spectrum looks like that of the aniline (see Section 7.6), except for a relatively small $RCH_2CO^+$ ion derived from α-cleavage of $M^{+\bullet}$ (*m/z* 43 in acetanilide).

**SAQ 7.5e**

Figure 7.5i shows the mass spectrum of an amide (*Unknown 11*) which is isomeric with butanamide. Suggest a structure for this compound.

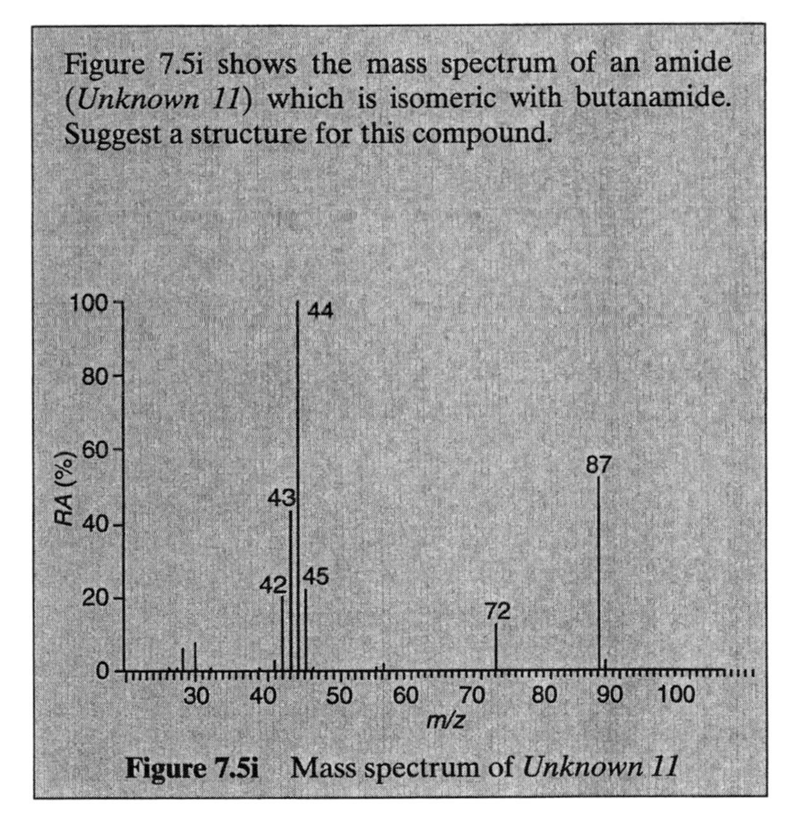


**Figure 7.5i** Mass spectrum of *Unknown 11*

**SAQ 7.5e**

## 7.6 FRAGMENTATIONS OF AMINES

The structures of amines are analogous to structures we have already encountered. For example, primary amines are analogous to alcohols ($CH_2\overset{+\bullet}{N}H_2$ and $CH_2\overset{+\bullet}{O}H$), secondary/tertiary amines to ethers ($CH_2N\overset{+\bullet}{N}R(Ar)$ and $CH_2{-}\overset{+\bullet}{O}{-}R(Ar)$), and aromatic amines to phenols ($C_6H_5\overset{+\bullet}{N}H$ and $C_6H_5\overset{+\bullet}{O}H$). The imonium ions, $>C{=}\overset{+}{N}H_2$ and $R{-}\overset{+}{N}H{=}CH_2$, are even more stable than oxonium ions; nitrogen is better able to carry a positive charge. Consequently, since the $M_r$ is odd, aliphatic-amine spectra are dominated by ions occurring in the *even* series, *m/z* 30, 44, 58, 72, 86, 100, etc., compared with the *odd* series, *m/z* 31, 45, 59, 73, 87, 111, etc. For example, in the spectrum of $CH_3CH_2NH_2$, (M–1) is 18% ($M^{+\bullet}$ = 17%) and *m/z* 30 ($CH_2{=}\overset{+}{N}H_2$) is the base peak.

$$H{-}\underset{CH_3}{\overset{H}{C}}{-}\overset{+}{N}H_2 \xrightarrow{-H^\bullet} CH_3CH{=}\overset{+}{N}H_2 \quad m/z\ 44$$

$$H{-}\underset{CH_3}{\overset{H}{C}}{-}\overset{+}{N}H_2 \xrightarrow{-CH_3^\bullet} CH_2{=}\overset{+}{N}H_2 \quad m/z\ 30$$

In fact, for most $RCH_2NH_2$ compounds, the base peak is *m/z* 30. The presence of even-mass ions in a spectrum may alert you to the presence of an amine, particularly in conjunction with the absence of a strong $M^{+\bullet}$ peak.

Hydrocarbon ions in the series $CH_3(CH_2)_n^+$, *m/z* 29, 43, 57, 71, etc., i.e. of *m/z* (14*n* + 1), where *n* is the number of carbon atoms in the ion, are quite weak in amine spectra, even when *n* is large, in contrast to higher alcohols and ethers. Amines do not fragment in analogy to the loss of $H_2O$ from alcohols. Instead, they undergo a fragmentation to form cyclic ammonium ions of variable ring sizes, in the series *m/z* 58, 72, 86, 100, etc.

$$R{-}H_2C\text{—}(CH_2)_n\text{—}\overset{+\bullet}{N}H_2 \longrightarrow R^\bullet + H_2C\text{—}NH_2^+ \ (\text{cyclic, with } (CH_2)_n)$$

*n* = 2, 3, 4, 5, 6, etc.
with *m/z* = 58, 72, 86, 100, 114, etc.

The *m/z* 86 ion is usually the most prominent of these because it has a stable six-membered-ring structure.

Secondary and tertiary aliphatic amines behave very much like ethers, as the example of *N*-ethyl-(1,3-dimethylbutyl)amine shows (Figure 7.6a).

The mass spectrum is dominated by the α-cleavage reaction which gives rise to the loss of the largest radical, namely $^{\bullet}CH_2CH(CH_3)_2$ (Stevenson's rule). The resulting ion further breaks down by loss of ethene and transfer of a β-hydrogen to the positive nitrogen, in a similar way to the oxonium ions in ethers (Section 7.3):

$$\begin{array}{c} H_2C-\overset{H}{\overset{|}{N}}{}^{+}=CHCH_3 \\ | \\ H_2C-H \\ m/z\ 72 \end{array} \longrightarrow \begin{array}{c} H_2N=\overset{+}{C}HCH_3 + H_2C=CH_2 \\ m/z\ 44 \end{array}$$

If the first α-cleavage reaction gives rise to an ion of the type, $RCH_2CH_2\overset{+}{N}H=CH_2$, the following elimination of $RCH=CH_2$ will lead to $H_2\overset{+}{N}=CH_2$ (*m/z* 30). This explains why some secondary and tertiary amines will give *m/z* 30, the expected base peak of a primary amine.

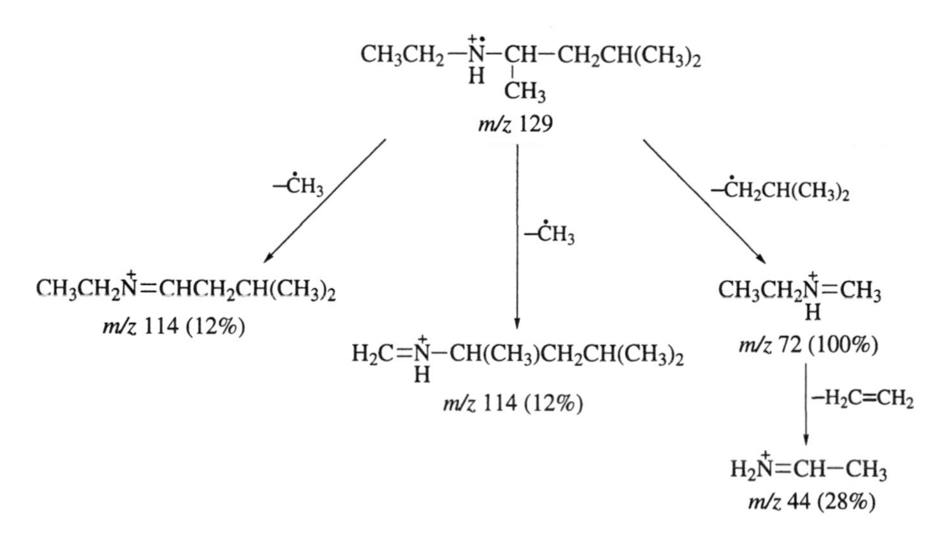


**Figure 7.6a** Fragmentation pattern of *N*-ethyl-(1,3-dimethylbutyl) amine

**SAQ 7.6a**

Figure 7.6b shows the mass spectrum of a tertiary amine (*Unknown 12*). Suggest a structure for this compound.

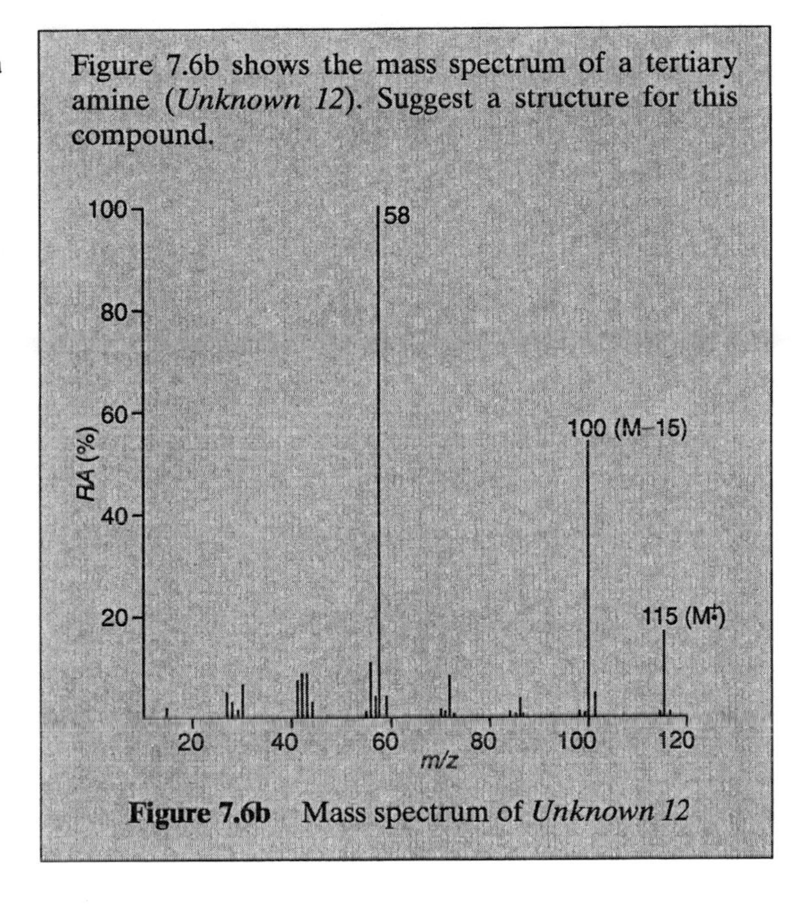


**Figure 7.6b** Mass spectrum of *Unknown 12*

**SAQ 7.6a**

Cycloalkylamines, such as substituted cyclopentyl- and cyclohexylamines, show fragmentations similar to these seen in cyclanols (α-cleavage), which lead to ions in the series, *m/z* 70, 84, 98, etc., by the process illustrated for *N*-methylcyclopentylamine in Figure 7.6c.

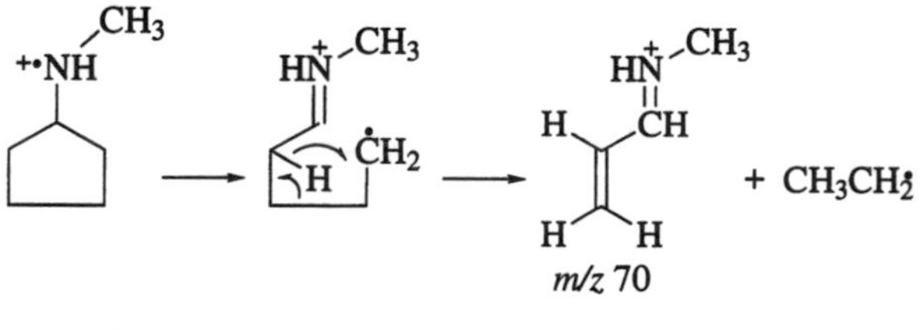

$$\text{+•NH(CH}_3\text{)-cyclopentyl} \longrightarrow \ldots \longrightarrow \text{HN}^+(\text{CH}_3)\text{=CH-CH=CH}_2 \; (m/z\ 70) + \text{CH}_3\text{CH}_2^{\bullet}$$

**Figure 7.6c** Fragmentation pattern of *N*-methylcyclopentylamine

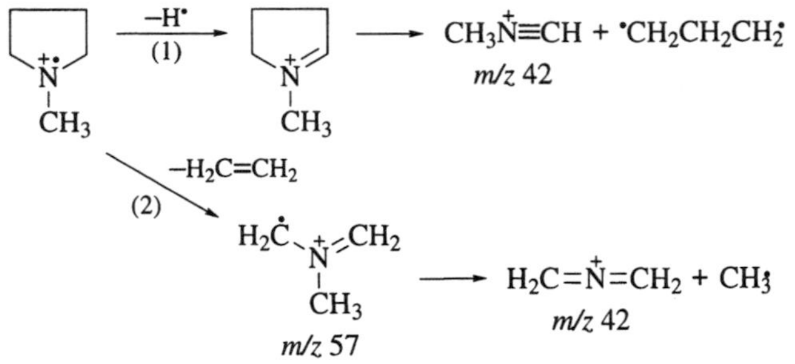

–H• (1)

$CH_3\overset{+}{N}{\equiv}CH + {}^{\bullet}CH_2CH_2CH_2^{\bullet}$

*m/z* 42

$CH_3$ $CH_3$

$-H_2C{=}CH_2$ (2)

$H_2\dot{C}\overset{+}{N}{=}CH_2$ $CH_3$

*m/z* 57

$H_2C{=}\overset{+}{N}{=}CH_2 + CH_3^{\bullet}$

*m/z* 42

**Figure 7.6d** Fragmentation pattern of *N*-methylpyrrolidone

When the nitrogen is actually in the ring, e.g. in pyrrolidines and piperidines, two main processes occur, as shown in Figure 7.6d for *N*-methylpyrrolidine. The first of these is the loss of an α-hydrogen from the ring, followed by the formation of an $R{-}\overset{+}{N}{\equiv}CH$ ion (route (1)), while the second is α-cleavage within the ring and then expulsion of the alkene to form $\overset{+}{N}{=}CH_2(CH_2^{\bullet})$ ions (route (2)).

Anilines show intense $M^{+\bullet}$ ion peaks and fragment relatively little by loss of HCN, either from the $M^{+\bullet}$ itself, or from $(M{-}H)^+$, which is usually a moderately intense peak:

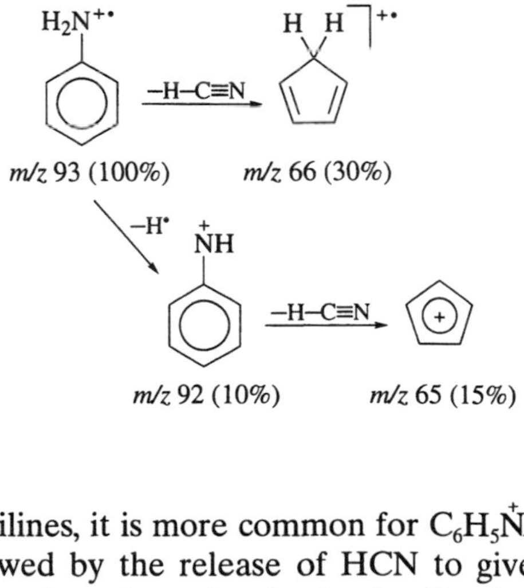

$H_2N^{+\bullet}$ –H–C≡N H H +•

*m/z* 93 (100%) *m/z* 66 (30%)

–H• $\overset{+}{N}H$

–H–C≡N

*m/z* 92 (10%) *m/z* 65 (15%)

In substituted anilines, it is more common for $C_6H_5\overset{+}{N}H_2$ or $C_6H_5\overset{+}{N}H$ to be formed, followed by the release of HCN to give *m/z* 66 and 65, respectively, rather than the molecular ion itself to lose HCN.

Alkylanilines, however, behave rather like alkylphenols. The three methylanilines lose hydrogen to give *m/z* 106 (the azotropylium ion, as the base peak), and then lose HCN to give *m/z* 79:

$$CH_3C_6H_4NH_2^{+\bullet}\ (m/z\ 107) \xrightarrow{-H^\bullet} \text{methylazatropylium}\ (m/z\ 106) \longleftrightarrow \text{etc.} \xrightarrow{-HCN} m/z\ 79$$

*N*-alkylanilines all give intense *m/z* 106 peaks, supposedly $CH_2{=}\overset{+}{N}HC_6H_5$, by α-cleavage, while the mass spectra below this value appear similar to the various methylanilines. Therefore, the presence of *m/z* 106, 79, 92, and 65 in a mass spectrum could indicate $—CH_2NHC_6H_5$, $CH_3NHC_6H_4$, or $H_2NC_6H_3CH_3$ in the parent molecule.

Pyridines have very intense $M^{+\bullet}$ peaks. Alkylpyridines lose hydrogen to form azatropylium ions and both these and the $M^{+\bullet}$ lose HCN.

Pyridyl compounds will often show $C_5H_4N^+$ ions, *m/z* 78, and both *m/z* 51 and *m/z* 52, caused by loss of HC≡N and HC≡CH, respectively, from this ion. If there is a side-chain possessing a hydrogen which γ is to the nitrogen, a 'McL'-type rearrangement can occur where the terminus of the hydrogen shift is $C{=}N^{+\bullet}$.

$$\text{2-}(CH_2CH_2CH_2R)\text{pyridine}^{+\bullet} \longrightarrow \text{2-methylene-1H-pyridinium}\ (m/z\ 93) + RCH{=}CH_2$$

**SAQ 7.6b**

Figure 7.6e shows the mass spectrum of an amine (*Unknown 13*). Suggest a structure for this compound.

**SAQ 7.7b**
(Contd)

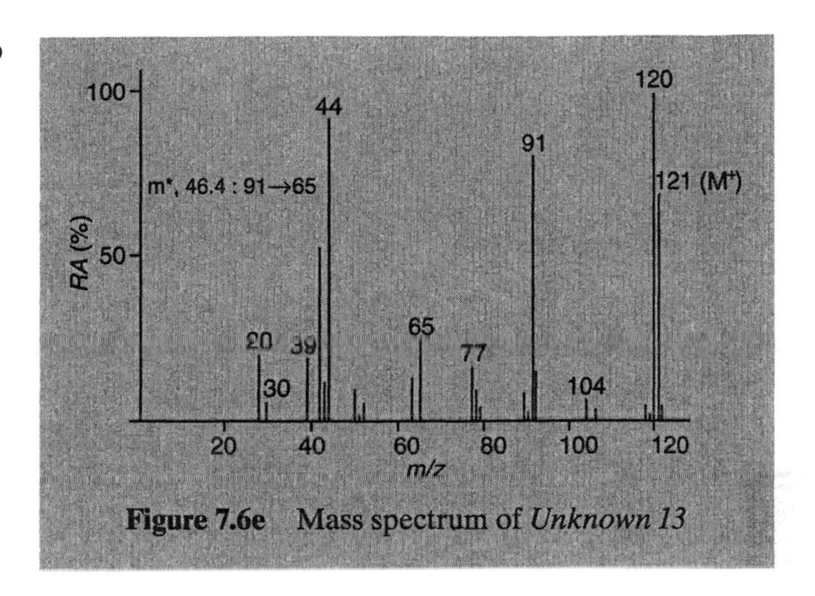


**Figure 7.6e** Mass spectrum of *Unknown 13*

**SAQ 7.6b**

## 7.7 FRAGMENTATIONS OF HYDROCARBONS

You may have wondered why the mass spectra of hydrocarbons were not considered first in this chapter, i.e. as they are often dealt with in many textbooks of organic chemistry. This is because their mass spectra are often quite complex with many possible fragmentation routes.

Many molecules contain substantial hydrocarbon rings or chains, as well as a functional group. Therefore, the mass spectrum of such a compound consists of two types of peaks, namely those arising from the presence of the functional group and those due to fragmentation of the hydrocarbon chain or ring. So, when you are trying to determine the compound that is present, it is helpful if you can eliminate some of the apparent peaks.

Saturated hydrocarbons exhibit quite distinctive mass spectra. However, their $M^{+\bullet}$ peaks are quite weak, thus making it very difficult to identify long-chain hydrocarbons, e.g. oils and greases. The main features of the fragmentations that occur are illustrated by the mass spectrum of dodecane, $M_r$ 170, which is shown in Figure 7.7a.

In the initial ionisation, an electron is removed from any carbon in the chain followed by cleavage of the adjacent C—C bond, with the loss of $R^\bullet$ and formation of $R'CH_2^+$ ions. These form an odd-numbered series $m/z$ $(14n + 1)$, from $CH_3^+$ and $CH_3CH_2^+$ to $CH_3(CH_2)_9^+$, where $n$ is the number of carbon atoms in the fragment. Therefore, the typical ions in a saturated-hydrocarbon spectrum are $m/z$ 15, 29, 43, 57, 71, 85, 99, ... ($M-CH_2CH_3$), differing by 14 amu.

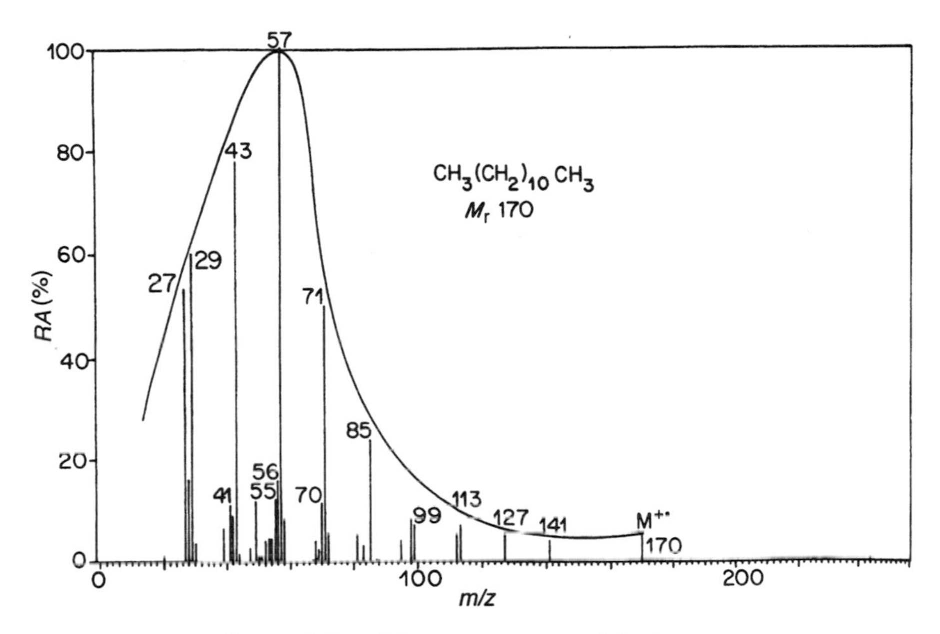


**Figure 7.7a** Mass spectrum of dodecane

The second characteristic feature is the distribution of the relative intensities of these ions. In a saturated, unbranched chain, the most intense ion is *m/z* 57 ($C_4$ fragment), while the $C_3$ and $C_5$ fragments (*m/z* 43 and 71, respectively) are also intense. Either side of this range, the ion intensities fall off rapidly. The whole spectrum falls within an envelope shown by the smooth curve in Figure 7.7a.

$$\underset{m/z\ 99}{CH_3(CH_2)_4{-}CH_2{-}\overset{+}{C}H_2} \xrightarrow[m^*\ 50.9]{-H_2C=CH_2} \underset{m/z\ 71}{CH_3(CH_2)_2{-}CH_2{-}\overset{+}{C}H_2} \xrightarrow[m^*\ 26.0]{-H_2C=CH_2} \underset{m/z\ 43}{CH_3(CH_2)_2^+}$$

There are successive losses of $H_2C{=}CH_2$ until their excess energy has been dissipated. Losses of $:CH_2$ are much less favoured and rarely occur. Repetition of this process with all of the initial fragments favours the formation of the $C_3$, $C_4$ and $C_5$ carbocations.

A third distinguishing feature of saturated-hydrocarbon spectra is the formation of two further series of ions, separated by 14 amu, which are due to loss of RH from $M^{+\bullet}$, with the formation of $^\bullet(CH_2)_n\overset{+}{C}H_2$ ions, and $H_2$ from the $(14n + 1)$-series ions.

$$R(CH_2)_nCH_3]^{+\bullet} \xrightarrow{-RH} {}^{\bullet}(CH_2)_nCH_2^{+} \longrightarrow \text{could cyclise or form alkene ions}$$
$$(A-1)^{+}$$

$$\underset{A^+}{RCH_2\overset{+}{C}H_2} \xrightarrow{-H_2} \underset{(A-2)^+}{RC{=}\overset{+}{C}H_2}$$

If the (14$n$ + 1) ions are termed $A^+$, these ions appear at $(A-1)^{+\bullet}$ and $(A-2)^+$, and are always of lower intensity than the $A^+$ ions.

Branching in hydrocarbon chains results in large changes in distribution of the (14$n$ + 1) ions, as secondary or tertiary carbocations are formed which are much more stable (see Section 6.3). The ion formed by the loss of the largest radical ($R^\bullet$) is usually the base peak (Stevenson's rule):

$$R{-}\underset{H}{\overset{R_1}{C}}{-}R_2]^{+\bullet} \longrightarrow R_2^{\bullet} + \underset{\text{base peak}}{(R)(R_1)CH^+} \qquad R_2 > R_1 > R$$

**SAQ 7.7a**

Figure 7.7b shows the mass spectrum of a hydrocarbon (*Unknown 14*). Try to identify this compound.

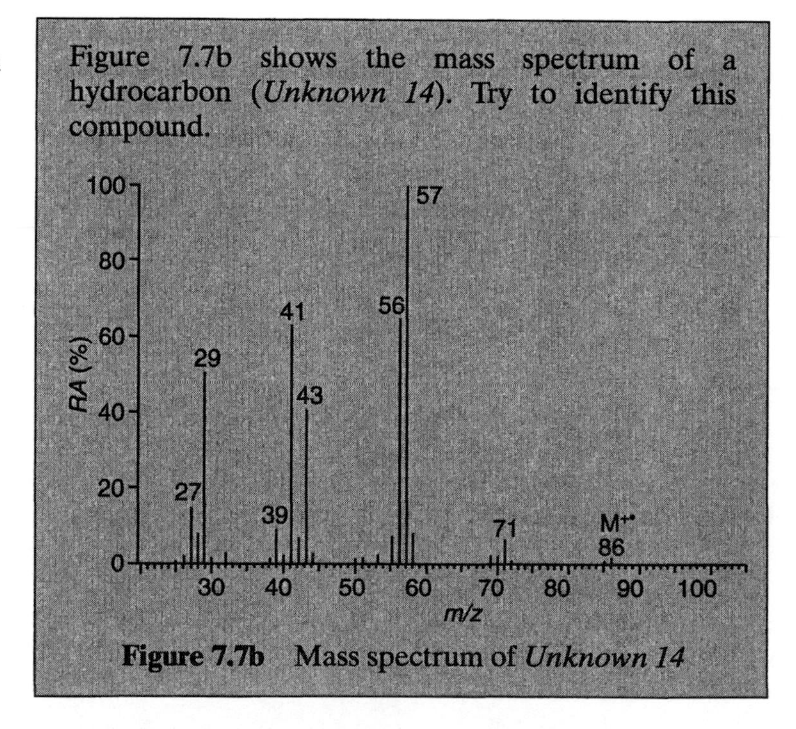


**Figure 7.7b** Mass spectrum of *Unknown 14*

**SAQ 7.7a**

In answering the last question, you have probably realised how difficult it is to be precise about the exact structure of a saturated hydrocarbon from its mass spectrum. It is often imperative to use other analytical techniques, such as NMR spectroscopy, to confirm an identification.

The mass spectra of cycloalkanes show prominent even-mass ions in addition to peaks formed from the loss of any R groups attached to the ring. This is because the rings fragment by successive losses of ethene. Their $M_r$ values are 2 amu less than the corresponding alkanes if monocyclic, 4 amu if bicyclic, etc. Figure 7.7c shows the spectrum of methylcyclopentane, which has prominent $(M–CH_3)^+$ (*m/z* 69) and $(69–H_2C{=}CH_2)^+$ (*m/z* 41) ions, while the base peak (*m/z* 56) is formed by loss of $H_2C{=}CH_2$ from $M^{+\bullet}$.

Alkenes also have molecular ions which are 2 amu less than the corresponding alkane, 4 amu less if dienes, etc. Their molecular ions tend to be quite intense relative to their saturated analogues because ionisation occurs on the double bond by removal of one of the π-electrons, thus leaving the carbon skeleton relatively undisturbed.

When alkenes fragment, they tend to give a series of ions corresponding to $C_nH_{2n}^{+\bullet}$ and $C_nH_{2n-1}^{+}$, i.e. $14n$ and $(14n–1)$, of which the

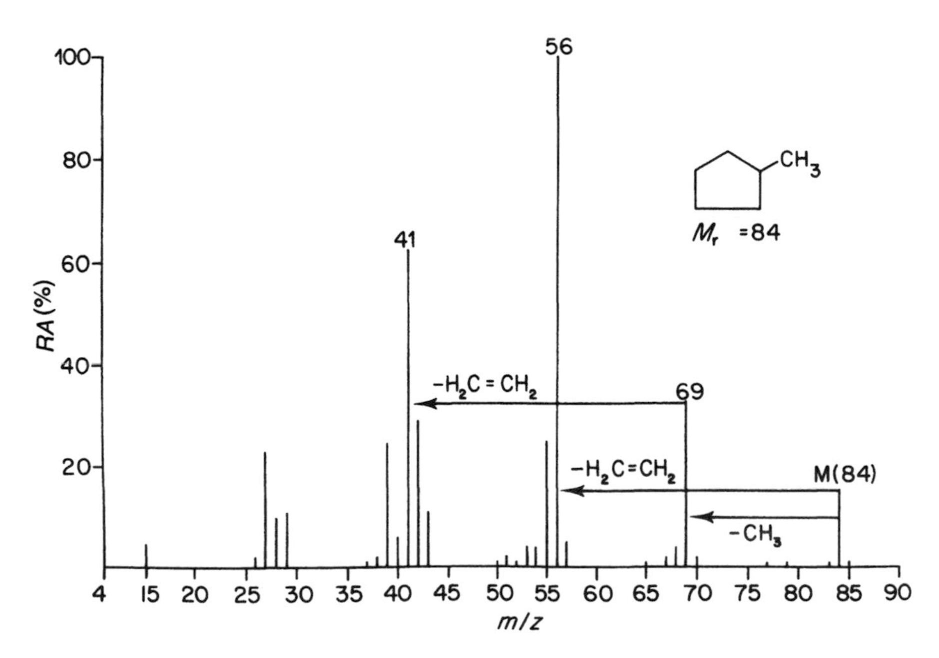


**Figure 7.7c** Mass spectrum of methylcyclopentane

latter are the more intense. These are formed by a cleavage which is β- to the double bond, known also as *allylic* cleavage:

$$\mathrm{R{-}CH_2{-}\dot{C}H{-}\overset{+}{C}H{-}R'} \longrightarrow \mathrm{R^{\bullet} + CH_2{=}CH{-}\overset{+}{C}H{-}R'} \longleftrightarrow \mathrm{\overset{+}{C}H_2{-}CH{=}CHR'}$$

$M^{+\bullet}$ $\qquad$ $(M{-}R^{\bullet})$, $C_nH_{2n-1}$

One problem in alkene mass spectra is the migration of double bonds along the chains by a series of hydrogen-shifts:

$$\mathrm{R{-}CH_2{-}\dot{C}H{-}\overset{+}{C}HCH_2R'} \longrightarrow \mathrm{R\dot{C}H{-}CH_2\overset{+}{C}H{-}CH_2{-}R'} \longrightarrow \mathrm{R\dot{C}H{-}CH_2{-}CH_2{-}\overset{+}{C}HR'}$$

This means that more $C_nH_{2n-1}^+$ ions are formed than expected, and more importantly, that isomeric alkenes (and many cycloalkenes) give virtually identical mass spectra. A further disadvantage is that once you have ionised a double bond, free rotation about this bond is possible, so *E*- and *Z*-isomers (cis- and trans-forms) also give nearly identical spectra.

Cyclohexenes do show a characteristic retro Diels–Alder fragmentation to a dienyl ion and ethene:

cyclohexene $\xrightarrow[70\ \mathrm{ev}]{e}$ radical cation $\longrightarrow$ butadiene radical cation + $CH_2{=}CH_2$

*m/z* 54

Alkynes also display quite intense molecular ions (4 amu less than the corresponding alkanes), and if terminal they often show quite marked $(M{-}H)^+$ ions. These are probably *propargyl ions* of the form:

$$\mathrm{H{-}C{\equiv}C{-}\overset{+}{C}HR} \rightleftarrows \mathrm{H\overset{+}{C}{=}C{=}CHR}$$

The *m/z* 39 ion is quite intense and is characteristic of alkynes:

$$\mathrm{H\overset{+}{C}{=}\dot{C}{-}CH(R){-}H} \longrightarrow \mathrm{H\overset{+}{C}{=}C{=}CH_2 + R^{\bullet}}$$

*m/z* 39 (propargyl ion)

Carbocations derived from the R group ($R^+$ itself and its decomposition products) are present but these are generally weaker.

If the R group possesses hydrogens which are γ- to the C=C bond, a 'McL'-type process results in the loss of alkenes:

$$\mathrm{R{-}CH(H){-}CH_2{-}C^{+}{=}\dot{C}H} \longrightarrow \mathrm{RCH{=}CH_2} + \mathrm{H_2C{=}C^{+}{-}\dot{C}H_2} \equiv \left[\mathrm{CH_2{=}C{=}CH_2}\right]^{+\bullet} \quad m/z\ 40$$

So, to sum up, ions at *m/z* 39, 40 and $(M\text{–}1)^+$ tend to be typical of alkynes.

The mass spectra of most aromatic hydrocarbons show very intense $M^{+\bullet}$ peaks, as the rings are so stable. When they do fragment, $(M\text{–}C_2H_2)$ and $(M\text{–}C_3H_3)$ ions are formed, and to a limited extent, there is also loss of $H^\bullet$, followed by loss of $C_2H_2$ and $C_3H_3$. $C_6H_5$ compounds, in addition, show characteristic ions at *m/z* 77, 52, and 39. When an alkyl group is attached to a benzene ring, preferential cleavage occurs β- to the ring (*benzylic* cleavage) to form the tropylium ion, *m/z* 91:

$$\left[\mathrm{C_6H_5{-}CH_2{-}R}\right]^{+\bullet} \xrightarrow{-R^\bullet} \mathrm{C_6H_5\overset{+}{C}H_2} \equiv \mathrm{C_7H_7^+}\ (\text{tropylium}) \quad m/z\ 91$$

The mass spectra of dimethylbenzene isomers are almost identical and show loss of $CH_3^\bullet$ to give $CH_3C_6H_4^+$ (*m/z* 91), plus loss of hydrogen from one of the $CH_3$ groups to give a peak at *m/z* 105, which is methyltropylium. This is an example of benzylic cleavage where R=H. Owing to the symmetry of the substituted tropylium ion, the mass spectra of alkyl-substituted aromatics are very similar and therefore patterns of substitution cannot be found by using mass spectrometry.

On the other hand, branching in the side-chain may give rise to considerable differences. For example, isopropylbenzene readily loses one of the $CH_3$ groups by benzylic cleavage to give *m/z* 105, while propylbenzene loses $CH_3CH_2$ by this process to give *m/z* 91:

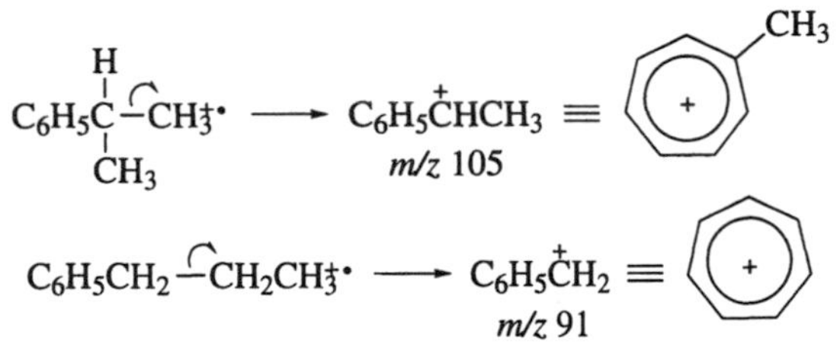

$$C_6H_5C(H)(CH_3){-}CH_3^{+\bullet} \longrightarrow C_6H_5\overset{+}{C}HCH_3 \equiv \text{(methyltropylium ion)}$$
$m/z$ 105

$$C_6H_5CH_2{-}CH_2CH_3^{+\bullet} \longrightarrow C_6H_5\overset{+}{C}H_2 \equiv \text{(tropylium ion)}$$
$m/z$ 91

Another useful fragmentation is the 'McL' rearrangement of γ-hydrogen from an alkyl side-chain to the aromatic ring, which gives rise to ions which are 1 amu higher than the tropylium ions (see Figure 7.7d).

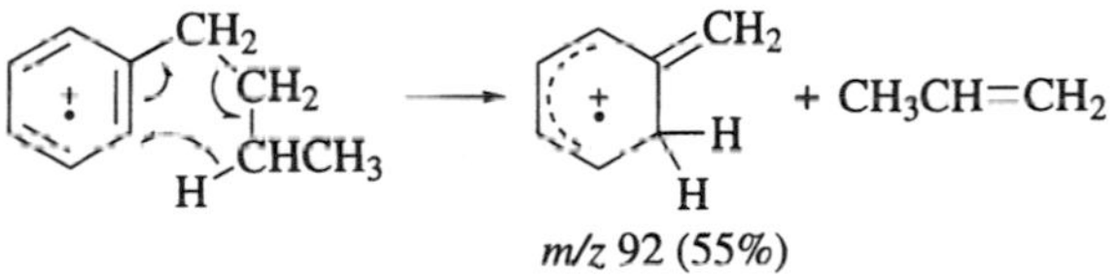

$$C_6H_5CH_2CH_2CHCH_3(H)^{+\bullet} \longrightarrow (C_6H_6{=}CH_2)^{+\bullet} + CH_3CH{=}CH_2$$
$m/z$ 92 (55%)

**Figure 7.7d** Illustration of the 'McL' rearrangement in butylbenzene

However, be careful, as isobutylbenzene, $C_6H_5CH_2CH(CH_3)_2$, also contains a γ-hydrogen and can give a similar $m/z$ 91 peak.

**SAQ 7.7b**

Figure 7.7e shows the mass spectrum of a hydrocarbon (*Unknown 15*). Suggest a structure for this compound and explain the main features of the spectrum.

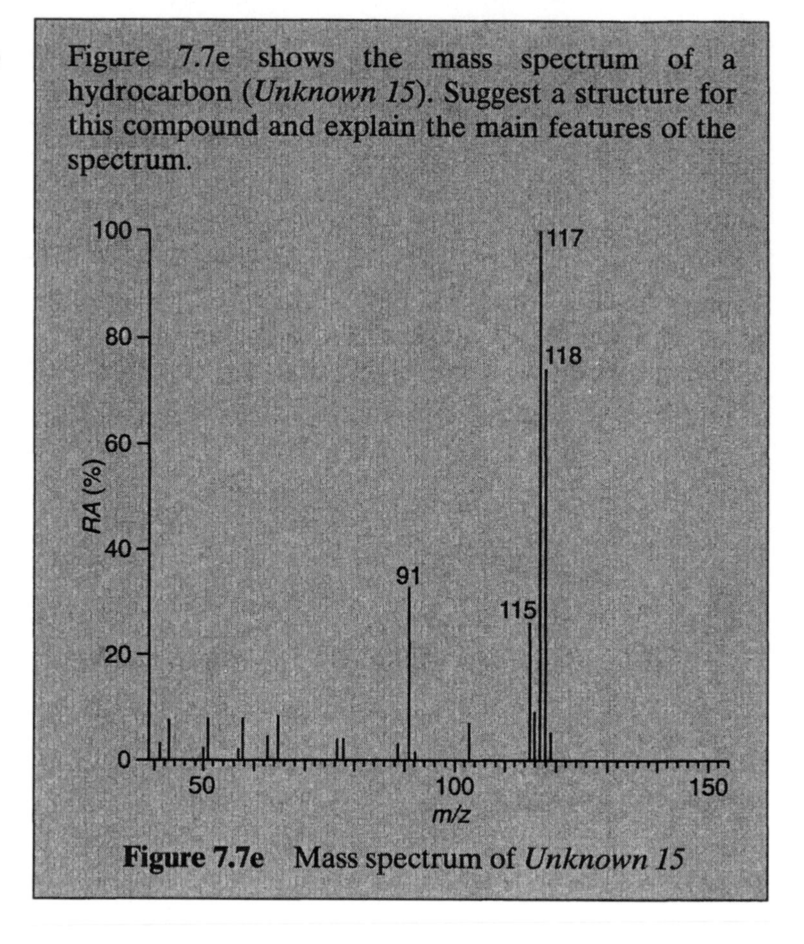


**Figure 7.7e** Mass spectrum of *Unknown 15*

**SAQ 7.7b**

## 7.8 FRAGMENTATIONS OF HALOCOMPOUNDS

In Sections 5.3 and 5.4, we considered methods for the calculation of relative abundances for the ion clusters produced by chlorine and bromine isotopes. These were summarised in Table 5.3b. Refer back to this table if you are unsure about these points. It was also stated that fluorine and iodine are monisotopic, and could be identified by typical increments in the $M_r$ values of the compounds in which they were substituted, i.e. 18 amu for F and 126 for I.

Haloaromatics can be divided into two categories, namely fluoroaromatics and the rest. The high bond strength of the C—F bond (Section 5.3.1), means that $F^{\bullet}$ is rarely lost from a fluorine compound. Other, weaker bonds cleave first, giving rise to the normal fragmentation products of other functional groups which are 18 amu higher than expected. Therefore, in the mass spectrum of 2,4-dinitrobenzene (Figure 5.3b), which has an intense $M^{+\bullet}$ ion, the loss of one nitro group leads to the small ion at *m/z* 140, followed by the loss of the second to the base peak at *m/z* 94 ($C_6H_3F^+$). The third prominent ion, i.e. *m/z* 30, is $NO^+$. Loss of HF (the even-electron rule) gives rise to $C_6H_2$ (*m/z* 74) and its daugther ion, $C_4H_2^+$ (*m/z* 50):

$$C_6H_3F^+ \longrightarrow HF + C_6H_2^+ \longrightarrow C{\equiv}C + C_4H_2^+$$

The other haloaromatics behave differently. They have quite intense $M^{+\bullet}$ peaks, and because the C—X bonds are weaker (in the order I < Br < Cl) they tend to lose the halogen atoms via successive fragmentations, thus leading to the formation of hydrocarbon ions, which being aromatic, are quite stable, for example:

$$C_6H_3X_3^{+\bullet} \longrightarrow X^{\bullet} + C_6H_3X_2^+ \longrightarrow X^{\bullet} + C_6H_3X^+ \longrightarrow X^{\bullet} + C_6H_3^+$$

Π Can you see something odd about the second fragmentation reaction?

It disobeys the even-electron rule (Section 6.3.5), namely that an even-electron ion, e.g. $C_6H_3X_2^+$, loses a radical $X^{\bullet}$ instead of an even-electron species. This could be due to the weakness of the C—X bond and also the stability of the halogen radicals. (However, HF is sometimes lost, even allowing for this!)

When a mixture of haloaromatics fragment, the most intense ions in the spectrum, apart from $M^{+\bullet}$, will be those with the weakest C—X bonds, i.e. I before Br, etc. Therefore, bromochloroiodobenzene will largely fragment as follows:

$$C_6H_3ClBrI^{+\bullet} \longrightarrow I^{\bullet} + C_6H_3ClBr^+ \longrightarrow Br^{\bullet} + C_6H_3Cl^{+\bullet}$$

The substitution pattern is not easily identifiable.

With benzyl halides, the molecular ion is usually observable, and the tropylium ion ($C_7H_7^+$) readily forms upon loss of the halogen atom:

$$C_6H_5CH_2{-}X^{+\bullet} \longrightarrow C_6H_5CH_2^+ \longrightarrow C_7H_7^+$$

When the aromatic ring carries substituents, a substituted tropylium ion is formed, with a substituted phenyl carbocation also being seen:

$$[YC_6H_4CH_2X]^{+\bullet} \longrightarrow {}^\bullet CH_2X + YC_6H_4^+$$

For aliphatic halogen compounds, there is a marked trend in the $M^{+\bullet}$ ion intensities (I(strongest) > Br > Cl > F). As the alkyl groups become longer, as the amount of α-branching increases, or the number of halogen atoms increases, the $M^{+\bullet}$ intensity falls. Therefore, it is often difficult to detect cluster intensities of the $M^{+\bullet}$ ions of aliphatic hydrocarbons and a knowledge of their fragmentation processes is therefore important.

The most important loss is that of the halogen itself, which leaves a carbocation. If this is a hydrocarbon, the rest of the spectrum will resemble that of the base hydrocarbon. This cleavage is most important when the halogen is a good leaving group. Therefore, the $(M–Cl)^+$ and $(M–F)^+$ ions would be weak in mass spectra of these halo compounds. In the mass spectrum of 1-bromohexane (Figure 7.8a), the peak at *m/z* 85 is due to the formation of the (hexyl)$^+$ ion. This then fragments by loss of propene to form a $C_3H_7^+$ ion, *m/z* 43.

Haloalkanes may also lose a molecule of hydrogen halide according to the following process:

$$[R{-}CH_2CH_2{-}X]^{+\bullet} \longrightarrow [RCH{=}CH_2]^{+\bullet} + HX$$

Bromo and iodo compounds tend to lose $Br^\bullet$ and $I^\bullet$.

A less important mode is α-cleavage:

$$R{-}CH_2{-}\overset{+\bullet}{X} \longrightarrow R^\bullet + CH_2{=}\overset{+}{X}$$

This mode is less frequently observed due to the halogen (high

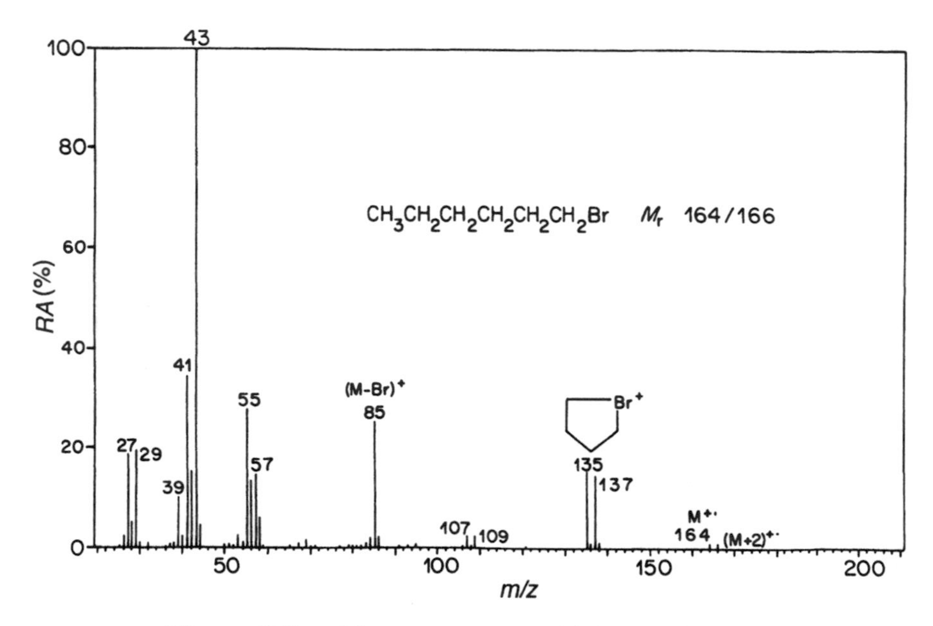


**Figure 7.8a** Mass spectrum of 1-bromohexane

electron affinity) preferring not to have any charge retained on itself.

Heterolytic loss of X• is favoured:

$$RCH_2\!-\!\overset{+\bullet}{X} \longrightarrow RCH_2^+ + X^\bullet$$

Where the α-position is branched, the largest alkyl group is lost preferentially (Stevenson's rule).

Another fragmentation (R—Cl, Br) involves the rearrangement to a cyclic halonium ion with loss of an alkyl radical. These have *m/z* 91/93 and 135/137, respectively:

$$R\text{–}(CH_2)_4\text{–}\overset{+\bullet}{X} \longrightarrow R^\bullet + \text{(cyclic)}\ \overset{+}{X},\ C_4H_8X^+ \quad X = Cl, Br$$

halonium ion

**SAQ 7.8**

Figure 7.8b shows the mass spectrum of a haloalkane (*Unknown 16*). Suggest a structure for this compound, and assign structures (as far as you can) to the ions $m/z$ 77/79, 63/65, 57, 56, and 41.

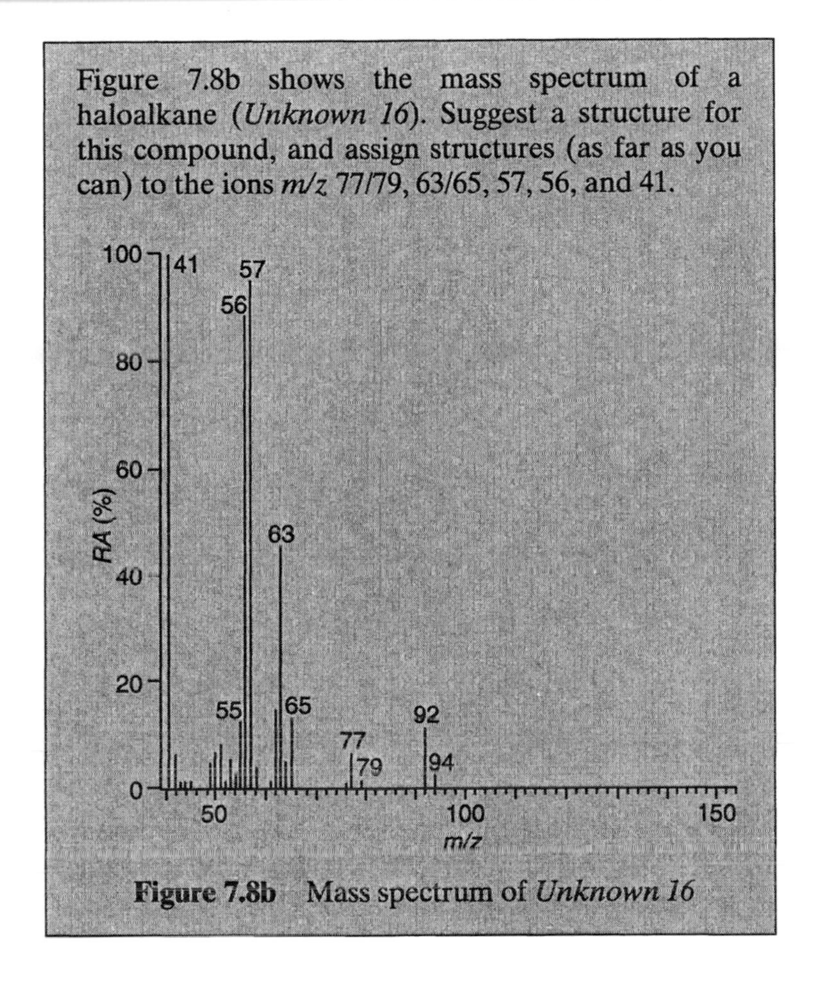


**Figure 7.8b** Mass spectrum of *Unknown 16*

**SAQ 7.8a**

## 7.9 FRAGMENTATIONS OF NITROCOMPOUNDS

Aliphatic nitrocompounds have very weak $M^{+\bullet}$ ions and fragment to give hydrocarbon ions:

$$R-\overset{+}{N}O_2 \longrightarrow NO_2 + R^+ \longrightarrow (\text{alkyl})^+ \text{ ions}$$

'McL' rearrangement does occur but only to a very limited extent, as is the case with β-cleavage. These give small peaks at *m/z* 61 and 60, respectively.

Therefore, (M–46) is typical for aliphatic compounds; $NO_2^+$, *m/z* 46, is not normally observed.

The isomeric nitrites, R—O—N=O, however, show easy β-cleavage to give an *m/z* 60 ion and $R^\bullet$:

$$R-CH_2-\overset{+\bullet}{O}-N=O \longrightarrow R^\bullet + CH_2=\overset{+}{O}-N=O \quad (m/z\ 60)$$

Nitrites also show strong $NO^+$, *m/z* 30, but $NO_2^+$ is not produced. There is no sign of the following isomerisation:

$$RNO_2^{+\bullet} \rightleftarrows RONO^{+\bullet}$$

which occurs in aromatic nitrocompounds, so these two forms can be easily distinguished.

Aromatic nitrocompounds have intense $M^{+\bullet}$ peaks, and loss of $NO_2^\bullet$ may give the $Ar^+$ as the base peak. Many $ArNO_2$ compounds show a small $(M\text{–}O)^+$ ion which is quite distinctive. An interesting fragmentation is the loss of NO, followed by CO, from $M^{+\bullet}$. This fragmentation, in the case of nitrobenzene, is shown in Figure 7.9a.

*Ortho*-compounds can lose OH by hydrogen transfer. For example, in 2-nitroaniline, *m/z* 121 (M–OH) is formed as well as *m/z* 122:

[2-nitroaniline radical cation → quinonoid imine + $HO^\bullet$]

*m/z* 121 (5%)

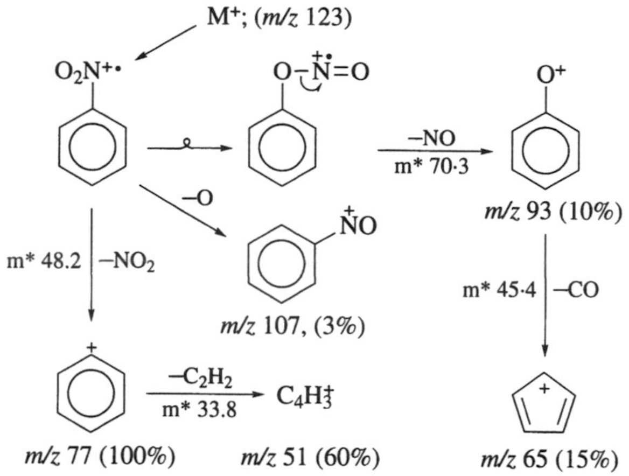

M⁺; (*m/z* 123)

$O_2N^{+\bullet}$ → $O-\overset{+\bullet}{N}=O$ —(−NO, m* 70·3)→ $O^+$ *m/z* 93 (10%)

−O → $\overset{+}{N}O$ *m/z* 107, (3%)

m* 48.2 | $-NO_2$

m* 45·4 | −CO

−$C_2H_2$, m* 33.8 → $C_4H_3^+$

*m/z* 77 (100%) *m/z* 51 (60%) *m/z* 65 (15%)

**Figure 7.9a** Fragmentation pattern of nitrobenzene

Therefore, the most general features of the mass spectra of aromatic nitrocompounds are M–16 (M–O), M–30 (M–NO) and M–46 (M–$NO_2$), but *not* $NO_2^+$, itself.

Figure 7.9b shows the mass spectrum of 4-nitroaniline.

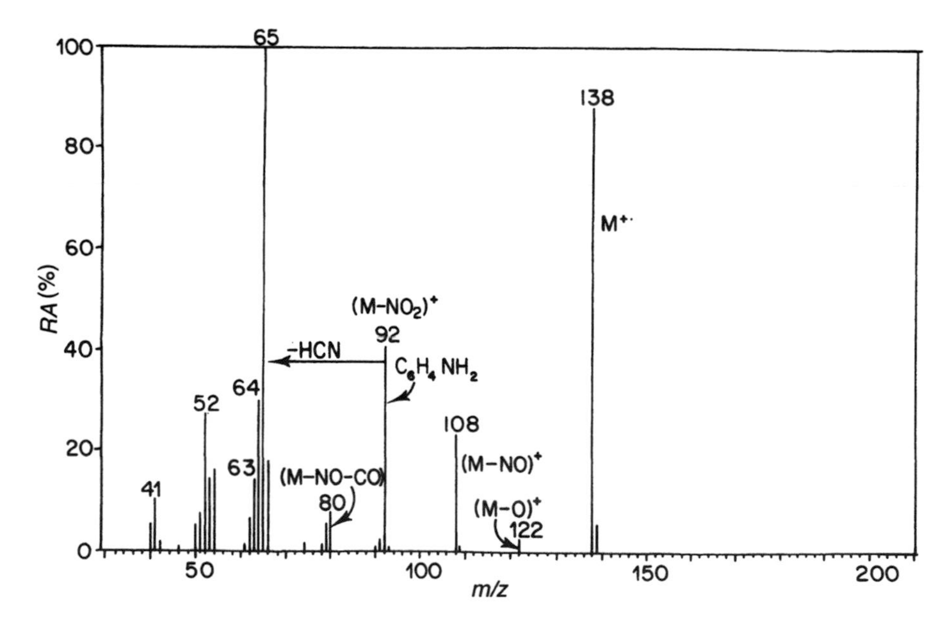


**Figure 7.9b** Mass spectrum of 4-nitroaniline

## 7.10 FRAGMENTATIONS OF HETEROCYCLIC COMPOUNDS AND SULPHUR COMPOUNDS

### 7.10.1 Heterocyclic Compounds

Figure 7.10a shows a number of common heterocyclic compounds that you may have come across before.

Such systems, in particular, benzo analogues such as quinoline, are very common in biological chemistry, since they form important components of living organisms and drugs.

We dealt with the fragmentation pattern of pyridine in Section 7.6, but let us now generalise this to include other heterocycles. For the most part, they give intense $M^{+\bullet}$, and the alkyl analogues lose H (if $CH_3$-substituted), or cleave the β-bond to lose $R^{\bullet}$ and give a tropylium species, usually as the base peak. The rings themselves break down by the loss of small and stable, neutral molecules, which are analogous to HCN, such as CH=NH, CO, HC=O, and HC=S, as well as HC=CH. In the substituted heterocycles, ArX-type compounds show varying amounts of $Ar^+$, while ArCOX-type compounds almost always have $ArCO^+$ as intense peaks (or $ArCOOH^+$ if X has a γ-hydrogen, which allows 'McL' rearrangements to occur).

Table 7.10a summarises the main features of the mass spectra of the heterocycles given in Figure 7.10a.

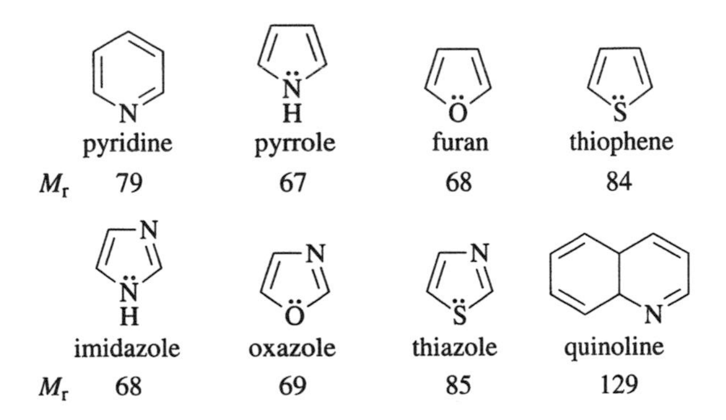


**Figure 7.10a** Some examples of common heterocyclic compounds

**Table 7.10a** Main features of the mass spectra of some common heterocyclic systems

| System | Neutrals lost (amu) | Ions formed ($m/z$) | | | |
|---|---|---|---|---|---|
| | | $Ar^+$ | Tropylium | 'McL' | $ArCO^+$ |
| Pyridines | HCN (27) | 78 | 92 | 93 | 106 |
| Quinolines | HCN (27)<br>HC≡CH (26) | 128 | 142 | 143 | 156 |
| Pyrroles | HCN (27)<br>CH=NH (28) | 66 | 80 | –[a] | 94 |
| Furans | HC≡CH (26)<br>CHO (29)[b] | 67 | 81 | –[a] | 95 |
| Thiophenes | HC≡CH (26)<br>HC≡S (45) | 83 | 97 | –[a] | 111 |
| Imidazoles | HCN (27) | 67 | 81 | –[a] | 95 |
| Oxazoles | HCN (27)<br>CO (28)<br>CHO (29)[b] | 68 | 82 | 83 | 96 |
| Thiazoles | HCN (27)[c]<br>HCN+H (28) | 84 | 112[d] | 99 | 112 |

[a]β-Cleavage to give the tropylium ion largely predominates
[b]Major-loss peak
[c]HCN loss occurs first, followed by loss of hydrogen; ions formed are similar in intensity
[d]Peak shown by dialkylthiazoles and methylthiazoles; tropylium ion not formed

It can be easily seen from Table 7.10a that apart from furans and imadazoles, all of the typical ions diagnostic of these heterocycles have different $m/z$ values and thus it should be possible to identify which system you have from these ions, remembering also that sulphur has a significant (M + 2) isotope.

Sometimes, it is possible to distinguish between compounds containing two substituents, if the side-chain at the 2-position is no longer than $CH_3$. 2-Methyl substituents lose an extra 14 amu in the neutral fragment, e.g. 2-methylfurans lose $CH_3CO$ (43 amu). Furans and thiophenes also show ions corresponding to the neutral species, e.g.

$HCO^+$ (*m/z* 29), $CH_3CO$ (*m/z* 43), $HCS^+$ (*m/z* 45), and $CH_3CS^+$ (*m/z* 59), which support this distinction.

*Ortho*-effect fragmentations occur in a similar manner to those found in benzene compounds.

**SAQ 7.10a**

The mass spectrum of an unknown compound shows the following features: $M^{+\bullet}$ *m/z* 126 (82%), 111 (100%), 83 (36%), 57 (16%), 45 (18%), and 39 (25%), plus $m^*$ 62.1 and 39.1. All of these ions have (M + 2) peaks of *ca.* 5%. Which of the following compounds do you think it is?

A $CH_3CO$–(thiazol-2-yl; ring with N and S)

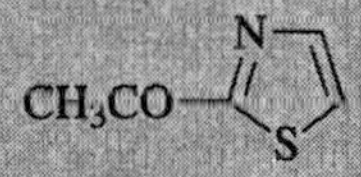

B $CH_3CO_2$–(furan-2-yl; ring with O)

C $CH_3CH_2CH_3$–(thiazol-2-yl; ring with N and S)

D $(CH_3)_2CH$–(thiazol-2-yl; ring with N and S)

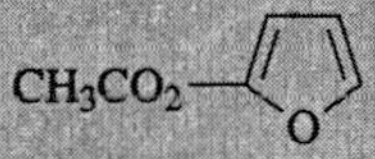

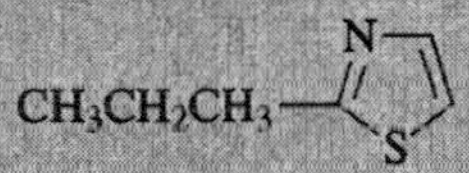

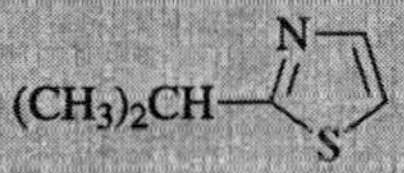

**SAQ 7.10a**

### 7.10.2 Thiols

The behaviour of thiols is similar to their oxygen analogues with just a few marked differences. Therefore, they should show $CH_2{=}\overset{+}{S}H$ (*m/z* 47), followed by *m/z* 61, 75, 89, etc., (M–$H_2S$), and (M–($H_2S$ + $CH_2{=}H_2$)), i.e. the (M–34) and (M–62) peaks, respectively.

All of these ions are, in fact, found, along with various hydrocarbon ions formed by cleavages at each point in the alkyl chain. In chains of five carbons or more, δ-cleavage is particularly favoured in thiols, as the ion, *m/z* 89, is cyclic:

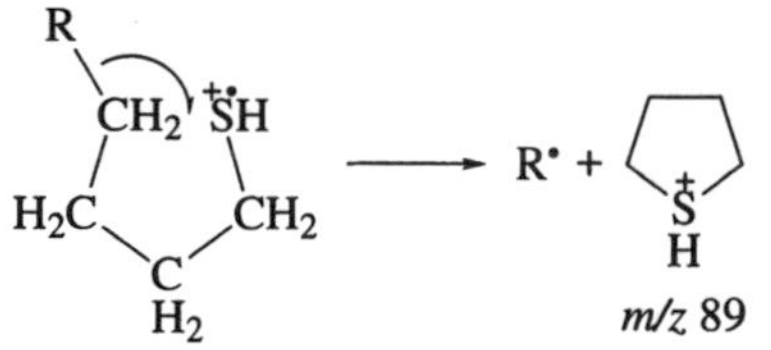

$$R{-}CH_2{-}CH_2{-}CH_2{-}CH_2{-}\overset{+\bullet}{S}H \longrightarrow R^{\bullet} + \text{(cyclic } \overset{+}{S}H\text{)}$$

*m/z* 89

This ion is analogous to the cyclic oxonium and halonium ions.

### 7.10.3 Thioethers (Sulphides)

In these compounds, you would expect to see α-cleavage of the longest chain, thus giving ions such as $R\overset{+}{\underset{R}{S}}{=}CH_2$, i.e. in the series *m/z* 61, 75, 89, etc. If the R group contained a β-hydrogen, transfer to the $\overset{+}{S}{=}CH_2$ would occur next with the loss of a neutral alkene, and therefore $H\overset{+}{S}{=}CH_2$ (*m/z* 47) would be expected. Furthermore, because of the extra stability of the $RS^+$ ion, cleavage occurs at the S—R links, with $RS^+$ ions of fairly high relative abundances appearing in the series *m/z* 47, 61, 75, 89, etc. Therefore, a thioether, R—S—R′, gives $RS^+$ and $R'S^+$ of unusual intensity. The second difference is that one of the alkyl groups cleaves off, with transfer of a β-hydrogen to the charged sulphur atom, thus giving an ion which is 1 amu higher than the $RS^+$ ion.

$$R{-}\overset{+\bullet}{S}{-}CH_2\overset{H}{\overset{|}{C}}H{\sim\sim\sim} \longrightarrow RS\overset{+\bullet}{H} + CH_2{=}CH_2{\sim\sim\sim}$$

The presence of such an even-mass ion is very useful in fixing the position of the sulphur in the chain.

### 7.10.4 Arylethers and Thiophenols

As you would expect, the $M^{+\bullet}$ is intense, and it loses $R^\bullet$ to give $ArS^+$, which would further expel C=S to form cyclopentadienyl ions. Methyl aryl ethers would expel $CH_2{=}S$ and transfer a hydrogen to the aromatic ring. The molecular ion would also fragment by loss of $RS^\bullet$ to give $Ar^+$ ions, while longer-chain R groups would lose an alkene and transfer a hydrogen to the sulphur to give $Ar{-}SH^+$. As an example, Figure 7.10b shows the fragmentation pathways and the corresponding mass spectrum of methylthiobenzene.

One further possible fragmentation is the formation of the tropylium ion, *m/z* 91, by expulsion of $HS^\bullet$. In higher-alkyl thioethers this is not as important as the formation of $Ar\overset{+}{S}{=}CH_2$ and $ArSH^{+\bullet}$. Therefore, (M–SH) is typical of $ArSCH_3$ compounds.

Thiophenols would be expected to show an intense $M^{+\bullet}$ ion, followed by loss of C=S to give cyclopentadienyl ions. Alkyl derivatives would

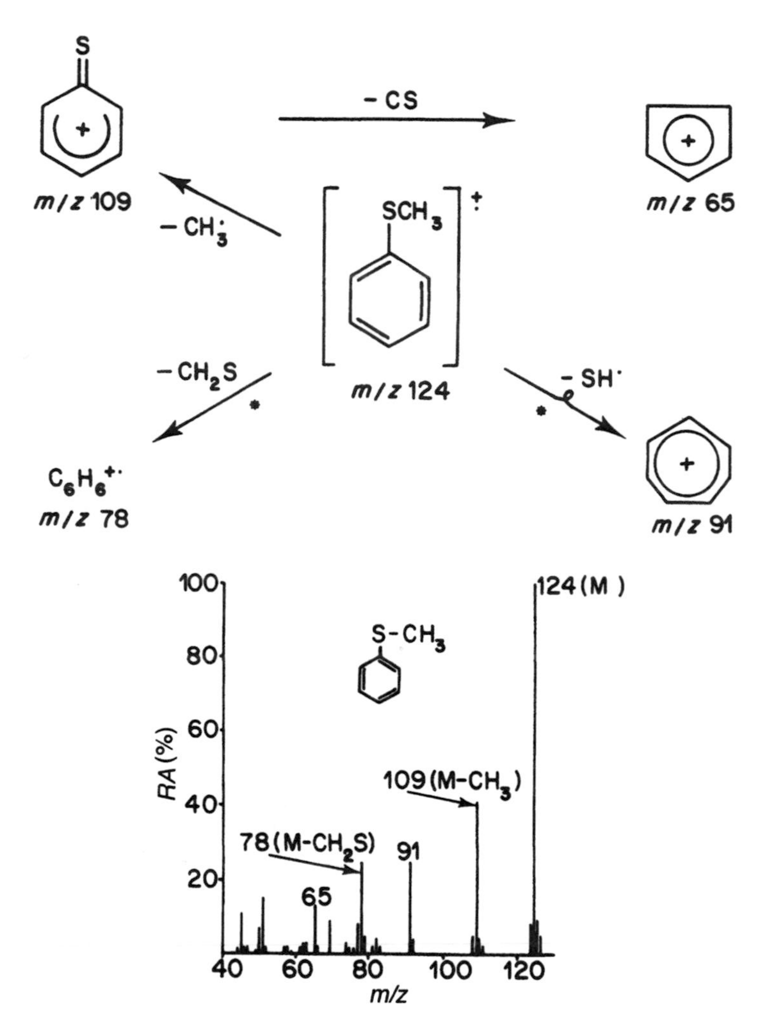


**Figure 7.10b** Fragmentation pattern and mass spectrum of methylthiobenzene

be expected to lose either H• or cleave off the β-R• group of the alkyl chain to give sulphhydryltropylium ions, which are analogous to the hydroxytropylium ions.

In practice, thiophenols differ significantly from these predictions, as can be seen in the mass spectrum of thiophenol itself in Figure 7.10c.

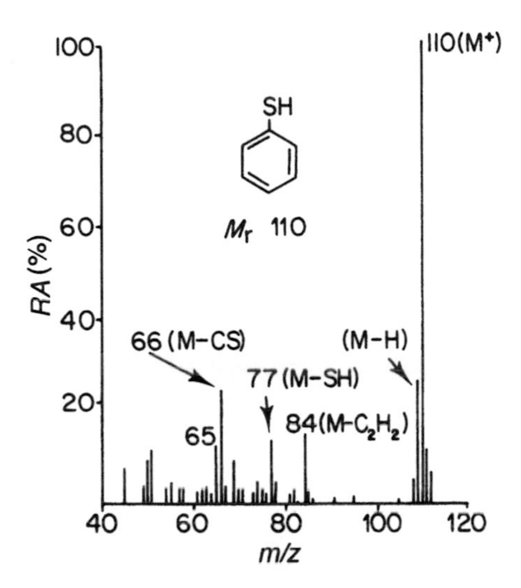


**Figure 7.10c** Mass spectrum of thiophenol

The two expected features *are* shown by thiophenol, e.g. an intense $M^{+\bullet}$ ion and (M–CS), *m/z* 66. However, there are also significant ions at (M–H), *m/z* 109, (M–$C_2H_2$), *m/z* 84, and (M–SH), *m/z* 77; these have no analogous features in the mass spectrum of phenol itself.

Presumably, (M–H) is a thiotropylium ion, which can then lose CS to give *m/z* 65 in the following way:

–CS

*m/z* 109 *m/z* 65

Alkyl thiophenols also show a different behaviour to that expected. They lose SH much more readily than groups from the α-carbons and give alkyl tropylium ions in the series *m/z* 91, 105, 119, etc. This may be due to the fact that the $SH^\bullet$ radical is more stable than the $OH^\bullet$ radical, and so it is more easily lost. As the alkyl chains become longer, α-cleavage competes with $SH^\bullet$ loss. Thiophenols also show the same sort of *ortho*-effects that 2-substituted phenols display, e.g. thiosalicyclic acid and its methyl ester eliminate $H_2O$ and $CH_3OH$, respectively, from $M^{+\bullet}$, whereas the 3- and 4-substituted isomers do not:

$$\text{[ring]}^{+\bullet}\text{-SH, -C(=O)OR} \longrightarrow \text{[ring]}^{+\bullet}(=S)(=CO) + ROH \xrightarrow{-CO} C_6H_4S^{+\bullet}$$

*m/z* 136 *m/z* 108

R = H (thiosalicyclic acid)
R = $CH_3$ (methyl thiosalicylate

**SAQ 7.10b**

An unknown compound contains one sulphur atom and its mass spectrum shows the following major ions (*m/z*): 138 (100%) ($M^{+\bullet}$); 123 (65%); 110 (66%); 109 (24%); 65 (21%); 51 (23%); 45 (32%). The sulphur atom is represented by peaks at *m/z* 138, 123, 110, 109 and 45.

The compound could be one of the following:

(a) $C_6H_5CH_2SCH_3$;

(b) 4-$CH_3C_6H_4SCH_3$;

(c) 4-$CH_3CH_2C_6H_4SH$;

(d) $C_6H_5SCH_2CH_3$.

Which do you think it is?

## 7.11 ANALYSIS OF BIOMOLECULES

Up until the end of the 1970s, the only techniques that were capable of characterising large biomolecules were electrophoretic, chromatographic or ultracentrifugation methods. The results were not very precise (10–100% relative error), with the only means of knowing the exact relative molecular mass ($M_r$) of a molecule being based on theoretical calculations of its assumed chemical structure. However, with the advent of plasma desorption (PD) and fast-atom-bombardment (FAB) techniques, followed by electrospray ionisation (ESI) and matrix-assisted laser desorption ionisation (MALDI) methods in the early 1990s, high-precision analysis of very-high-molecular-weight biomolecules can now be made. The characteristics performances of these various ionisation techniques are presented in Table 7.11a. All of these methods are of the soft-ionisation type.

Π What does this imply that their general spectra will look like?

The spectra will be characterised by the general absence of fragment-ion peaks.

As the resolution required to resolve the isotopic cluster increases with the mass (or with the charge carried by the ion), and because the resolution of most analysers is limited, the various peaks in the isotopic cluster combine and form a single peak that spreads over

**Table 7.11a** Characteristics of the various ionisation methods used in the mass-spectrometric analysis of peptides and proteins

| Ionisation method | Detection limit (pmol) | Common application field (Da) | Precision (%) | Analyser |
|---|---|---|---|---|
| Fast-atom-bombardment | 1–50 | 6000 | 0.05 | Magnetic or quadrupole |
| Electrospray | 0.01–5 | >130 000 | 0.01 | Magnetic or quadrupole |
| Matrix-assisted laser desorption | 0.001–1 | >300 000 | 0.05 | Time-of-flight |

several masses (15 Da (amu) at 10 000 Da and 45 Da at 100 000 Da). Therefore, there will be a slight difference in the mass measured, depending on how well the analyser can resolve the peaks (see Figure 7.11a).

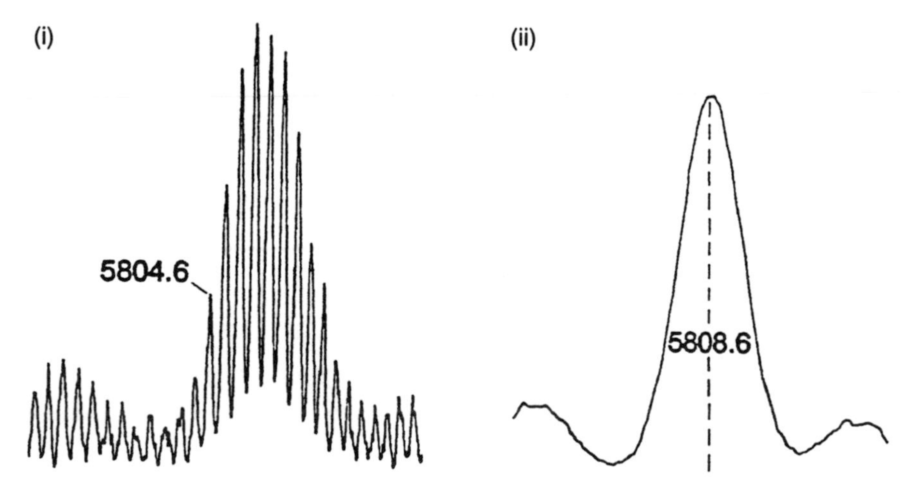


**Figure 7.11a** FAB mass spectra of human insulin isotopic cluster ($C_{257}H_{383}N_{65}O_{77}S_6$) measured at resolutions of (i) 6000, and (ii) 500

The absence of fragment ions means that complex mixtures can be analysed. However, this, of course, again depends on the resolution of the mass analyser.

A detailed discussion of the increasingly important role of mass spectrometry in biomolecule analysis is beyond the scope of this present text. We shall therefore only cover one important field, namely that of protein analysis.

### 7.11.1 Protein Analysis

Proteins can often by directly identified by their unique molecular weight. However, it is often best to break them down by the use of enzymes (e.g. trypsin) prior to mass-spectrometic investigations. A mixture of peptides that are characteristic of the precursor protein are obtained. MALDI is especially useful in this case as a preliminary chromatographic separation is not always needed.

**SAQ 7.11a**

> Describe the principles of operation of a MALDI ion source? Refer back to Chapter 2, if you are unsure.

Mass spectrometry also allows the detection of mutations within proteins, whether they are natural or are derived through mutagenesis. The strategy that is used for the natural mutants consists in comparing the molecular weights of a species of peptides obtained through enzymatic cleavage with those obtained for the native protein. A change in a peptide molecular weight indicates the position of the mutation, while the difference between the native-peptide molecular weight and that of the mutant peptide allows a possible determination of the nature of the amino acid that has mutated.

Disulphide bridges play an important role in establishing and preserving the three-dimensional structure of proteins. Mass spectrometry can be used to determine which cysteines are linked by these bridges, again by using enzymatic digestion. This is carried out by analysing the peptides produced by enzymatic cleavage before and after reduction of all of the disulphide bridges, as shown in Figure 7.11b.

The peptides containing an intermolecular disulphide bridge disappear, yielding two new peptides containing the cysteine residues responsible for the disulphide bridge.

Figure 7.11c shows the two FAB mass spectra of 'Paim I', a 73-amino-acid protease inhibitor of α-amylase. This protein was digested using protease V8, and spectra were obtained both before and after high performance liquid chromatography (HPLC) separation.

In the FAB spectrum of the mixture before separation, two ions with masses 2523 and 3066 Da appear. These do not correspond to any peptide generated from the protein sequence. Therefore, they could be peptides containing a disulphide bridge. Based on their mass, and on the specificity of the protease used, these ions characterise the

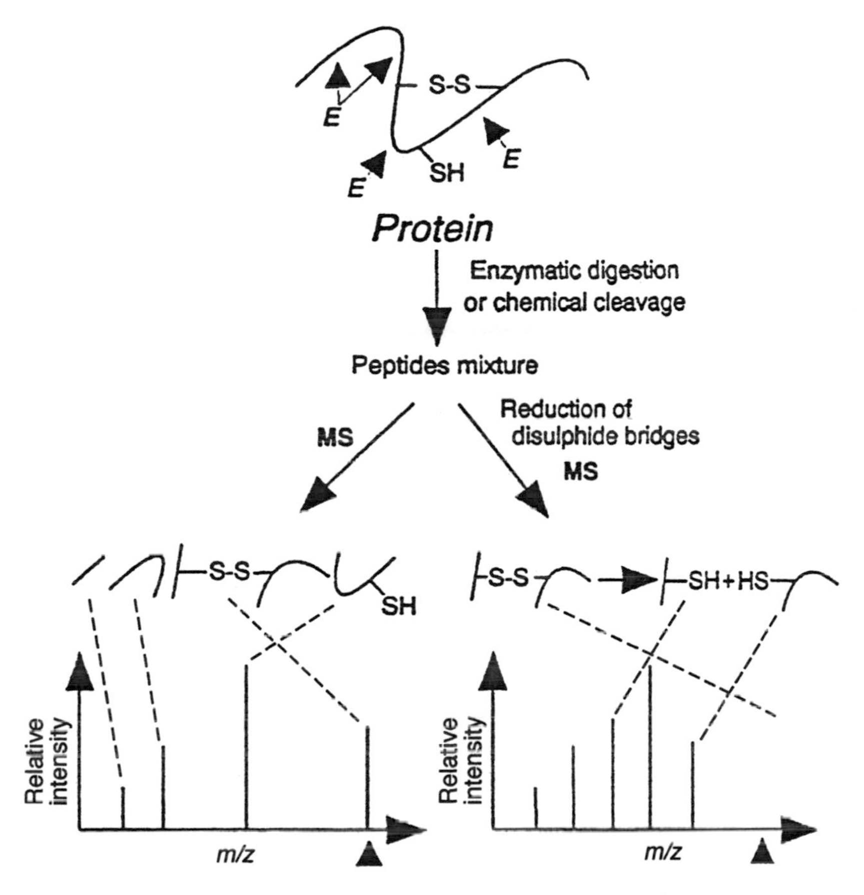


**Figure 7.11b** Illustration of the strategy used to locate cysteines linked by a sulphide bridge

bridged peptides Gly 37–Glu 55 + His 66–Ser 73 and Gly 37–Glu 55 + His 61–Ser 73. The FAB spectrum after HPLC separation allows this hypothesis to be confirmed. The peptides resulting from the cleavage of these bonds are easily reduced *in situ* during the FAB-MS analysis. This suggests that Cys-42 is linked with Cys-70. The spectrum of a *fraction* of the mixture shows ions with masses 1366 and 1472 Da, which correspond to the peptides Ala 1–Glu 13 and Ser 14–Asp 26, respectively, and an ion with mass 2836 Da, which corresponds to the bridged peptide Ala 1–Glu 13 + Ser 14–Asp 26. This suggests that the second disulphide bridge that is possible between Cys 8 and Cys 24 is the actual one.

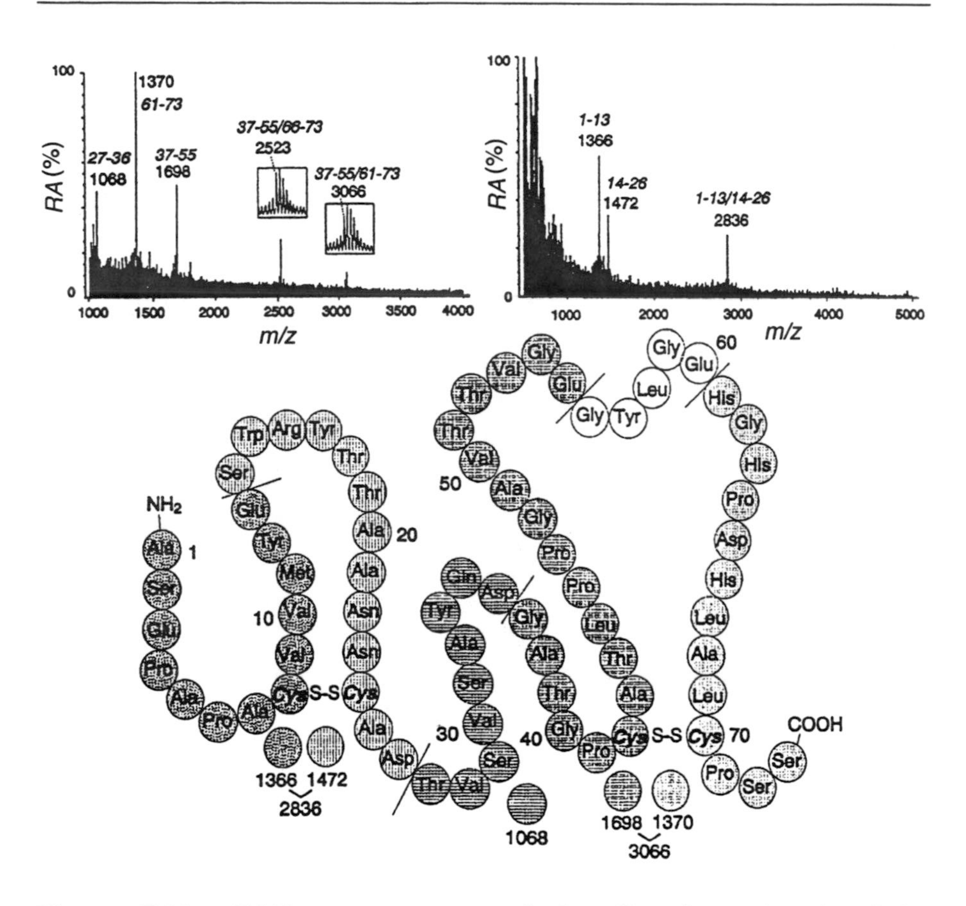


**Figure 7.11c** FAB mass spectra of the digestion of Paim I by protease V8 and of a peptide (HPLC-fractionated) present in this mixture

The strategy for sequencing a protein by using mass spectrometry starts with the precise determination of the molecular weight of that protein by using MALDi or ESI. The results allow both a verification of the sequence that is ultimately determined, and an assessment of the homogeneity of the sample.

In order to generate two different peptide series, the protein is digested with trypsin, which specifically cleaves on the C-terminal side of Lys and Arg, and by protease V8, which specifically cleaves on the C-terminal side of Glu and Asp. After a reversed-phase HPLC fractionation, the molecular weight of each peptide is determined.

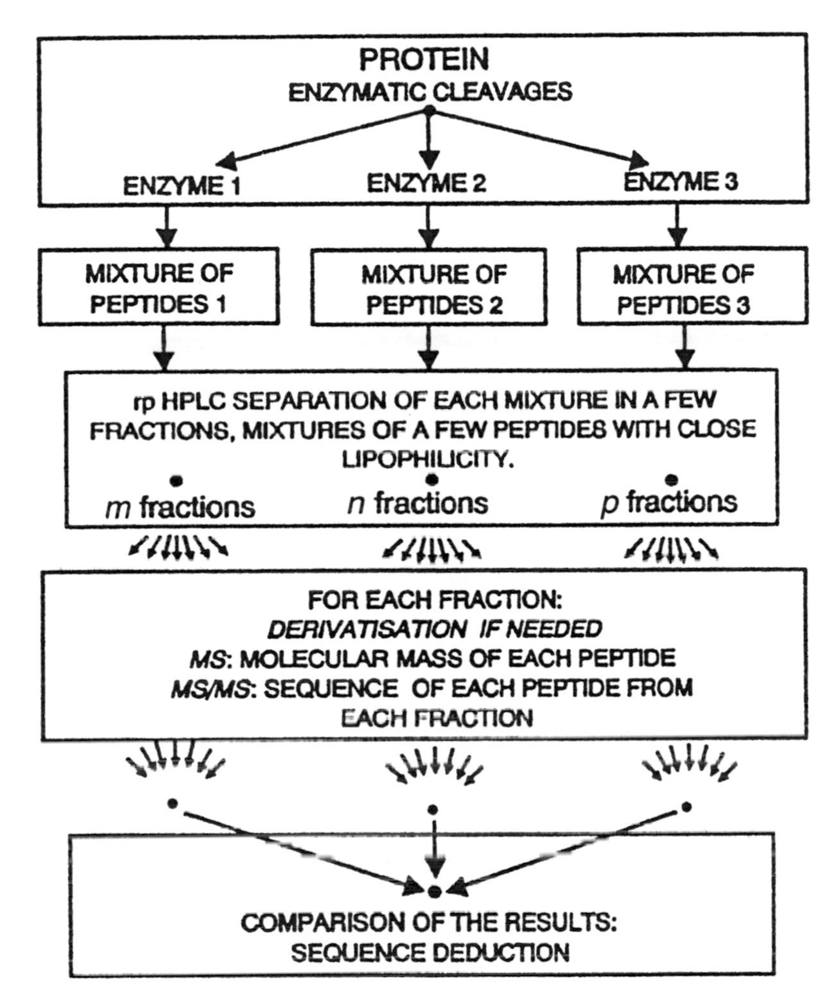


**Figure 7.11d** Illustration of the strategy followed in sequencing proteins by using mass spectrometry

Peptides with masses lower than 3000 Da which yield important signals are sequenced by using *Collision-Induced Dissociation* (CID), coupled with tandem mass spectrometry (MS/MS). CID-MS/MS is a technique enabling the stable ions produced from, e.g. FAB sources, to fragment, and therefore we can glean some additional information on the structure of the biomolecule in question, i.e. protein sequencing. Fragmentation is achieved by selecting the ion to be fragmented by first using a mass analyser, and then sending it to a collision cell, where it collides with uncharged gas atoms. Therefore, the kinetic energy is partly transformed into vibrational energy and the resulting fragments are analysed by a second mass spectrometer. By selecting only the predominant isotope peak, a fragmentation pattern which is free from complex isotopic clusters (particularly at high masses) can be achieved.

Peptides sequenced by using CID-MS/MS are then assembled by matching the peptides in both series, while considering the molecular weights of the large unsequenced peptides that contain several smaller sequenced peptides or the known sequence of a homologous protein. Figure 7.11d shows the strategy used in sequencing a protein by using mass spectrometry.

**SAQ 7.11b**

Can you think of any other classes of biomolecules that could be analysed by mass spectrometry?

**Summary**

This chapter has introduced you to a detailed system of analysing an unknown mass spectrum in order to gain as much information as possible about the structure of the compound in question. In Section 7.1, two correlation tables were presented, namely an (M–X) table and a mass composition table, which will help you to decide which class or classes an unknown compound might belong to, as well as likely structures for the main ions in the spectrum.

Once this has been decided, the typical fragmentations of the indicated class(es) of organic compounds can be referred to in Sections 7.2–7.10, as required. You should not expect to learn all of this detail at first, as it is far better in practice to build up your experience and confidence with interpretation of mass spectra over a period of time.

It is also important to realise that mass spectrometry is not always able to distinguish structural- and stereoisomers particularly well. The final structural determination is often not possible without reference to both NMR data and reference collections of mass spectra, etc.

**Objectives**

Now that you have studied this chapter, you should be able to:

- analyse a mass spectrum by a logical procedure, applying (M–X) and mass composition tables as appropriate, in order to assign the compound to its class or a limited range of classes;
- identify simple compounds from their mass spectra;
- recognise when the mass-spectral information is not sufficient or precise for a clear structural deduction to be made;
- present your conclusions in a clear and concise manner, indicating likely structures for the important ions in the mass spectrum.

# 8. Analysis of Mixtures by Hyphenated Mass Spectral Techniques

Mass spectrometry has been closely associated with chromatography since the 1960s when volatile organic mixtures were first analysed with the aid of a coupled gas chromatograph and GC/MS was born. By the mid-1970s, high performance liquid chromatography (HPLC/MS) had also been used to analyse involatile organic mixtures, and with the development of new ionisation methods such as field desorption and fast-atom-bombardment, LC/MS has now grown to become a valuable analytical technique. This chapter aims to introduce the reader to these two coupled chromatographic techniques, namely GC/MS (Section 8.1) and LC/MS, (Section 8.2). We shall concentrate on the methods of efficiently linking together the various types of instruments, and will also give some examples of their uses and the methodology of linked chromatography–mass spectroscopic analyses.

Tandem mass spectrometric techniques (MS/MS) will also be discussed later in the chapter.

## 8.1 GAS CHROMATOGRAPHY–MASS SPECTROMETRY (GC/MS)

### 8.1.1 Instrumentation

Discussion of the processes used to separate mixtures by chromatography is beyond the scope of this present text. It will be assumed that a good separation has been obtained on a conventional gas (or liquid) chromatograph and the need established for the identification or quantitation of some of the components of a complex mixture. These chromatographic techniques owe their usefulness in analytical chemistry to a combination of versatility, wide range of applicability, and sensitivity. Where materials are too involatile or too unstable to pass through a GC column, LC takes over. It is also important that both the mass spectrometer and the interface between the chromatograph and the mass spectrometer do not thermally degrade the compound and are sufficiently sensitive for the analysis being undertaken.

Of course, it is possible to have a separate chromatograph and mass spectrometer and merely trap (collect) the compounds as they are eluted from the column. Their mass spectra can be obtained afterwards by using a direct-insertion probe or all-glass heated inlet system (AGHIS).

Π Can you think of some advantages and disadvantages of combining a chromatograph with a spectrometer?

(a) Advantages:

1. It is possible to split the sample up into aliquots and conduct other tests (e.g. IR or NMR spectroscopy) to confirm the compound's identity.

2. A broad peak, or a peak with a shoulder, obtained from an initial separation would be best run through a second column with a different polarity to check if there is one or more components in the sample.

(b) Disadvantages:

1. There may be difficulties with efficient trapping, particularly of minor components.

2. It can be a very time-consuming method.

3. The technique is impossible to use with capillary columns owing to the minute amounts of sample being analysed.

Therefore, the modern approach is often to link chromatographs to dedicated mass spectrometers.

Let us now consider some vital questions we must ask ourselves if we are to conduct efficient GC/MS measurements.

It is often conventional to inject 1 microlitre (1 μl) of a volatile organic solution of concentration 1 part per million by weight (1 ppm) on to the GC column.

Π Would it be possible to obtain a mass spectrum with a reasonable signal-to-noise ratio from such a solution, assuming that all of it is delivered to the ion source?

1 ppm by weight is $10^{-6}$ g in 1 g of solution.

1 μl is $10^{-3}$ g (assuming unit density).

Therefore, the amount of the solute in 1 μl is $10^{-9}$ g, or 1 nanogram (1 ng). Modern mass spectrometers are well capable of determining useable mass spectra on ng quantities, and sensitivities can be as low as the femtogram (fg) level ($10^{-15}$ g). So, it is actually possible to obtain useful data at parts per billion (ppb) levels and lower!

The next parameter to look at is temperature. Gas chromatographs operate at temperatures between ambient and *ca.* 300 °C, depending on the nature of the sample and the stationary phase used on the column.

Π Is this temperature compatible with the mass spectrometer?

It is usual to operate mass-spectrometer inlet systems and source chambers at *ca.* 200 °C. This ensures the efficient transfer of samples through the system and rapid removal after ionisation, but not so high that it will decompose the compound itself. Therefore, both the chromatograph and the mass spectrometer are quite compatible with respect to temperature as long as the interface and transfer lines between them are heated to at least the temperature of the GC oven. Heating of these regions is also needed to prevent 'cold spots', which could result in some components of the mixture partially condensing on the walls. If this happens, the next component to emerge from the GC column would 'catch up with them' and the separation would be degraded. It is usual to heat all of the transfer lines and interface devices to 20–30 °C above the oven temperature used in the chromatograph and also line them with inert glass surfaces to avoid metal-catalysed thermal decompositions and rearrangements.

The next point to consider is the time factor. In GC/MS it is vital that a good mass scan of each component is obtained before it has finished eluting. Ideally, we would like to obtain at least ten scans across a GC peak while it is still eluting.

Π Why might one scan not be good enough?

1. The concentration of the component may vary over several orders of magnitude as it elutes off the column. It may be fairly difficult and time-consuming (when many components are being analysed) to choose the scan to coincide when the peak is at its maximum.

2. The mass spectrum may change considerably as the component is eluted and it is therefore desirable to obtain at least one scan where the concentration is changing as little as possible.

**SAQ 8.1a**

What would be the effect on the apparent intensity of the molecular ion and fragment-ion peaks if the sample concentration of the component was rapidly increasing as the scan was made, i.e. from low-mass to high-mass?

Therefore, it is important to scan many times as the component is eluted so as not to produce a biased mass spectrum.

For a packed column, the elution of a GC peak may take 1 min. However, for capillary-GC/MS it could be *ca.* 1 s. This means that the scan rate must be approximately $5 \times 10^{-2}$ s decade$^{-1}$, allowing for resetting time.

Π Which types of mass spectrometers can easily achieve such high-speed scanning requirements?

Quadrupole, time-of-flight and Fourier-transform instruments can scan quite easily at these speeds. Modern magnetic-sector instruments can now also carry out these tasks without magnet eddy-current problems occurring at high scan rates.

Another requirement for a GC detector is that it produces a chromatogram of concentration versus time of elution

Π Why is this an important requirement?

Retention indices are also an important means of component identification.

We can quite easily produce a chromatogram in a number of ways. The data system can sum up the intensities of all of the ions recorded by the electron multiplier at the end of each mass scan. The total intensity is then outputted as a voltage–time plot. The total ion current (TIC) trace is updated with every scan.

---

Another way of obtaining a TIC trace is to place a small negatively biased collector plate at the edge of the ion-source beam outlet in order to tap off some of the positive-ion beam. This is then amplified and outputted to a chart recorder.

A further way of recording packed-column chromatograms is to send some of the signal to a flame-ionisation detector or other conventional detector.

Π It appears that a gas chromatograph and a mass spectrometer are very compatible indeed, but what problem will there be with an interface that is linked to a conventional ion source?

This is the pressure difference. A gas chromatograph operates at an outlet pressure of 1 atmosphere and a mass-spectrometer ionisation chamber at *ca.* $10^{-6}$ torr. This is a pressure difference of about nine orders of magnitude. Therefore, the interface must be capable of sustaining this pressure gradient without any loss of sample or chromatographic-resolution problems. The carrier gas (often He) must be removed from the sample in order to concentrate the sample as much as possible. A popular interface is the *jet separator*, which is shown in Figure 8.1a.

The jet separator relies on the differential diffusion of the lighter carrier-gas molecules away from a jet created by passing the effluent

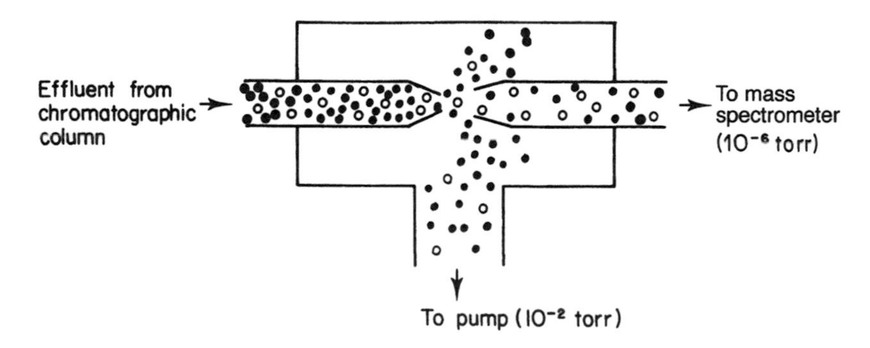


**Figure 8.1a** Schematic representation of a jet separator showing enrichment of the sample (○) by the removal of carrier gas molecules (●)

stream from the chromatograph into a small vacuum chamber. The gas stream expands as it enters the vacuum chamber via a small orifice and, according to Graham's law, the lighter carrier-gas molecules diffuse at a rate which is inversely proportional to their density. Therefore, as the jet expands into a cone the outer areas become richer in helium. The gas stream is directed towards a small orifice which leads to the mass spectrometer (about 1 mm away). When it reaches the latter, most of the He has diffused outwards and therefore misses the hole, while the heavier solute molecules have deviated very little from the straight path between the two jets and so enter the mass spectrometer, although clearly some will be lost. This design of separator, which is operated at oil-pump pressures (*ca.* $10^{-2}$ torr), removes about 90% of the carrier gas. Approximately 60% of the sample reaches the mass spectrometer with virtually no delay occurring. There is also a very small contact surface for the passing sample.

Π Why is it important that there should be a small contact surface for the passing sample?

This minimises any decomposition and rearrangements of the components caused by contact with hot surfaces and also reduces the sample dead volume, which helps to maintain chromatographic quality.

A jet separator can only be optimised for one particular density ratio of carrier gas to sample for a particular flow rate and a fixed jet gap.

Jet separators become more efficient as the relative molecular-mass difference between carrier gas and sample is increased; this is why He, a very light gas, is often used.

Further refinements of the jet separator include provision of a 'dump valve'. This is a thin strip of tungsten that can be switched into and out of the atom path between the jets by means of an external magnetic link. The GC effluent stream is thus deflected down the inlet backing-system pump to waste.

Π Why may we want to deflect the GC effluent stream from the mass spectrometer?

There are two reasons for this:

1. GC samples are often made up as dilute solutions in a solvent. This will be present in a large amount and might overload the mass spectrometer. In addition, we do not need to analyse the effluent stream because we know what its spectrum will look like.

2. Some samples may contain large amounts of known components which are best selectively dumped so as to avoid excessive pressure rises in the ion source and possible filament damage.

Having removed the carrier gas, the sample reaching the ion source from the jet separator is not very different from that entering from the conventional inlet system, and so the usual methods of ionisation can be used singly or in tandem.

The backing pumps in the inlet system and the diffusion or turbomolecular pumps in the ion source maintain the pressure at $10^{-2}$ and $10^{-6}$ torr, respectively, at these locations. The high pumping capacity of the pumps means that it is possible to couple a capillary GC column directly to the ion-source chamber, without needing a separator, since the gas flows used in capillary columns are *ca.* 10% of those used in packed columns. This means that no sample is lost at the interface. Full advantage can therefore be taken of the tremendous separations achievable by using capillary columns threaded directly through the interfacial glass-lined heated steel tubing.

There are limitations on the kind of GC column we can use for GC/MS, and also the column temperatures which can be employed. This is because the stationary phases, being organic polymers, will depolymerise as the temperature is raised and give characteristic decomposition products known as *bleed*. The latter continuously elute from the column through the interface into the mass spectrometer, where they contaminate the source and can give rise to ions which are as intense as the sample ions. The bleed is difficult to correct for, since it increases in a non-linear way with rises in temperature.

Π Can you suggest some practical measures we could take to minimise bleed problems?

1. Use longer columns with low stationary-phase loadings in packed column work, e.g. 5% phase or less.

2. Be careful not to exceed the decomposition temperature of the stationary phase.

3. Use the stable bonded-phase columns in capillary work, where mixtures of components of widely differing volatilities are to be analysed.

Two commonly used GC stationary phases are poly(ethylene glycol) (PEG, Carbowax) and dimethylsiloxane (OVI) polymers. These are available as bonded phases with the tradenames BP20 and BP1, respectively. BP20 is useful for natural oil and flavour work as it is a polar phase, while BP1 is non-polar and is widely used for polar molecules such as biochemicals, drugs, and metabolites. BP20 is relatively bleed-free, but BP1 and other silicone phases give silicon-containing bleed peaks at *m/z* 75, 147, 207, and 281, which the operator should be aware of. Most data systems have routines that can help eliminate bleed peaks from GC/MS spectra by subtracting component-free data from that obtained during the scan.

Π It is particularly important to use such a routine if the GC column is being temperature-programmed. Can you suggest a reason for this?

It is because the rate of decomposition of the polymer phase will give a non-linear increase in the bleed peaks as the temperature rises. Therefore, a background scan needs to be made as close as possible to the appearance of the chromatogram peak on the trace.

### 8.1.2 Derivatisation

Many types of compound need to be derivatised before GC separation because of their involatility. Trimethylsilyl (TMS) and *tert*-butyldimethylsilyl derivatives are often used, for example:

$$ROH \longrightarrow R{-}OSi(CH_3)_2Bu^t$$

However, such silyl derivatives do not always yield good structural information from their mass spectra, so acetyl, trifluoroacetyl and methyl esters are also used for derivatising alcohols and amines:

$$ROH + (CH_3CO)_2O \longrightarrow R{-}OCOCH_3 + CH_3CO_2H$$

$$ROH + (CF_3CO)_2O \longrightarrow R{-}OCOCF_3 + CF_3CO_2H$$

$$ROH + CH_2N_2 \longrightarrow R{-}OCH_3 + N_2$$

Acids are frequently converted into their methyl esters, and aldehydes and ketones into their *O*-methyloximes; these yield stable molecular ions and structurally informative fragment ions:

$$>{=}O + NH_2OCH_3 \longrightarrow >{=}NHOCH_3$$

Vicinal and 1,3-diols, and related systems with neighbouring active hydrogens, e.g. aminoalcohols and hydroxyamides, are easy to derivatise with alkyl boronic acids such as butyl-, *t*-butyl-, and phenylboronic acids, to give volatile cyclic boronates:

$$R(XH)(YH) + R'B(OH)_2 \longrightarrow R\langle X, Y \rangle B{-}R'$$

$R' = Bu, Bu^t, Ph;\ X,Y = O, N$

These derivatives are only formed if 5- or 6-membered rings are possible, and are widely used in the study of natural products related to carbohydrates.

Π What are the important factors in the choice of a derivative for GC/MS analysis?

1. Efficient GC separation must be achieved.

2. The derivative group should have a fairly small mass increment.

3. The derivative group must give abundant high-mass ions.

4. Sometimes, a larger-derivative group can be chosen so as to move the molecular ion of the compound away from any interferent bleed peaks.

Table 8.1a lists some common derivatives and their associated mass increments (per derivatised group).

**Table 8.1a** Some commonly used derivatives for GC/MS and their corresponding mass increments

| Typical substrate | Derivative | Mass increment (amu) |
|---|---|---|
| Alcohols<br>Phenols<br>Enols<br>Thiols<br>Amines | $-OCH_3$<br>$-OSi(CH_3)_3$<br>$-OSi(CH_3)_2Bu^t$<br>$-OCOCH_3$<br>$-OCOCF_3$ | 14<br>72<br>114<br>42<br>96 |
| Acids | $-OCH_3$<br>$-OSi(CH_3)_3$ | 14<br>72 |
| Aldehydes<br>Ketones | $=NOCH_3$ | 29 |
| $>C(OH)-C(OH)<$<br>$>C(OH)-C-C(OH)<$<br>$>C(NH_2)-C(OH)<$<br>$>C(NH_2)-C-C(OH)<$ | $BuB-$<br>$Bu^tB-$<br>$PhB-$<br>$-CH(CH_3)_2$ | 66<br>66<br>86<br>40 |

**SAQ 8.1b**

Suggest derivatives which would be suitable for GC/MS analysis of each of the following classes of compound:

(a) fatty acids;

(b) amino acids and peptides;

(c) carbohydrates;

(d) phenols;

(e) 2-hydroxyacids;

(f) steroid hormones (containing C=O and OH groups);

(g) fruit essences (containing alcohols, carbonyl compounds and esters).

In some cases there will be a choice — select which derivative you think would be best from the mass spectrometry point of view.

### 8.1.3 Quantitative Work

Quantitative work can be performed by GC/MS at very low levels, e.g. ng $cm^{-3}$ of blood plasma for drugs and metabolites, providing that suitable internal standards are available. Figure 8.1b illustrates the analytical methodology employed in quantitative GC/MS. Purification of the crude biological extract may not always be necessary. Mass-spectrometric sensitivity and selectivity are greatly aided by a data system capable of monitoring selected ions from the mass spectra, i.e. selected ion-monitoring (SIM).

The standard and the compound to be analysed must have similar GC and MS behaviour. For this reason, deuterated drugs are often used in combination with the 'natural' drug itself in pharmaceutical analysis. There will be no need for a chromatograph separation of the isotopes if its mass fragments, such as the $M^{+\bullet}$ and $(M+3)^{+\bullet}$ ions (if it contains a $CD_3$ group) can be measured by SIM. The ratio of the drug to the deuterated standard can therefore be calculated, with other background peaks in the spectrum being ignored.

Π (i) Why would we need at least three deuterium atoms in the standard compound?

(ii) What other advantages does SIM have over ordinary mass scanning?

(i) The (M–1) and (M–2) peaks, due to $^{13}C$, $^{2}H$, $^{15}N$, $^{18}O$, and other isotopes, would seriously interfere with the determination of the ratio of drug to standard if only one or two deuterium atoms were incorporated. An increment of 3 amu is the minimum required to shift the mass of the standard away from the common isotope region.

(ii) When using SIM, the sensitivity is much increased, i.e. down to pg and even lower levels, because we are only trying to measure two *m/z* values. The 'dwell-time' on each ion can be 100 times greater than that possible in a full scan of the mass spectrum, and therefore an increase of two orders of magnitude can be achieved.

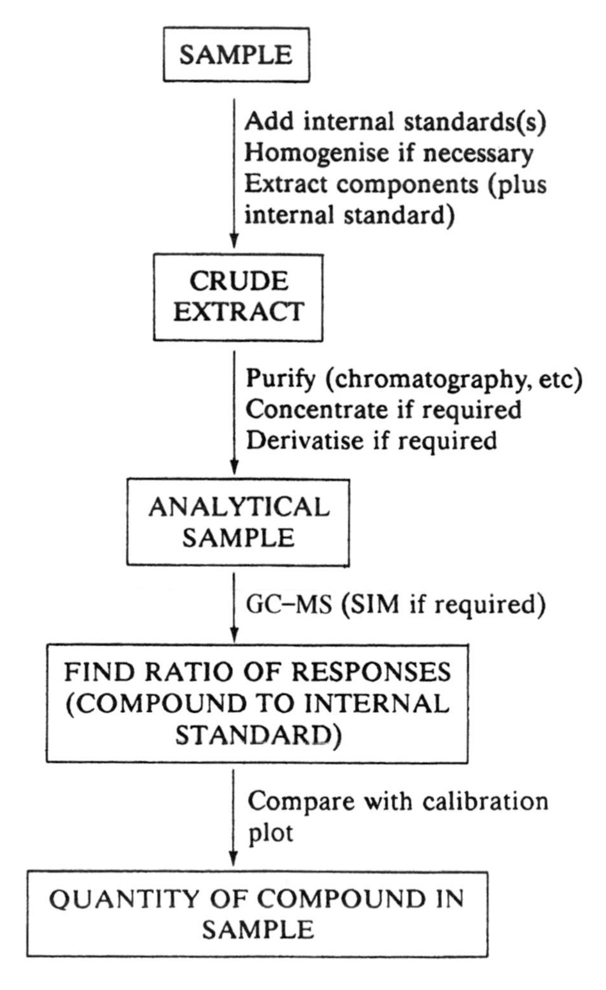


**Figure 8.1b** Illustration of the typical analytical methodology employed in GC/MS

**SAQ 8.1c**

Discuss the ways in which you think GC and MS are (i) compatible, and (ii) incompatible techniques.

**SAQ 8.1d**

Think back to how a magnetic-sector mass spectrometer is scanned. What do you think is the best way of carrying out SIM? Are quadrupole analysers suitable for SIM work?

## 8.2 LIQUID CHROMATOGRAPHY–MASS SPECTROMETRY (LC/MS)

### 8.2.1 Instrumentation

Early work on LC/MS (if we could call it that) concentrated on the collection of fractions from a high performance liquid chromatography (HPLC) separation. These fractions were evaporated and subjected to EI, CI, FD or FAB methods of ionisation by using direct-probe insertion techniques.

The increased interest in bioanalytical chemistry in the last decade or so has stimulated the development of LC to make it capable of handling aqueous buffers at flow rates of at least 1 $cm^3$ $min^{-1}$ without degrading the performance of either the chromatograph or the spectrometer, and its ability to obtain $M_r$ values and structural information on large, non-volatile and thermally labile molecules.

At the present time, GC/MS works for neutral molecules with $M_r$ values up to about 1500, and LC/MS for charged substrates with $M_r$ values in the mass range 1000–20 000. With the development of soft ionisation techniques, such as FAB, and new desorption processes, substances having molecular weights of 5000 and above, which are present in living organisms, have now been separated and characterised.

The combination of LC and MS presents a major challenge. The mobile liquid phases used in LC range from low-boiling-point organic solvents to aqueous mixtures, often modified with a variety of acids, bases and organic and inorganic salts to buffer them and improve chromatographic performance. The problem of removing the high gas volumes produced by on-line evaporation of the mobile phases, in order to reach the vacuum required for good mass spectra, is much more acute than in GC/MS. Typical flow rates for LC are 0.5–5 $cm^3$ $min^{-1}$, which translates into gas flow rates in the range 100–3000 $cm^3$ $min^{-1}$. Interfaces for LC/MS can be broadly divided into those which use the mobile phase to assist in ionisation, e.g. as a CI reagent gas, and those where the mobile phase is removed before ionisation.

One of the first approaches tried was the direct-liquid-introduction (DLI) method, which was developed in the 1970s. In this technique, a small portion of the eluent from the chromatograph is fed into the MS ion source via a capillary inlet, where the vaporised solvent becomes a CI reactant gas, thus giving $MH^+$ ions. Unfortunately, there are a number of disadvantages of this procedure, namely structural information from CI spectra are limited, flow rates of only 10–20 μl $min^{-1}$ can be tolerated, and only a limited range of compounds give $MH^+$ ions readily.

### 8.2.2 Thermospray (TS) Interface

An alternative approach which overcomes some of these problems is the *thermospray* technique. The spectra produced are typical of those produced by field-desorption ion sources. The mechanism for ion production centres around charge exchange between salt ions and the eluent, when the liquid flow from the LC is heated to *ca.* 150 °C in a capillary before spraying into a vacuum chamber. The solvent contains buffer ions, e.g. ammonium ethanoate, and as the solvent evaporates charged solid particles, containing a core of the solute molecules, form in the gas phase.

Because of the very small size of these particles, the field gradient created across them is *ca.* $10^2$ V $m^{-1}$, and this increases rapidly as the particles dissociate in the vacuum. The substrate is usually present as a $MH^+$ or $MX^+$ species, where X is $NH_4$, Na, or K, depending on the

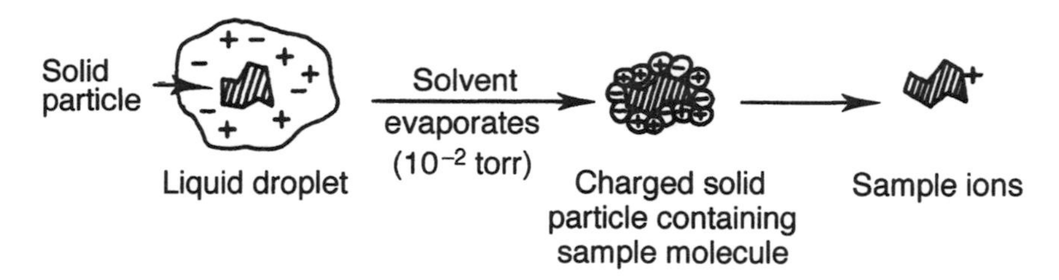


**Figure 8.2a** Schematic representation of ion formation during the thermospray process

nature of the buffer. Figure 8.2a shows a schematic representation of the mechanism of ion formation during the thermospray process.

This technique is widely applicable to virtually any molecule present as either a *positive or negative ion in aqueous solution*. Solvents such as methanol and acetonitrile do not give rise to ions even when dissolved salts (*ca.* 0.1 mol dm$^{-3}$) are used, but provided that at least 10% water is present in reversed-phase LC efficient direct-ion evaporation occurs. The ionised substrate molecules need to be at concentrations of *ca.* 10$^{-3}$ mol dm$^{-3}$, which is clearly a signiciant limitation on sensitivity.

Ammonium ethanoate buffers, which are commonly used as LC solvents, give rise to ammonium ions. These will bring about ammonia chemical ionisation, but many samples will have proton affinities less than ammonia and so will not be observed as $MH^+$, although polar solvents may yield $(M + NH_4)^+$ ions. Relatively non-polar molecules may not produce any detectable positive ions under these conditions.

In order to turn thermospray into a universal LC/MS 'interface' an electron beam is placed between the end of the capillary and the ion sampling aperture leading into the MS analyser. The thoriated irridium filament which is used is more stable to burnout than the usual tungsten or rhenium filament. It is then possible to obtain ammonia CI, by using the $NH_4^+$ ions produced in the spray of an ammonium-ethanoate buffered LC effluent, on those non-polar molecules which do not respond to direct ion evaporation. The whole interface can be designed to be inserted into an ion source through the direct-probe insertion lock, and is applicable to any mass

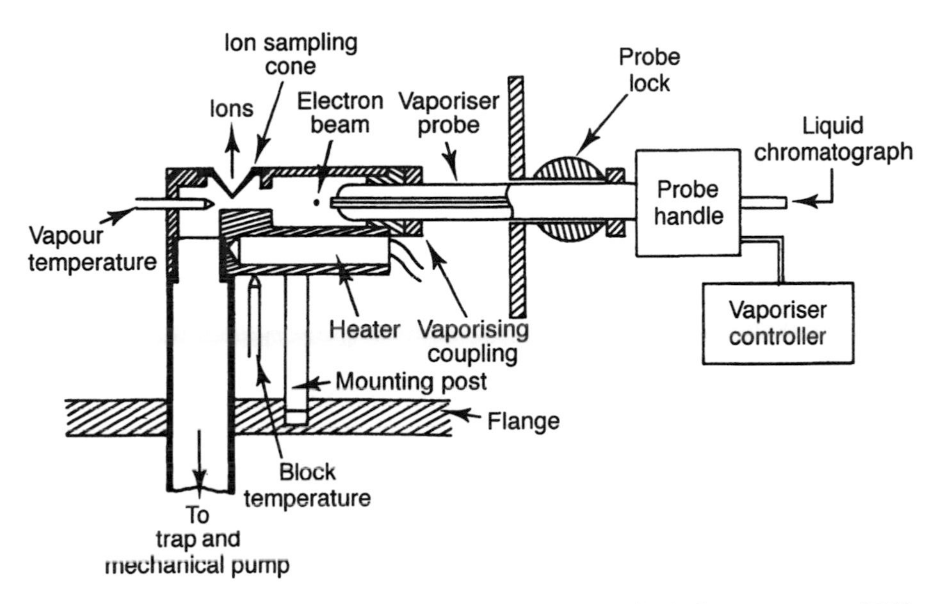


**Figure 8.2b** Schematic representation of a typical thermospray (TS) interface

spectrometer capable of CI operation. Figure 8.2b shows a typical design for a thermospray interface.

This is essentially similar to a CI source, with the addition of a pumping line from the ion source which is connected through a cold trap to an oil pump. This is necessary to remove the large amounts of vapour resulting from flow rates of 1 $cm^3$ $min^{-1}$ (or more) of solvent. The ion-exit aperture to the mass-spectrometer analyser is at the apex of a cone which is perpendicular to the TS jet. The use of such a 'sampling cone' is not essential, but it does appear to provide a better sample of directly vaporised and ionised molecules, rather than those which are deposited on surfaces and then revaporised or pyrolysed.

Figure 8.2c shows a thermospray mass spectrum for cyanocobalamin (vitamin $B_{12}$, $M_r$ 1354), obtained by using ammonium ethanoate (0.1 mol $dm^{-3}$) as the solvent.

In this case, the intact $M^{+\bullet}$ or $MH^+$ ions are not obtained, but rather the cation $C^+$ (produced by loss of $CN^-$) is the only peak in the molecular-ion region. The major peaks in the spectrum are the doubly

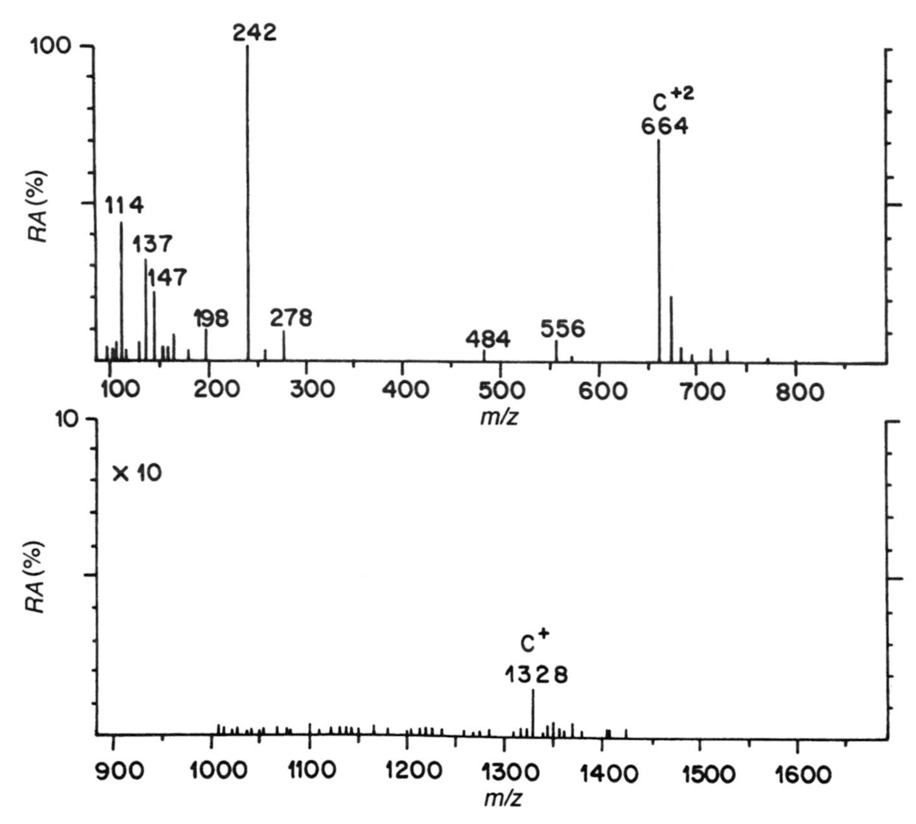


**Figure 8.2c** Thermospray mass spectrum of vitamin $B_{12}$ (cyanocobalamin) (10 nmol)

charged cation $C^{2+}$ (*m/z* 664) and satellites corresponding to the replacement of a proton by an alkali ion. At the low-mass end, some structurally significant ions are also produced. Multiply protonated molecular ions are often found when complex polar molecules are subjected to the TS process, particularly in buffers of low pH. For example, the peptide glucagon ($M_r$ 3483) has been detected as the triply and quadruply protonated molecular ions by using a quadrupole mass spectrometer which only has a mass range of 1300 amu.

The mass-spectral analysis of a series of alkaloids and drugs is shown in Figure 8.2d. In this study, aliquots containing 5 μg of each of the listed compounds were injected at 1.5 min intervals on to an LC

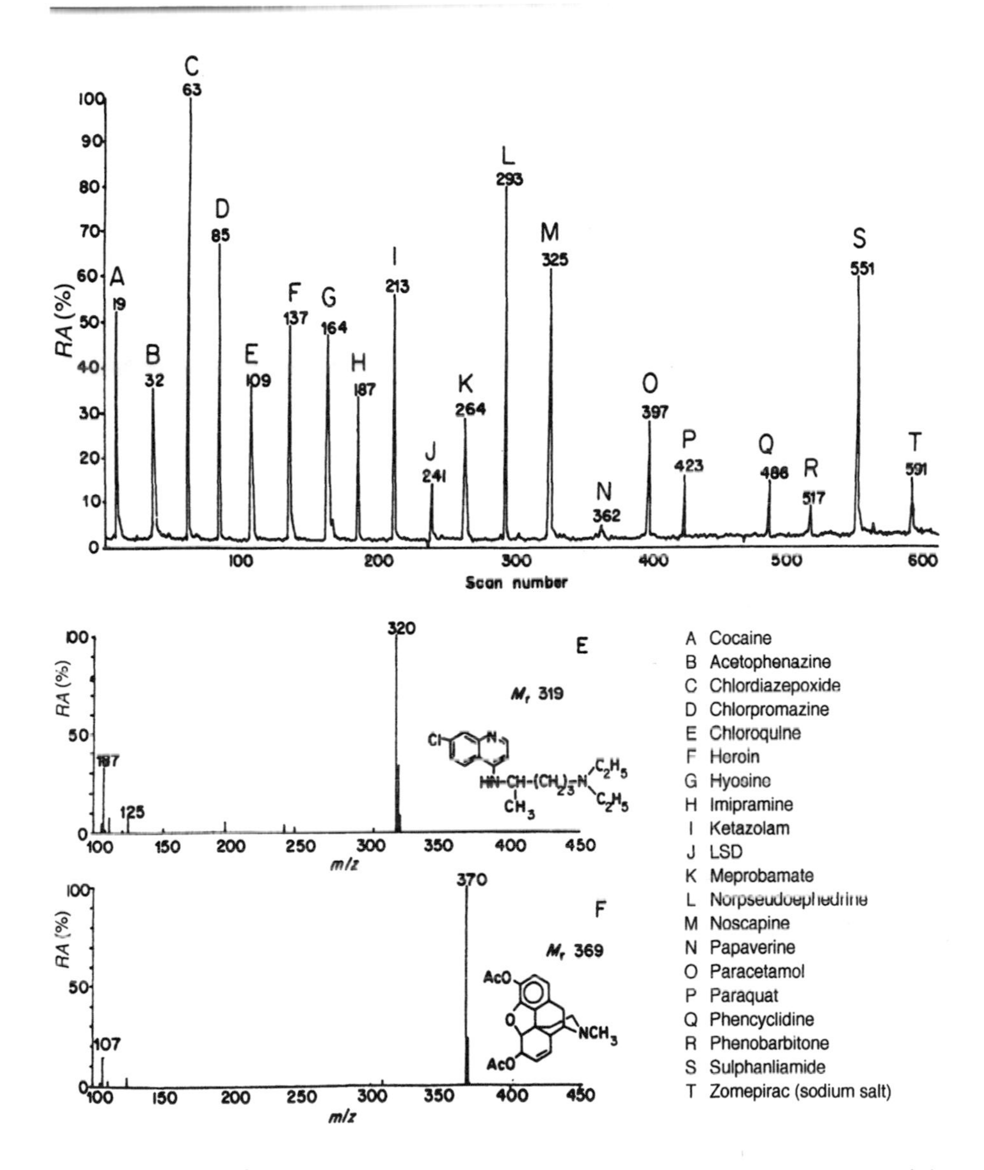


**Figure 8.2d** (i) LC/MS trace of a series of alkaloids and drugs; (ii) corresponding mass spectra of chloroquine (E) and heroin (F)

column and eluted with $CH_3OH$ (0.2 mol dm$^{-3}$)/$NH_4OCOCH_3$ (80/20), flowing at a rate of 0.8 cm$^3$ min$^{-1}$. A quadrupole mass spectrometer was used, recording over the range 40–450 amu.

This figure shows that all of the drugs were detected, although there were considerable differences in sensitivity, i.e. over a 20-fold range, with papaverine (N) being the least sensitive. TS ionisation had yielded $MH^+$ ions, which were often quite intense, thus providing $M_r$ information for chloroquine and heroin (E and F, respectively, in the figure).

Π Which compound appears to have the most fragment ions, E or F? Can you say why?

Compound E has more fragment ions. The amount of such ions produced in E and F may be related to whether the molecule has a stable fused ring system (as in F) or a more linear structure.

### 8.2.3 Moving-belt Interface Coupling

An alternative approach to the LC/MS interface is to remove the solvent prior to entry into the mass spectrometer. The most popular method used to achieve this is the moving-belt interface (Figure 8.2e).

In this system, the LC eluent is sprayed on to an endless moving belt in a low-vacuum chamber. The solvent evaporates readily and this is further helped by means of a small infrared heater. The sample

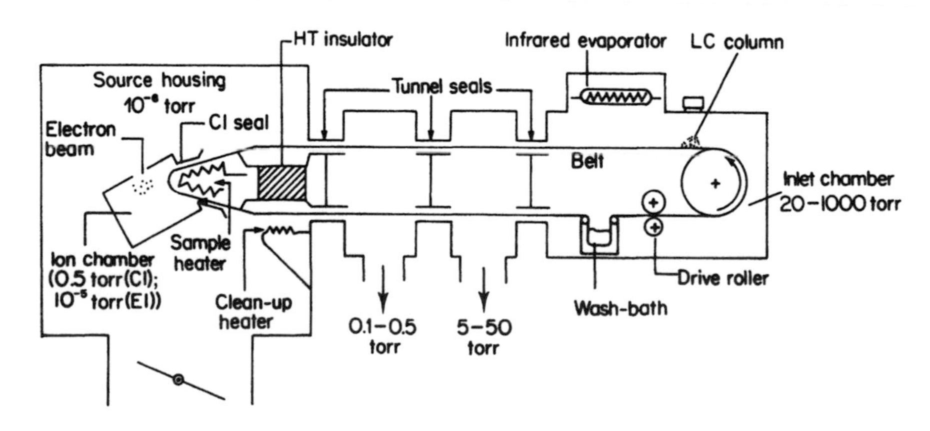


**Figure 8.2e** Schematic representation of a moving-belt interface

deposits are carried into the ion source through a system of seals and vacuum chambers, where any remaining solvent is progressively removed. Transfer efficiencies of 40–80% can be achieved in this way. EI, CI and, most importantly, FAB methods of ionisation can be used directly from the belt surface, which means that large biomolecules can be readily examined. A wash-bath is placed at the exit from the vacuum system in order to clean the belt. This also has the advantage that the matrix required for FAB can be dissolved in the wash-bath and renewed each cycle. Microbore columns, combined with angled spray deposition on to the moving belt improves the sensitivity, so that moving-belt devices can be up to ten times more sensitive than thermospray interfaces, although the belts have shorter working lives.

### 8.2.4 Particle-beam Interface Coupling

This interface allows a very efficient and rapid separation of the solvent and eluted molecules from a liquid chromatograph column. The eluate is pumped through the capillary up to a glass concentric nebuliser (see Figure 8.2f). The eluate is then transformed into a cloud of droplets, which are scattered by a helium concentric flow. The spray passes through a desolvation chamber, the walls of which are heated, and where the pressure is slightly lower than atmospheric. During their transport, the droplets undergo partial desolvation, thus yielding partially solvated droplets of the eluted compound.

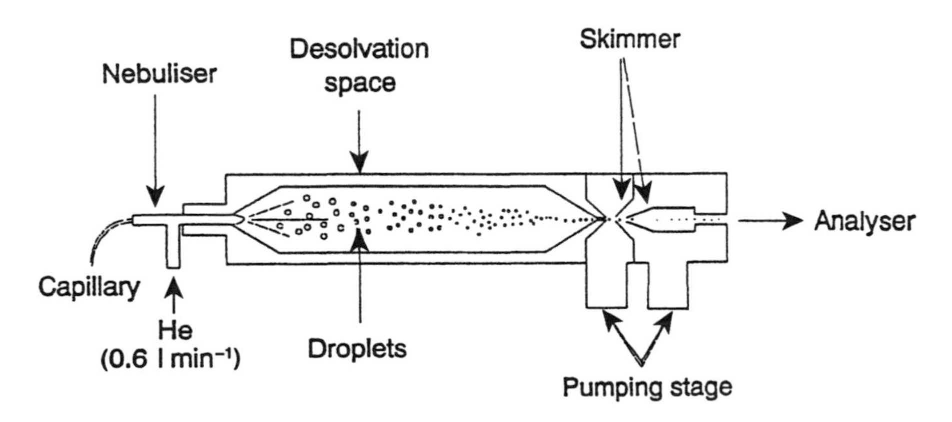


**Figure 8.2f** Schematic representation of a particle-beam interface

The desolvation chamber is linked with a double-stage-pumping, molecular-beam separator. When it leaves this evaporation chamber, the mixture of helium, solvent vapour and particles undergoes a supersonic expansion into the first pumping stage (*ca.* 10 torr). A high-speed gas beam containing particles of the eluted molecules thus results, in which the particles diffuse less rapidly from the centre of the beam towards the periphery. Separation is achieved by skimming the peripheral layers of the beam with μ. The skimmed solvent vapour and the helium are pumped out in two stages (*ca.* 500 mtorr). The narrow particle beam, now bound for the ion source, is typically smaller than 100 nm.

### 8.2.5 Continuous-flow FAB Coupling

This coupling device, which was developed in the late 1980s, operates by linking the end of a chromatographic capillary column with the end of a FAB nozzle by a capillary passing through the introduction nozzle. Glycerol (1–5%) is added to the chromatographic solvent as

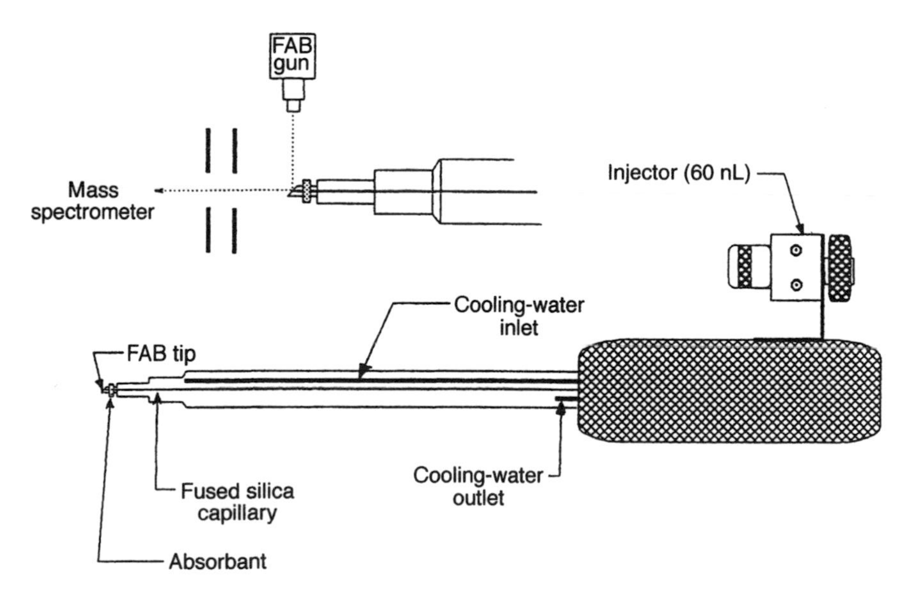


**Figure 8.2g** Schematic representation of a continuous-flow FAB coupling system

the FAB matrix, and remains as the solvent evaporates. The flow rate is 1–5 μl min$^{-1}$. Figure 8.2g shows a diagram of such a coupling.

### 8.2.6 Atmospheric-pressure Chemical Ionisation (APCI)

This is a chemical-ionisation technique which uses gas-phase ion–molecule reactions at atmospheric pressure. Primary ions are produced by corona discharge on a solvent spray (see Figure 8.2h). The chromatographic eluate (2 ml min$^{-1}$) is directly introduced into a pneumatic nebuliser, where it is converted into a thin fog by a high-speed air or nitrogen beam. Droplets are then displaced by the gas flow through a carefully controlled heated quartz tube, known as a desolvation/vaporisation chamber. The hot gas (120 °C) and the compounds leaving this tube and arriving at the reaction area of the source are chemically ionised (under atmospheric pressure) via proton transfer. The evaporated mobile phase acts as the ionising gas and yields $(M+H)^+$ ions. The electrons needed for primary ionisation result from corona discharges or negatron emitters. Thermal decomposition is kept at a minimum and ionisation of the substrate is very efficient. Differential pumping, over several stages separated by skimmers, maintains a good vacuum and allows entry of as many ions as possible to the analyser. Figure 8.2h shows a schematic diagram of an APCI source while Figure 8.2i displays an example of an application of HPLC–APCI coupling, in this case the separation of a mixture of five benzodiazapines.

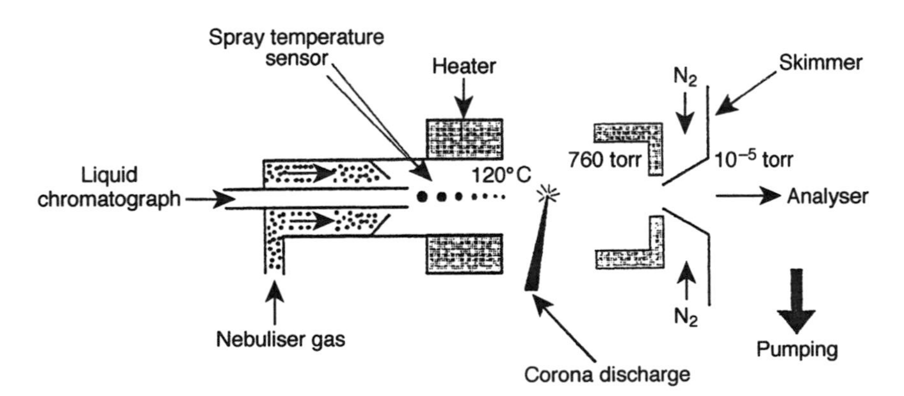


**Figure 8.2h** Schematic representation of an APCI source

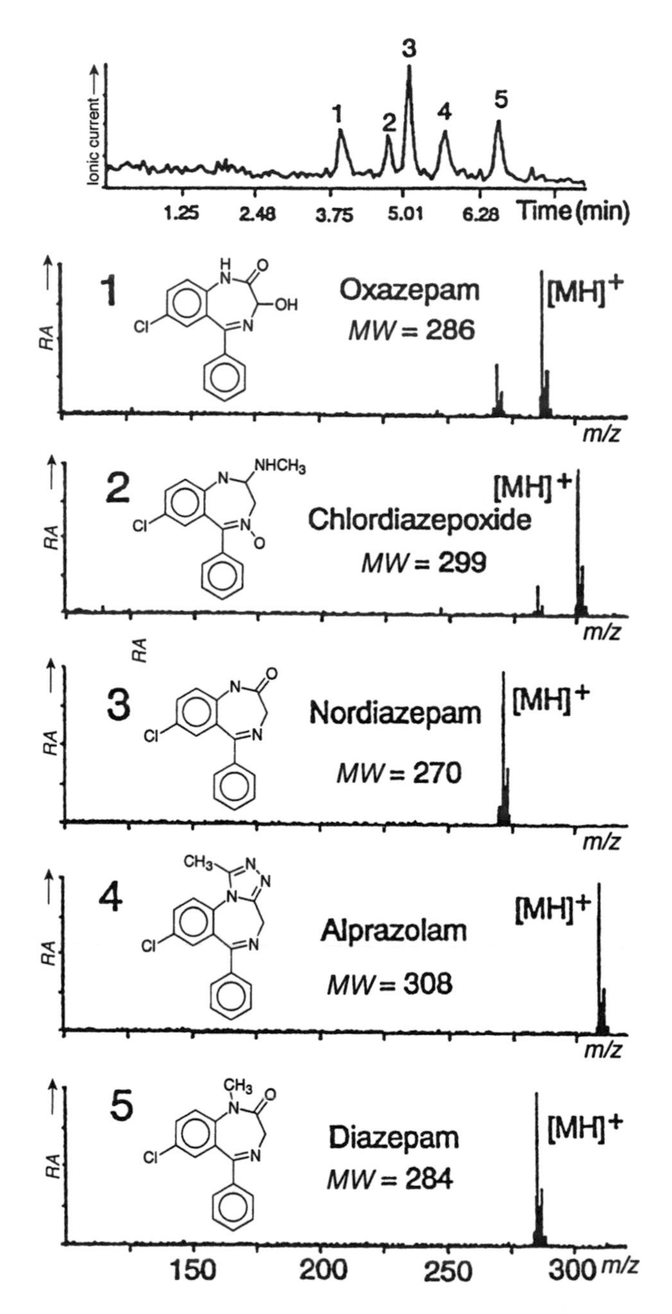


**Figure 8.2i** HPLC–APCI separation of a mixture of five benzodiazepines using $CH_3CN/CH_3OH/H_2O$ (40/25/35) (wt/vol) as the eluate, with a flow rate of 1.2 ml min$^{-1}$

### 8.2.7 Electrospray (ES) and Ionspray

An electrospray (ES) is produced by applying a strong electric field, under atmospheric pressure, to a liquid passing through a capillary tube with a weak flux (1–10 $\mu$l min$^{-1}$). The electric field is obtained by applying a potential difference of 3–6 kV between this capillary and the counter electrode, which are separated by 0.3–2.0 cm (see Figure 8.2j). This field induces a charge accumulation at the liquid surface located at the end of the capillary, which will then break up to form highly charged droplets. As the solvent contained in these droplets evaporates, they shrink to the point where the repelling coulombic forces come close to their cohesion forces, thereby causing their 'explosion'. Eventually, after a cascade of further 'explosions', the highly charged ions desorb. Electrospray can also be used in the case of molecules without any ionisable sites, through the formation of sodium, potassium, ammonium, or other adducts. A variation of this

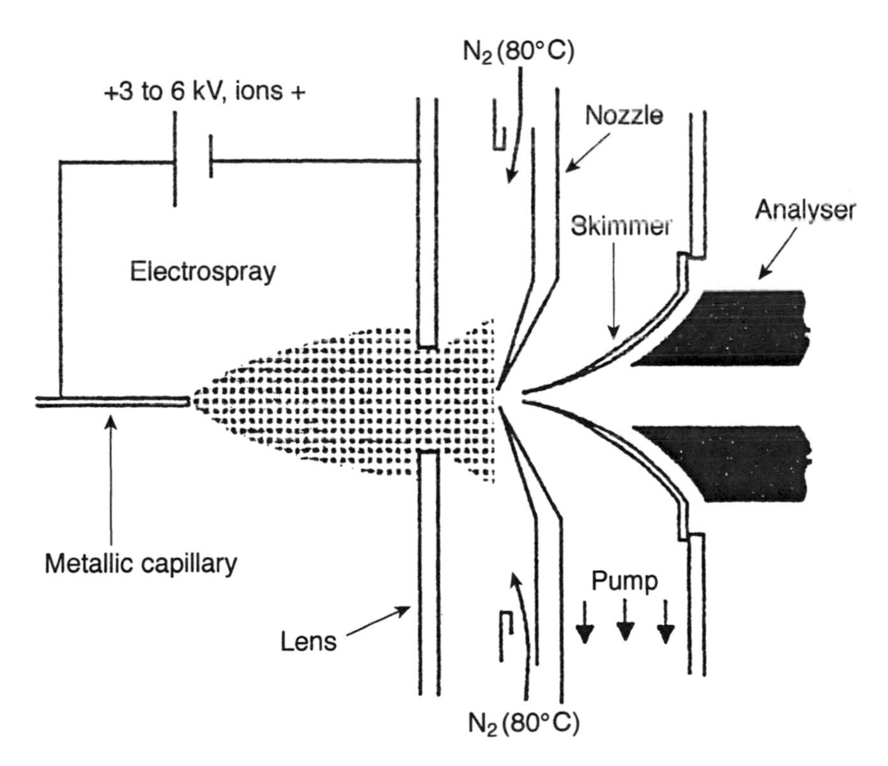


**Figure 8.2j** Schematic representation of an electrospray source

method, known as 'micro-electrospray', uses flow rates of a few nl min$^{-1}$. This enables sensitivities in the range of a few hundred attomoles to be achieved. Electrospray mass spectra normally correspond to a statistical distribution of multiply charged molecular ions obtained from protonation $(M + nH)^{n+}$, while avoiding the contributions from dissociations or from fragmentations. Computer-based algorithms have been developed to allow the determination of the molecular mass through the transformation of the multiply charged peaks in the ES spectrum into singly charged peaks. Even simple mixtures can be deconvoluted (see Figure 8.2k).

Ionspray, which is another form of electrospray, creates charged droplets by using a potential difference and a pneumatic nebuliser in

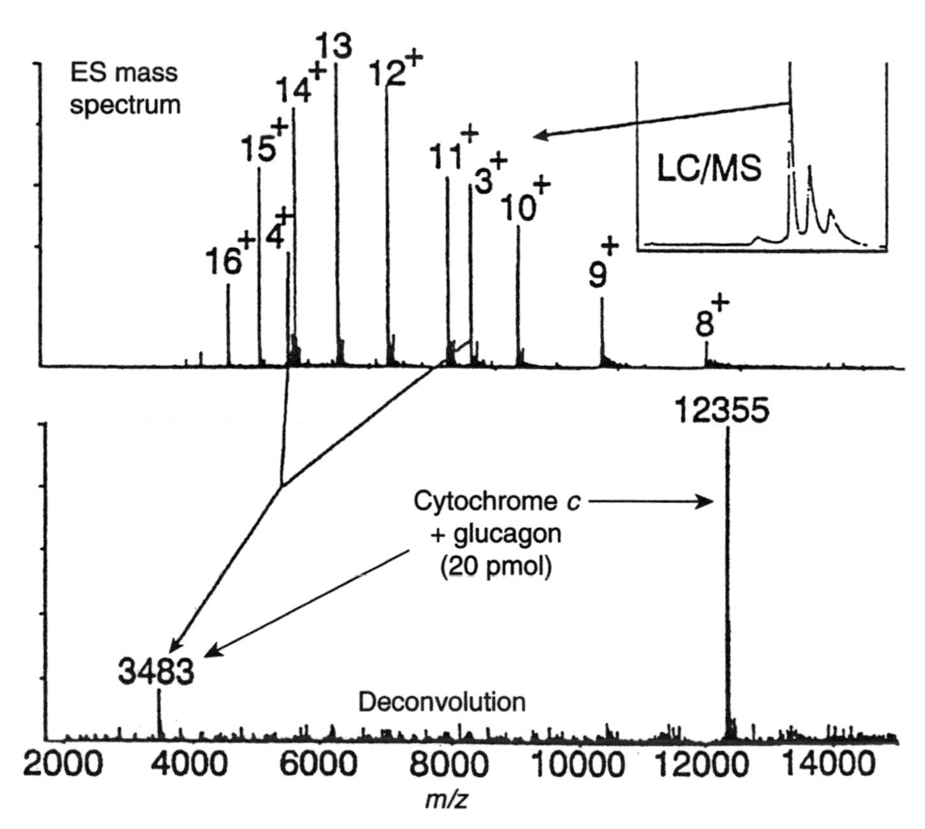


**Figure 8.2k** Deconvolution of an ES-mass spectrum as applied to the analysis of a protein

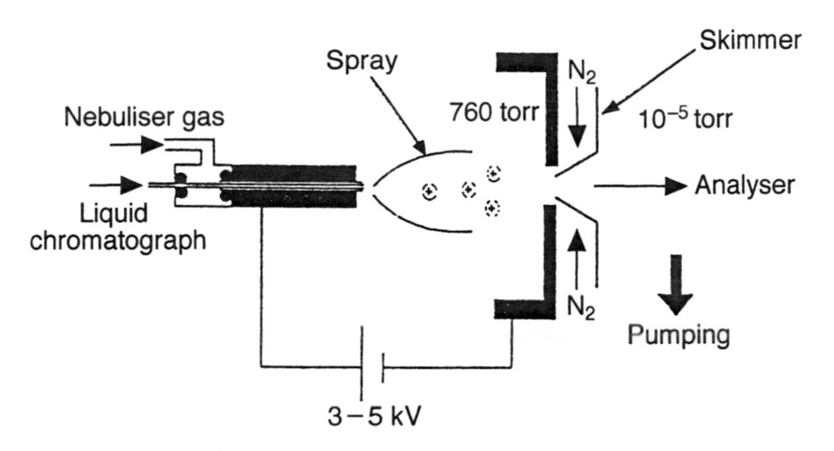


**Figure 8.21** Schematic representation of an ionspray interface

order to accept higher flow rates and to produce a spray that is less dependent on the nature of the liquid phase. The ionspray technique (see Figure 8.21) consists of pumping the sample solution through a pneumatic nebuliser, which is maintained at a high voltage so as to form a fog of charged droplets, even if the flow rate being used is high. The ions are then ejected from the droplets which evaporate, via a low-energy process that does not produce fragments. Flow rates are *ca.* 50–200 μl $min^{-1}$. A grounded screen lens acts as a spray splitter and thus inhibits the condensation of droplets that make the spray unstable at high flow rates. In this way, rates of up to 2 ml $min^{-1}$ are then possible.

Figure 8.2m illustrates the application ranges of different mass spectrometric/chromatographic coupling methods as a function of mass and of the nature of the compound being eluted.

**SAQ 8.2a**

What are the fundamental requirements for a LC/MS system?

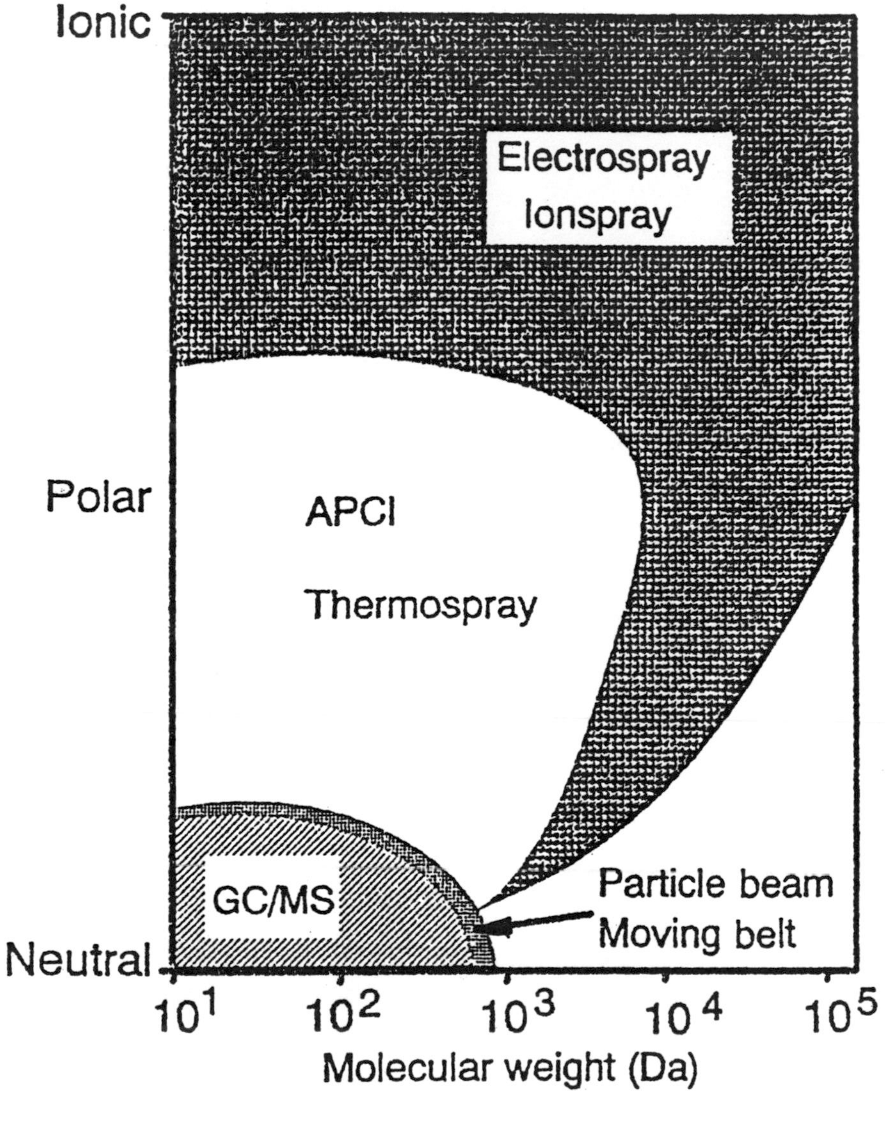


**Figure 8.2m** Illustration of the application ranges of different mass spectrometric/chromatographic coupling methods

**SAQ 8.2b**

What is the major problem in coupling a liquid chromatograph directly to a mass spectrometer and how has this problem been overcome?

## 8.3 TANDEM MASS SPECTROMETRY (MS/MS)

This is a technique where we couple one mass spectrometer to a second spectrometer in series. Here, the first mass spectrometer serves to isolate the molecular ions of the mixture's various components; these ions are then taken, one at a time, and introduced into the second machine, where they are fragmented to give a series of mass spectra, one for each molecular ion produced in the first spectrometer. This technique is called *tandem mass spectrometry* (often abbreviated as MS/MS) and was first developed in the 1970s. MS/MS was briefly outlined in Chapter 3.

The first spectrometer often uses a chemical ionisation source or another soft technique in which molecular ions (or protonated molecular ions) are predominantly produced. These ions then pass to the second spectrometer, where the ions source is normally a field-free collision chamber through which helium is pumped. The high-velocity *parent* ions and helium atoms collide, which results in further fragmentation of the parent ions to form a multitude of *daughter* ions. The spectrum of these daughter ions is then scanned by the second spectrometer. Three methods of scanning are commonly adopted.

Π Can you recall from Section 3.9 what these methods are?

The first is *fragment (daughter)-ion* MS/MS. Here, the analyte may consist of two markedly different compounds that have identical nominal molecular masses, i.e. they are isomers. In order to discriminate between the two, the first spectrometer is set to the mass of the protonated parent ions. The spectra of the daughter ions are quite different, and therefore structural identification is possible.

The second technique is called *precursor (parent)-ion* MS/MS. Here, the first spectrometer is scanned while the second spectrometer is set to the mass of one of the daughter ions. Similarly structured compounds usually give several daughter ions of a similar mass, so this method is particularly useful where the identities and concentrations of members of a homologous series of compounds are required. An example of an application of this type of tandem mass spectrometry involves the determination of alkylphenols ($HOC_6H_4CH_2R$) in solvent-refined coal. The second spectrometer is set at an *m/z* value of 107, which corresponds to the ion $HOC_6H_4CH_2^{+}$. The first spectrometer is then used to scan the sample. All of the alkyl phenols in the sample give ions of *m/z* 107, whatever the nature of the substituent R.

The third method, namely *neutral-loss scanning*, was covered in detail in Section 3.9.

Tandem mass spectrometers can be made up of various combinations of magnetic-sector, electrostatic-sector and quadrupole-filter separators, which are often represented as B, E and Q, respectively. Therefore, an EBEB instrument consists of two double-focusing mass spectrometers with each of these containing an electrostatic- and a magnetic-sector separator. A popular tandem mass spectrometer which is commonly used has the configuration QQQ; this system is shown in Figure 8.3a.

In this particular set-up, the sample is first introduced into the chemical-ionisation source, where molecular ions are formed. These ions are then accelerated into the first stage, or parent-ion separator, which is a normal quadrupole filter. The high-velocity ions pass into quadrupole 2 (Figure 8.3a), where they collide with He atoms ($10^{-3}$–$10^{-4}$ torr) and further fragment. This quadrupole is operated on RF mode only (i.e. no DC potential), so that the ions are focused, but not filtered. The resulting daughter ions then pass into quadrupole 3, where they are scanned and recorded.

Tandem mass spectrometry offers similar advantages to those of GC/MS and LC/MS, but in this case with the specific benefit of speed.

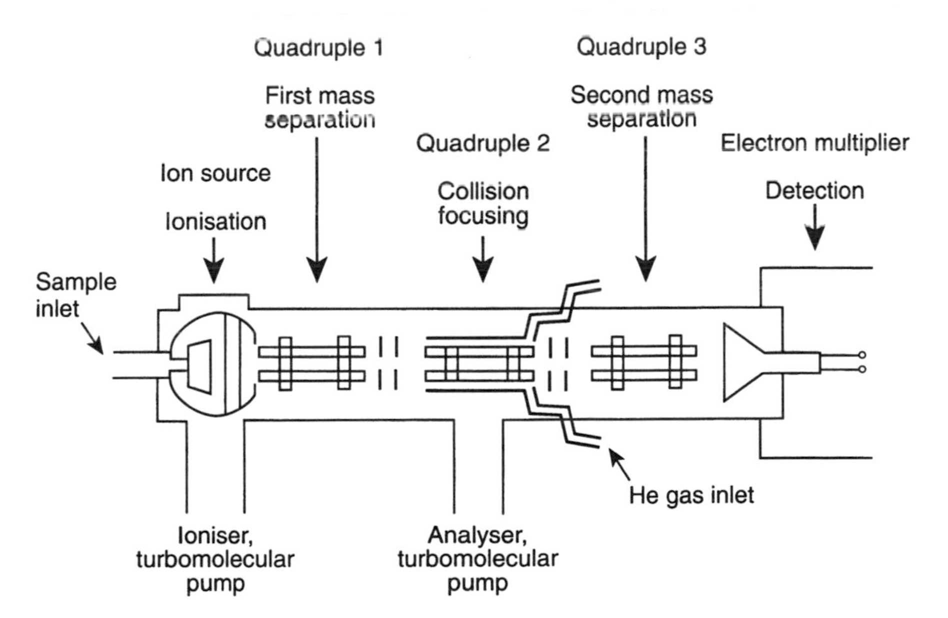


**Figure 8.3a** Schematic representation of a tandem quadrupole mass spectrometer system

Π Why do you think an MS/MS instrument is quicker?

Separations on a chromatographic column may take several minutes or even hours, while Mass spectrometers can scan in milliseconds.

Π Can you think of any other advantages that MS/MS has over GC/MS or LC/MS?

The chromatographic techniques require dilution of the sample with large excesses of a mobile phase (and the later removal of this phase), which adds to the problem of the introduction of potential chemical interferences. Therefore, tandem mass spectrometry is potentially much more sensitive than these other hyphenated techniques. A disadvantage of MS/MS, however, is the greater cost of the equipment that is required.

Tandem mass spectrometry has sought wide application in both qualitative and quantitative analysis of various components in the complex materials that are found in both natural and synthetic systems. Coupled with new soft-ionisation techniques such as laser desorption, MS/MS shows much promise for future research and application.

## 8.4 CHROMATOGRAPHY DATA: ACQUISITION, RECORDING AND TREATMENT

### 8.4.1 Data Acquisition

There are three data-acquisition modes that can be considered, namely scanning, selected-ion monitoring (SIM, not to be confused with SIMs, i.e. 'secondary ion mass spectrometry' and selected-reaction monitoring (SRM). We have dealt with SIM earlier in this chapter (see Section 8.1.3) and saw then the gain in sensitivity for two extreme masses that can be achieved when compared to conventional scanning.

Selected-reaction monitoring achieves an even better sensitivity and selectivity gain than SIM. SRM is based on the decomposition

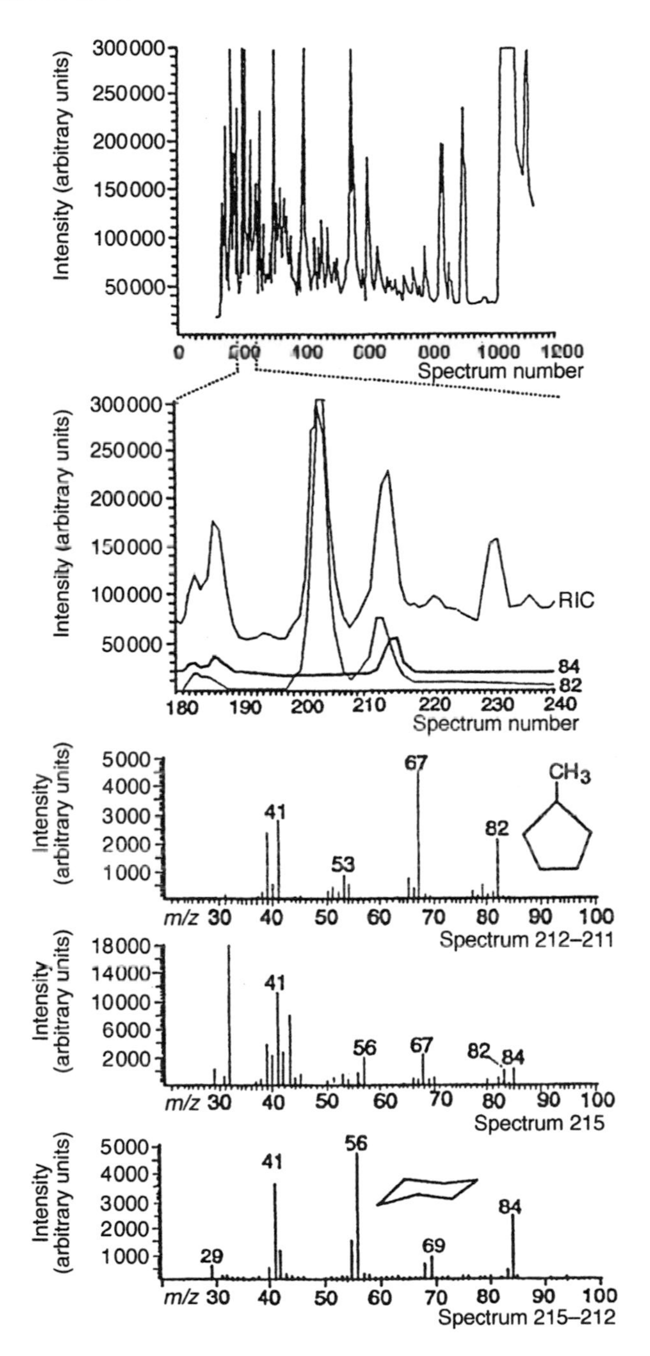


**Figure 8.4** Illustration of the deconvolution of a chromatographic peak representing two compounds

reactions of ions that are characteristic of the compounds to be analysed and requires the use of tandem mass spectrometry. The spectrometer is set so as to let through only the ions produced by a decomposition reaction in the chosen reaction region between the two analysers. Selectivity gains arise from the fact that the fragmentation reaction implies two different characteristics of the compound under study, i.e. the *m/z* ratios of both the product ions and the precursor ions must satisfy the selected values. Advanced systems now allow the instrument tuning parameters to by fully programed via compouter investigation of the data.

### 8.4.2 Data Recording and Treatment

The spectrometer provides two series of data as a function of time, namely the number of detected ions and a physical parameter (e.g. the Hall-probe reading) which indicates the individual masses of these ions. (Refer back to Chapter 4 for a more detailed discussion of these points.) Interactive programs allow the operator to compare the spectra measured at the beginning of the elution of a chromatographic peak and at the end. If these are identical, the probability that the peak contains only one compound is very high.

The spectra shown in Figure 8.4 indicate the presence of two compounds, with *m/z* values of 82 and 84. A spectrum can be selected within the chromatographic-peak region and others, representing the background noise, at the peak edge. The second spectrum is then subtracted from the first. In the figure, RIC corresponds to the reconstructed ion spectrum.

### Summary

This chapter has considered the linking together of both gas and liquid chromatographs with mass spectrometers. Interfacing is a much easier task in GC/MS than in LC/MS.

In the former technique, this is effectively achieved by the use of a jet separator and direct introduction of the capillary column into the ion source. Sensitivity down to the pg region of the components injected on to the columns can be realised. A wide range of derivatising reagents are available which enhance the volatility of polar samples and also their mass-spectrometric behaviour. Relative-molecular-mass values in the 1000–2000 amu range are routinely measurable, but ionic compounds are not directly analysed by this method.

Interfacing for LC/MS is carried out via either direct-liquid introduction, thermospray, electrospray, ionspray, atmospheric-pressure chemical ionisation, particle-beam, continuous-flow FAB, or belt interfaces. Ionisation of charged substrates is readily achieved in thermospray inlets, with some useful fragmentation often occurring, while belt systems are quite compatible with all of the current ionisation methods, particularly FAB. Therefore, LC/MS and GC/MS can be considered as complementary techniques, because LC is more suited to ionic compounds, such as acid salts, and base hydrochlorides of biomolecules. These can be separated by liquid chromatography using buffered eluents, but are unsuited to gas chromatography without extensive extraction and derivatisation to render them volatile.

We have also discussed the technique of tandem mass spectrometry (MS/MS) and described three methods of component analysis, namely fragment (daughter)-ion, precursor (parent)-ion and neutral-loss MS/MS. In addition, we have looked briefly at the acquisition, recording and treatment of chromatographic data.

## Objectives

Now that you have completed this chapter, you should be able to:

- appreciate the practical problems involved in coupling gas and liquid chromatographs to mass spectrometers;
- appreciate the reasons why mass spectrometers are good detectors for GC and LC;
- describe the design and principles of operation of jet separators for GC/MS;

- describe the design and principles of operation of direct-liquid introduction, thermospray, ionspray, electrospray, atmospheric-pressure chemical-ionisation, continuous-flow FAB, particle-beam and belt interfaces for LC/MS;

- appreciate the range of samples and concentrations which can be examined by each of the separators;

- understand the role of derivatisation in GC/MS;

- suggest derivatives for use with given functional groups in GC/MS;

- understand the methodologies involved in quantitative analysis in GC/MS and LC/MS, and the importance of selecting a suitable internal standard;

- suggest suitable methodologies for the GC/MS or LC/MS analysis of mixtures of various sorts of compounds, including the basic choice of a GC or LC approach;

- describe the design and name three principles of operation of a tandem mass spectrometer;

- recognise how data is acquired, recorded and treated in chromatography.

# 9. Inorganic Mass Spectrometry

Inorganic mass spectrometry, although often omitted from many textbooks on mass spectrometry, has become increasingly widely used for determination of the elemental composition of samples. This present chapter will deal with some of the many different techniques in common use today. For convenience these methods have been grouped into three areas of study, namely surface methods, analysis of bulk solids and elemental analysis in solution.

Π Can you think of any technique you may have come across before in atomic spectroscopy that uses a mass spectrometer?

Inductively coupled plasma mass spectrometry (ICPMS) is such a technique.

## 9.1 SURFACE METHODS INVOLVING MASS SPECTROMETRY

### 9.1.1. Secondary-ion Mass Spectrometry (SIMS)

You may recall (from Section 2.4) that if a beam of ions (e.g. $Cs^+$) of a sufficiently high energy are directed on to a surface, then secondary ions could be generated, i.e. *sputtered*, from this surface. The technique of a secondary-ion mass spectrometry (LSIMS) for example, is powerful way of looking at large species, particularly biomolecules. However, LSIMS and other similar techniques are actually modifications of techniques which have been used to investigate inorganic surfaces for a number of years. Often these techniques are discussed under the blanket term of *secondary-ion mass spectrometry* (SIMS), because in many cases, but not all, the primary beam is an ion beam.

The basic process is as follows. A high-energy (5–20 keV) beam of ions such as $Ar^+$ and $Cs^+$, or occasionally $O_2^+$ or $N_2^+$, is directed on to a surface such as a metal film or semiconductor, or even a mineral (see Figure 9.1a).

This results in a number of secondary particles being ejected, including both positive and negative ions of the elements making up the surface. These ions can be passed into a magnetic quadruple or other type of analyser for separation into individual *m/z* values and then detected. The resulting mass spectrum is normally very simple, containing just one peak for each isotope of the elements that are present on the surface and can give measurable signals from as little as $10^{-15}$ g of material. All elements from hydrogen upwards can be determined in this way, but truly quantitative data is difficult to obtain. Since the primary ion beam produces a small crater at the point of impact it is also possible to perform depth–profile studies in the surface layers of the sample.

### 9.1.2 Ion-microprobe Mass Analysis (IMPMA) and Ion-beam Lithography/Modification

This is a modification of normal SIMS which uses a much more sophisticated (and expensive) ion beam; this passes through two or more electrostatic lenses which focus it down from the 0.2–5.0 mm diameter spot used in SIMS to a spot with a diameter of 1–2 μm which can be directed, with the aid of an ordinary optical microscope, on to specific areas on a surface.

Again, secondary ions are produced and are normally directed through

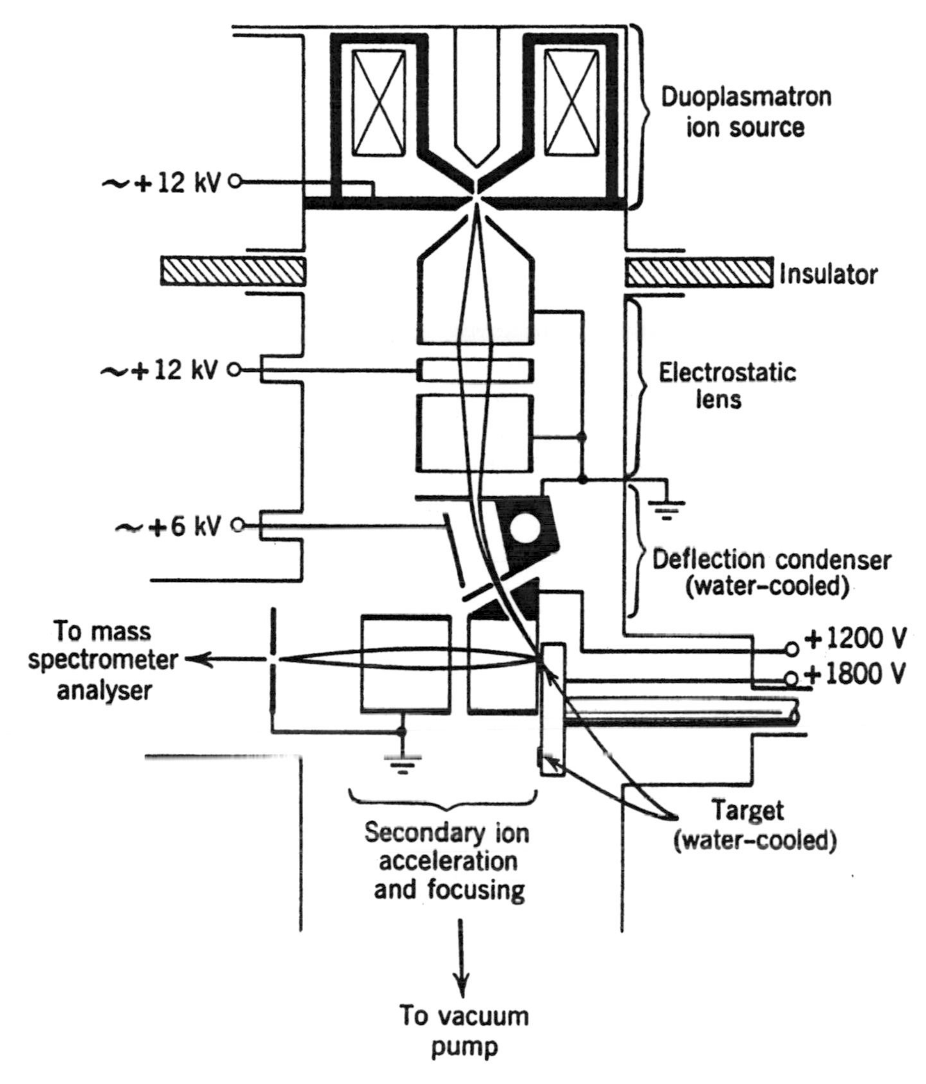


**Figure 9.1a** Schematic representation of a SIMS source

a double-focusing analyser to give high resolution and high sensitivity ($10^{-15}$ g). Using this technique, changes in the surface composition can be mapped as the beam is traversed across the surface.

IMPMA has become a powerful research tool for surface studies of semiconductors or deposited metal films.

Π Samples for IMPMA are often stored under high vacuum before analysis. Why is this?

The high sensitivity of the technique means that great care must be exercised in sample preparation and presentation and consequently very high vacuums are normally used to ensure that the surface is not contaminated by a surface film of 'impurities' (often oxidation products) prior to analysis.

The high capital cost of such analytical systems has, however, prevented them from wide usage. Ion beams are less subject to scattering than electrons and produce only secondary electrons of low energies and thus high spatial resolution is maintained. Microstructures can also be fabricated. Figure 9.1b shows such an arrangement, used in this case for the chemical etching of a gallium arsenide sample.

### 9.1.3 Laser-assisted Microprobe Mass Analysis (LAMMA)

Rather than employ a beam of primary ions, it is now possible to use a focused beam of light from a pulsed Nd–YAG laser, which can provide extremely high power levels at the surface, giving up to $10^{11}$ W $cm^{-2}$ on a spot of a few microns in diameter. This can achieve detection limits down to $10^{-20}$ g; in this way, minute samples such as individual

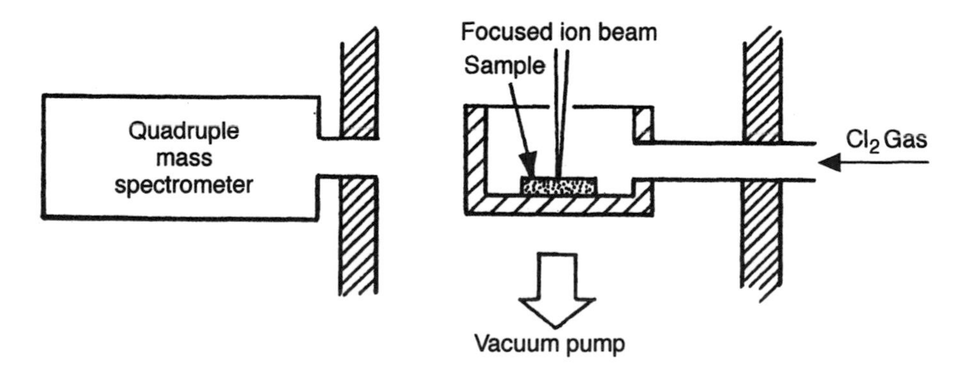


**Figure 9.1b** Schematic representation of an IMPMA arrangement used for the chemical etching of GaAs

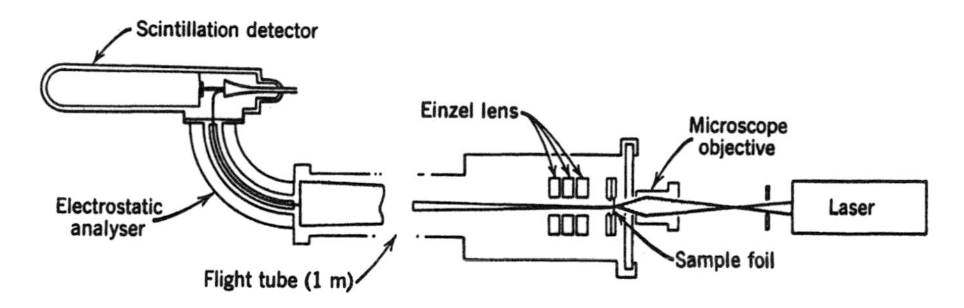


**Figure 9.1c** Schematic representation of a LAMMA system

dust particles can be studied, or specific areas on a surface can be spatially differentiated with a resolution of *ca.* 1 mm (see Figure 9.1c).

The production of ions may involve the following processes:

(a) direct ionisation of the solid sample by the laser beam;

(b) surface ionisation.

**SAQ 9.1a** Can you think of any more processes that might be involved?

At low-laser-power densities ($<10^8$ W cm$^{-2}$), the ionisation efficiency is *ca.* $10^{-5}$; but at higher power densities ($10^9$–$10^{10}$ W cm$^{-2}$) the ionisation efficiency may reach 0.1%. A limitation of the use of the laser microprobe technique for analytical work on bulk solids is the large variation in the probability of ionisation of vaporised samples and the differences in the ablations rates for various materials.

This technique is now beginning to find applications in the life sciences as well as more traditional areas, where, for example, element variation across small parts of tissue samples can be mapped, thus providing an insight into the distribution and function of common electrolyte ions, such as Na, K, Ca, Mg, etc., in systems such as neurones (nerve cells).

### 9.1.4 Particle-induced X-ray/Gamma Emission (PIXE/PIGE) and Rutherford Backscattering (RBS)

Analysis by PIXE involves the use of a particle beam, which is directed on to a sample surface; protons are often used, but $^4He^+$ or other 'projectiles' with energies of a few MeV can also be employed in this technique. Figure 9.1d shows schematically the radiation processes that are induced when energetic ion beams are directed on to a solid material. The emitted radiation is measured by the use of an energy-dispersive X-ray detector or gamma counter.

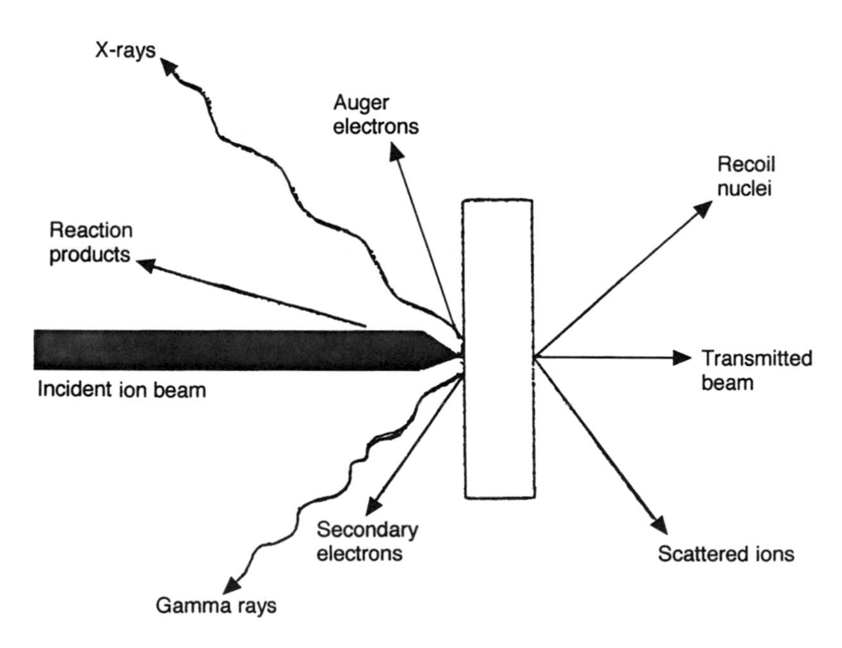


**Figure 9.1d** Schematic illustration of the radiation processes induced when energetic ion beams are directed on to a solid material

Modern machines use beam sizes down to a few microns in diameter (micro-PIXE), with beam currents of 100 pA; these external beams have been used (in air) to probe samples of archeological or biological interest. Standard reference samples are required for quantitative work. Figure 9.1e shows a schematic diagram of a typical PIXE system, used in this case for measuring aerosols emitted from aircraft engines.

The samples are deposited on Mylar film, characteristic X-rays are detected by a liquid-nitrogen-cooled Si(Li) detector. Recent applications have included the determination of the elemental composition of histopathological lesions in the brains of Alzheimer patients and the analysis of paints in works of art. Sensitivities can approach IPPB and $10^{-15}$ g.

Rutherford backscattering (RBS) again principally uses particle beams, but in this technique reflected ions are detected from the

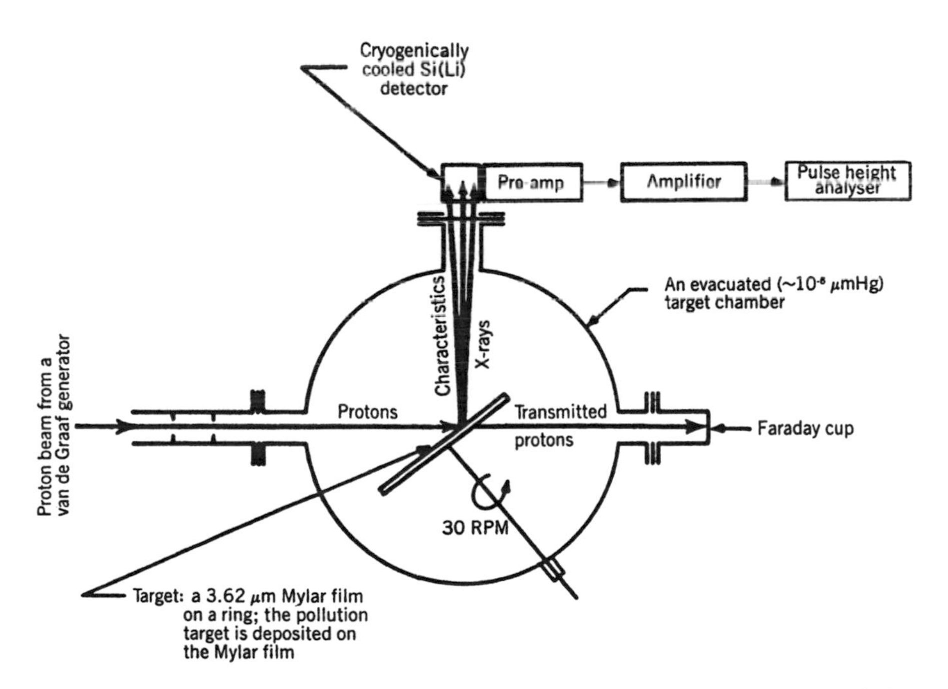


**Figure 9.1e** Schematic representation of a PIXE system used for measuring aerosols emitted from aircraft engines

radiation incident on the atoms of the sample, instead of measuring X-rays, gamma rays, etc. For sample atoms below the surface of a solid, the additional energy losses of the backscattered ions provide depth information. The technique is non-destructive and quantitative, and can be applied to all elements. Unlike PIXE, this form of analysis can detect ions up to a few mm below the surface of a solid sample.

**SAQ 9.1b** Why can't PIXE be used to measure ions which are located a few mm below the surface of a sample?

There are some facilities, such as the Isotope Separation On-Line (ISOLDE) at The European Laboratory for Particle Physics (CERN), Geneva, which utilise high-energy protons to generate radioactive nuclear-reaction products in the sample material for further use in particle-physics studies.

### 9.1.5 Glow-discharge Mass Spectrometry (GDMS)

Glow discharge usually occurs when a DC current is applied to a solid, conducting sample. If the sample is non-conducting, then prior mixing with a conducting substrate e.g. copper powder, is employed.

A flow of electric current through the glow-discharge source produces electrons which are attracted to the anode. These electrons hit argon (the 'fill-gas') atoms and some of these are then ionised. A plasma is subsequently formed and the positively charged argon ions are attracted towards the cathode (the sample). Metal atoms are thus

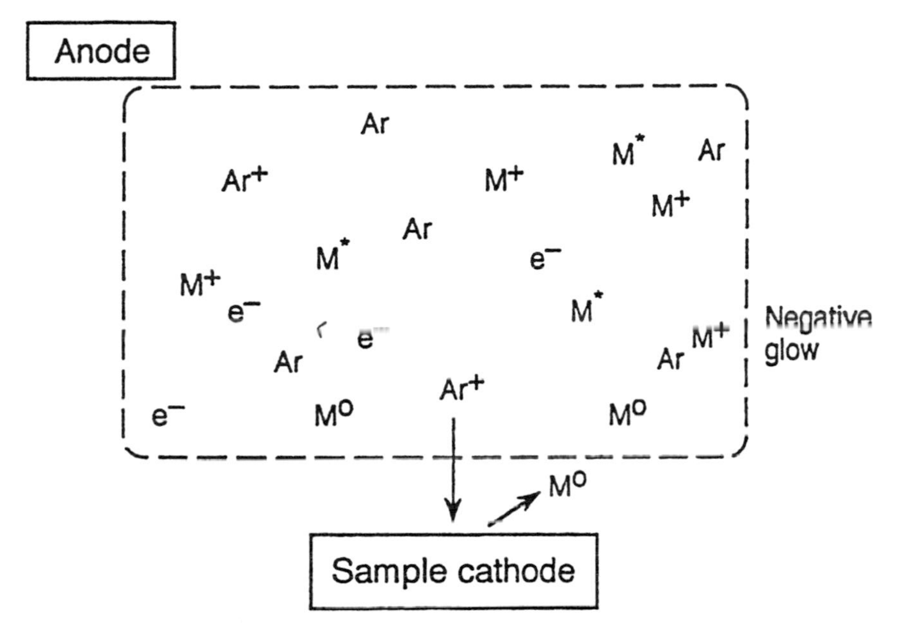


**Figure 9.1f** Principles of a glow-discharge source

sputtered off the surface and become ionised after colliding with further excited argon atoms. Figure 9.1f illustrates the principles involved in the glow-discharge process.

The sample ions are extracted from the source and are passed through a slit into a double-focusing mass analyser. Typical operating conditions employ currents of *ca.* 4 mA, voltages of *ca.* 1.2 kV and pressures of $2 \times 10^{-4}$ mbar (0.02 Pa).

## 9.2 ANALYSIS OF BULK SOLIDS

### 9.2.1 Accelerator Mass Spectrometry (AMS)

Accelerator mass spectrometry has now become an important analytical tool for elemental determinations at ultratrace concentrations. Detection limits approaching a few attograms ($10^{-18}$ g) have been achieved. This technique has proved to be particularly important for the detection of radionuclides with medium and long half-lives. In fact, the determination of the age of ancient antiquities is often carried out by using AMS, i.e. by measuring the $^{14}C/^{12}C$ ratio. A recent example of this is The Turin Shroud , which was shown to be a medieval reproduction by using this technique. The fundamental advantage of AMS is that it makes use of several atomic parameters in order to resolve nuclides of nearly identical mass. In fact, of the elements only natural indium cannot be uniquely identified by using AMS.

Π Why cannot indium be uniquely identified by AMS?

Indium has no stable isotope with a unique mass number.

As before, accelerator mass spectrometry utilises SIMS and a $Cs^+$ sputter beam. However, in this case negative ions at a potential of *ca.* 20 keV are extracted, and we tend to be more interested in the bulk properties of a sample rather than just its surface. In the case of $C^-$ ions, about 5% ionisation efficiencies are achieved. Another advantage in using negative ions is that some interfering isobars (isotopes with the same atomic mass) are relatively unstable and thus can be totally discriminated against. For example, $N^-$ ions are unstable, thus allowing $^{14}N$ and $^{14}C$ to be totally discriminated in 'carbon-14' dating. The stable negative-ion beams are then focused by means of electrostatic lenses, passed through a magnet (inflector-type) and enter a tandem van de Graaff accelerator (see Figure 9.2).

Here, the ions are accelerated towards a positively charged central terminal (made out of graphite foil) which is usually held at a potential of 2–3 MV. Any molecular ions that are present will dissociate and some electrons will be removed. A charge-state separator selects the ions of interest, e.g. $^{14}C^{3+}$, which are now accelerated away from the positive centre terminal, focused into an analysing magnet, and finally passed through to a magnetic separator and detector. A measure of the beam current of the stable $^{12}C$ isotope can be made in a movable Faraday cup placed in the beamline, after adjusting the magnets for mass 12.

Π Why do you think the van de Graaff accelerator is sometimes referred to here as a tandem accelerator?

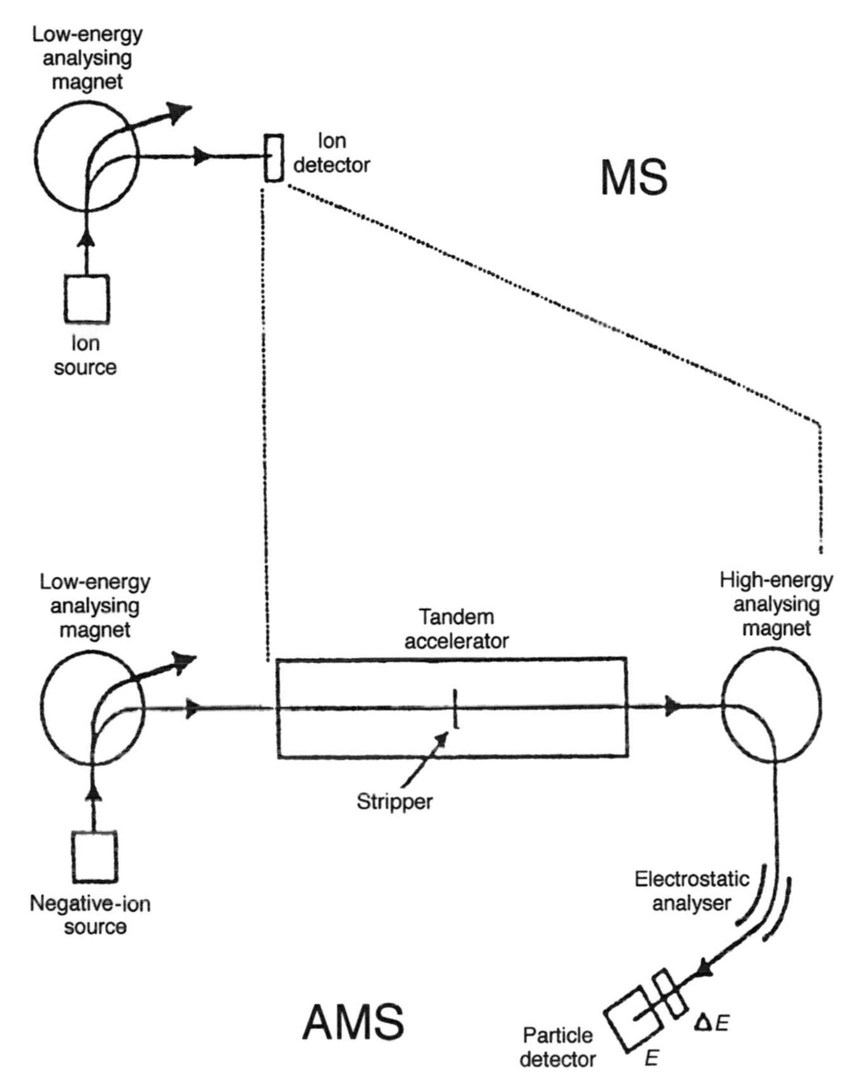


**Figure 9.2** Schematic representation of an accelerator mass spectrometry system; a conventional mass spectrometry arrangement is included for comparison purposes

It is called a tandem accelerator because there are two areas for acceleration, namely pre- and post-terminal.

Π Why do we need to measure isotopc ratios?

If the ion-source efficiency or the focusing of any electrostatic device alters, thus resulting in a reduction in the counts observed in the detector for a rare isotope, then this will be paralleled by a reduction in the stable ion current. Therefore, the ratio will remain the same. If the stable ion current is not measured, any reduction in the number of counts observed in the detector could be due to the sample itself or to the ion-source efficiency, etc.

**SAQ 9.2a** What is a possible weakness of this method of analysis?

Electrostatic analysers and velocity filters can be placed pre-detector to further aid particle identification. Isotope ratios approaching $10^{-15}$ ($^{14}C/^{12}C$) can be measured, which are a few orders of magnitude less than conventional radiochemical counting techniques. AMS has been applied to many different areas of research, including meteorite analysis, polar ice cores, and as a tracer in monitoring the fate of toxic metals in the body.

### 9.2.2 Spark-source Mass Spectrometry

Although the mass spectra obtained by using the above beam methods give elemental compositions, the data relate only to the surface of the sample, with not all materials being suitable for this type of surface analysis. Spark-source mass spectrometry provides a way of using mass spectrometry to give the elemental compositions of a wider range of sample types.

In this technique, an ion plasma is produced by passing a 30 kV RF spark between two electrodes in a vacuum chamber. If the sample material is a conducting solid it can form one of the electrodes.

**SAQ 9.2b** How could you increase the conductivity of a solid?

The ions thus produced are accelerated by a high DC potential into a double-focusing system, which often employs the Mattauch–Herzog geometry (see Section 3.3) and, until recently, generally used photographic-plate detection methods. The double-focusing system is required because a wide range of energies are imparted to the ions by the spark source, and multiply charged ions are produced. This requires the analyser to have a resolving power of at least 5000 in order to avoid inter-element interferences in the mass spectra. Spark-source mass spectrometry can provide the same type of multi-element determination as the previously mentioned beam techniques, and with similar sensitivities, but can be used on a wider range of materials and with less sample preparation. With this technique, a wide dynamic range can be investigated, providing data on compositions ranging from high percentage values down to parts per billion (i.e. a dynamic range of $\geqslant 10^6$). However, obtaining reliable quantitative information requires much skill and patience because the source of ions, i.e. the RF spark, is not reproducible even over short periods of time, and even when the integrating properties of photographic detection are utilised, quite large (i.e. 20%) relative standard deviations are obtained.

Various improvements to this technique, such as replacing the spark by a glow-discharge source, similar in action to the hollow-cathode lamp used in atomic absorption spectroscopy but with the sample forming the cathode, or by using modern electronic detectors with photon-counting capabilities and computerised data storage, have helped to improve the quantitative performance of these instruments, as has the technique of using ratio measurements between the analyte and an internal standard material. However, it seems likely that the technique of inductively coupled plasma mass spectrometry (ICP–MS), which will be described later in this chapter (see Section 9.3.1), will largely supersede spark-source spectrometry in the near future.

### 9.2.3 Resonance Ionisation Mass Spectrometry

A further development in the use of lasers has been resonance ionisation mass spectrometry (RIMS). In this method, the sample to be analysed is deposited on to a heated solid support. The gaseous atoms that are produced are bombarded with a Nd–YAG laser in pulsed mode. When the frequency of the laser pulse is equivalent to the energy between the ground state and the first excited state of the element in question, the photons that are produced promote the absorption of energy. After a few pulses, the element can become ionised; these ions are accelerated, passed through a magnet and then detected and measured. Isobaric separation is achieved and detection limits can approach $10^{-18}$ g.

## 9.3 ANALYSIS OF SOLUTIONS

### 9.3.1 Inductively Coupled Plasma Mass Spectrometry (ICP–MS)

This is a powerful and rapidly developing technique for multi-element determinations at very low levels in solution or in liquid samples.

A conventional inductively coupled plasma (ICP) source (positioned horizontally) is used to produce the primary ions, which are then analysed in a quadrupole mass spectrometer with a mass range up to 300 and unit resolution across the range. The preferred method for sample introduction is via a pneumatic nebuliser/spray chamber assembly. The interface allows the coupling of the ICP source (at atmospheric pressure) with the mass spectrometer (under high

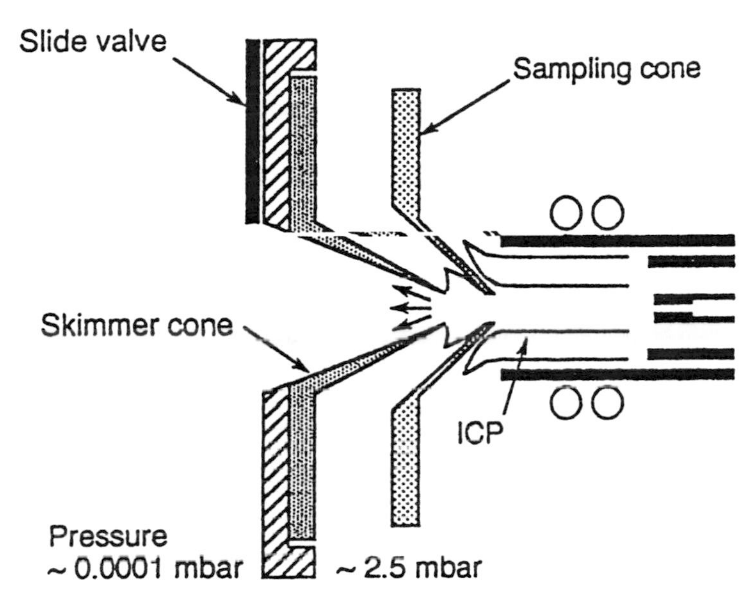


**Figure 9.3** Schematic representation of the inductively coupled plasma–mass spectrometer interface

vacuum). This consists of a water-cooled outer sampling cone which is positioned in close proximity to the plasma source (see Figure 9.3). The sampling cone is made of nickel because of its high thermal conductivity, relative resistance to corrosion, and robust nature. The pressure differential created by the sampling cone is such that ions from the plasma and the plasma gas itself are drawn into the region of lower pressure through the small orifice (1.0 mm). The region behind the sampling cone is maintained at a moderate pressure (*ca.* 2.5 mbar (250 Pa)) by the use of a rotary vacuum pump. Since the gas flowing through the sampling cone is large, a second cone is placed close enough behind it to allow the central portion of the expanding jet of plasma gas and ions to pass through the skimmer cone. The latter, which is also made of nickel, has an orifice diameter of *ca.* 0.75 mm. The pressure behind the skimmer cone is maintained at *ca.* $10^{-4}$ mbar (0.02 Pa). The extracted ions are then focused by a series of electrostatic lenses into the mass spectrometer.

Mass analysers can be of the double-focusing or quadrupole type, as outlined previously in Chapter 3 (Sections 3.3 and 3.4, respectively).

The transmitted ions are then detected with a channel electron multiplier (see Section 4.1.1). Extensive use of computer control and optimised geometry means that a full elemental analysis can be made in one scan, with measurement times of *ca.* 10 s per element, even for concentrations as low as the ppb level.

ICP–MS can be carried out in two distinctly different modes, i.e. with the mass filter transmitting only one single mass/charge ratio, or with the DC and RF voltages being changed continuously. The former case would allow single-ion monitoring, while the latter allows multi-element analysis. In multi-channel scanning, all $m/z$ ratios are repeatedly scanned, thus providing a complete fingerprint of the unknown sample composition. The scanning mode can also be useful for the quick qualitative analysis of unknown samples. Typical dwell times per $m/z$ ratio are 0.1–0.5 ms.

Interferences in ICP–MS can be broadly classified according to their origins, namely isobaric, molecular and matrix-dependent. An example of the first interference would be mass 58, where nickel (67.8% abundance) and iron (0.31% abundance) both occur. Table 9.3a shows the isobaric interferences for elements from period 4 of the periodic table.

**SAQ 9.3a**

Why is there a problem analysing for Ca in ICP–MS?

**Table 9.3a** Isobaric interferences for elements from period 4 of the periodic table

| Atomic mass | Element of interest[a] | Interfering element[a] |
|---|---|---|
| 39 | K (93.10) | – |
| 40 | Ca (96.97) | Ar (99.6)[b]; K (0.01) |
| 41 | K (6.88) | – |
| 42 | Ca (0.64) | – |
| 43 | Ca (0.14) | – |
| 44 | Ca (2.06) | – |
| 45 | Sc (100) | – |
| 46 | – | Ca (0.003); Ti (7.93) |
| 47 | – | Ti (7.28) |
| 48 | Ti (73.94) | Ca (0.19) |
| 49 | – | Ti (5.51) |
| 50 | – | Ti (5.34); V (0.24); Cr (4.31) |
| 51 | V (99.76) | – |
| 52 | Cr (83.76) | – |
| 53 | – | Cr (9.55) |
| 54 | – | Cr (2.38); Fe (5.82) |
| 55 | Mn (100) | – |
| 56 | Fe (91.66) | – |
| 57 | – | Fe (2.19) |
| 58 | Ni (67.88) | Fe (0.33) |
| 59 | Co (100) | – |
| 60 | Ni (26.23) | – |
| 61 | – | Ni (1.19) |
| 62 | – | Ni (3.66) |
| 63 | Cu (69.09) | – |
| 64 | Zn (48.89) | Ni (1.08) |
| 65 | Cu (30.91) | – |
| 66 | Zn (27.81) | – |
| 67 | – | Zn (4.11) |
| 68 | – | Zn (18.57) |
| 69 | Ga (60.40) | – |
| 70 | – | Zn (0.62); Ge (20.52) |
| 71 | Ga (39.60) | – |
| 72 | Ge (27.43) | – |

*(continued overleaf)*

**Table 9.3a** *(continued)*

| Atomic mass | Element of interest[a] | Interfering element[a] |
|---|---|---|
| 73 | – | Ge (7.76) |
| 74 | Ge (36.54) | Se (0.87) |
| 75 | As (100) | – |
| 76 | – | Ge (7.76); Se (9.02) |
| 77 | – | Se (7.58) |
| 78 | Se (23.52) | Kr (0.35) |
| 79 | Br (50.54) | – |
| 80 | Se (49.82) | Kr (2.27) |
| 81 | Br (49.46) | – |
| 82 | – | Se (9.19); Kr (11.56) |
| 83 | Kr (11.55) | – |
| 84 | Kr (56.90) | Sr (0.56)[c] |
| 85 | – | – |
| 86 | – | Kr (17.37); Sr (9.86)[c] |

[a]Percentage abundance is shown in parentheses
[b]Not in the 4th period of the periodic table but included because of its origin in the plasma source
[c]Not in the 4th period of the period table, but included for the sake of completeness

The second type of interference stems from polyatomic and doubly charged species, which can arise from the analyte's associated aqueous solution, the plasma gas itself or the acid used for digestion. For example, $^{56}Fe$ (91.66% abundance) is occluded by $^{40}Ar^{16}O^{+}$. Doubly charged species arising from cerium, lanthanum, strontium, thorium, and barium are of particular note. Table 9.3b gives details of these potential polyatomic interferences.

Matrix-dependent effects result from the presence of excess salts in the plasma source and lead to a loss in sensitivity. In addition, blockages in the nebuliser can result in erratic signal generation from high-solid-content samples, while mass discrimination can also occur.

**Table 9.3b** Potential polyatomic interferences

| Atomic mass | Element of interest[a] | Polyatomic interference |
|---|---|---|
| 39 | K (93.10) | $^{38}Ar^{1}H^{+}$ |
| 40 | Ca (96.97) | $^{40}Ar^{+}$ |
| 41 | K (6.88)q | $^{40}Ar^{1}H^{+}$ |
| 42 | Ca (0.64) | $^{40}Ar^{2}H^{+}$ |
| 43 | Ca (0.14) | – |
| 44 | Ca (2.06) | $^{12}C^{16}O^{16}O^{+}$ |
| 45 | Sc (100) | $^{12}C^{16}O^{16}O^{1}H^{+}$ |
| 46 | – | $^{14}N^{16}O^{16}O^{+}$; $^{32}S^{14}N^{+}$ |
| 47 | – | $^{31}P^{16}O^{+}$; $^{33}S^{14}N^{+}$ |
| 48 | Ti (73.94) | $^{31}P^{16}O^{1}H^{+}$; $^{32}S^{16}O^{+}$; $^{34}S^{14}N^{+}$ |
| 49 | – | $^{32}S^{16}O^{1}H^{+}$; $^{33}S^{16}O^{+}$; $^{14}N^{35}Cl^{+}$ |
| 50 | – | $^{34}S^{16}O^{+}$; $^{36}Ar^{14}N^{+}$ |
| 51 | V (99.76) | $^{35}Cl^{16}O^{+}$; $^{34}S^{16}O^{1}H^{+}$; $^{14}N^{37}Cl^{+}$; $^{35}Cl^{16}O^{+}$ |
| 52 | Cr (83.76) | $^{40}Ar^{12}C^{+}$; $^{36}Ar^{16}O^{+}$; $^{36}S^{16}O^{+}$; $^{35}Cl^{16}O^{1}H^{+}$ |
| 53 | – | $^{37}Cl^{16}O^{+}$ |
| 54 | – | $^{40}Ar^{14}N^{+}$; $^{37}Cl^{16}O^{1}H^{+}$ |
| 55 | Mn (100) | $^{40}Ar^{14}N^{1}H^{+}$ |
| 56 | Fc (91.66) | $^{40}Ar^{16}O^{+}$ |
| 57 | – | $^{40}Ar^{16}O^{1}H^{+}$ |
| 58 | Ni (67.88) | – |
| 59 | Co (100) | – |
| 60 | Ni (26.23) | – |
| 61 | – | – |
| 62 | – | – |
| 63 | Cu (69.09) | $^{31}P^{16}O_2^{+}$ |
| 64 | Zn (48.89) | $^{31}P^{16}O_2\,^{1}H^{+}$; $^{32}S^{16}O^{16}O^{+}$; $^{32}S^{32}S^{+}$ |
| 65 | Cu (30.91) | $^{33}S^{16}O^{16}O^{+}$; $^{32}S^{33}S^{+}$ |
| 66 | Zn (27.81) | $^{34}S^{16}O^{16}O^{+}$; $^{32}S^{34}S^{+}$ |
| 67 | – | $^{35}Cl^{16}O^{16}O^{+}$ |
| 68 | – | $^{40}Ar^{14}N^{14}N^{+}$; $^{36}S^{16}O^{16}O^{+}$; $^{32}S^{36}S^{+}$ |
| 69 | Ga (60.40) | $^{37}Cl^{16}O^{16}O^{+}$ |
| 70 | – | $^{35}Cl_2^{+}$; $^{40}Ar^{14}N^{16}O^{+}$ |
| 71 | Ga (39.60) | $^{40}Ar^{31}P^{+}$; $^{36}Ar^{35}Cl^{+}$ |

*(continued overleaf)*

**Table 9.3b** *(continued)*

| Atomic mass | Element of interest[a] | Polyatomic interference |
|---|---|---|
| 72 | Ge (27.43) | $^{37}Cl^{35}Cl^+$; $^{36}Ar^{36}Ar^+$; $^{40}Ar^{32}S^+$ |
| 73 | – | $^{40}Ar^{33}S^+$; $^{36}Ar^{37}Cl^+$ |
| 74 | Ge (36.54) | $^{37}Cl^{37}Cl^+$; $^{36}Ar^{38}Ar^+$; $^{40}Ar^{34}S^+$ |
| 75 | As (100) | $^{40}Ar^{35}Cl^+$ |
| 76 | – | $^{40}Ar^{36}Ar^+$; $^{40}Ar^{36}S^+$ |
| 77 | – | $^{40}Ar^{37}Cl^+$; $^{36}Ar^{40}Ar^1H^+$ |
| 78 | Se (23.52) | $^{40}Ar^{38}Ar^+$ |
| 79 | Br (50.54) | $^{40}Ar^{38}Ar^1H^+$ |
| 80 | Se (49.82) | $^{40}Ar^{40}Ar^+$ |
| 81 | Br (49.46) | $^{40}Ar^{40}Ar^1H^+$ |
| 82 | – | $^{40}Ar^{40}Ar^1H^1H^+$ |
| 83 | Kr (11.55) | – |
| 84 | Kr (56.90) | – |
| 85 | – | – |
| 86 | – | – |

[a]Percentage abundance is shown in parentheses

## Summary

In this chapter we have covered some common methods of inorganic analysis involving mass spectrometry which are in use today. You should now be aware of the vast differences in instrumentation that are possible when using these techniques. Given a sample for elemental determination, you should be able to suggest appropriate methods of analysis.

## Objectives

On completion of this chapter you should now be able to:

- describe and explain the principles of operation of some common methods of analysis in inorganic mass spectroscopy;
- give the advantages and limitations of these techniques;
- describe some applications of such techniques.

# Appendix

## Appendix 1

TABLE OF ISOTOPES IN ASCENDING MASS ORDER

| $Z$ | Symbol | $M$ | % | % (relative) | $M$ (exact) | $M$ (chemical) |
|---|---|---|---|---|---|---|
| 0 | n | 1 | – | – | 1.008 665 | – |
| 1 | H | 1 | 99.985 | 100 | 1.007 825 | 1.00794 |
| | D | 2 | 0.015 | 0.015 | 2.014 | |
| 2 | He | 3 | 0.000137 | 0.000137 | 3.016 030 | 4.0026 |
| | | 4 | ≈100 | 100 | 4.002 60 | |
| 3 | Li | 6 | 7.5 | 8.0108 | 6.015 121 | 6.941 |
| | | 7 | 92.5 | 100 | 7.016 003 | |
| 4 | Be | 9 | 100 | 100 | 9.012182 | 9.012182 |
| 5 | B | 10 | 19.9 | 24.84 | 10.012 937 | 10.811 |
| | | 11 | 80.1 | 100 | 11.009 305 | |
| 6 | C | 12 | 98.90 | 100 | 12.000 000 | 12.011 |
| | | 13 | 1.10 | 1.112 | 13.003 355 | |
| 7 | N | 14 | 99.63 | 100 | 14.003 074 | 14.00674 |

| | | | | | | |
|---|---|---|---|---|---|---|
| | | 15 | 0.37 | 0.37 | 15.000 108 | |
| 8 | O | 16 | 99.76 | 100 | 15.994 915 | 15.9994 |
| | | 17 | 0.04 | 0.04 | 16.999 133 | |
| | | 18 | 0.20 | 0.20 | 17.999 160 | |
| 9 | F | 19 | 100 | 100 | 18.998 403 | 18.9984 |
| 10 | Ne | 20 | 90.48 | 100 | 19.992 435 | 20.1797 |
| | | 21 | 0.27 | 0.298 | 20.993 943 | |
| | | 22 | 9.25 | 10.22 | 21.991 264 | |
| 11 | Na | 23 | 100 | 100 | 22.989 768 | 22.9898 |
| 12 | Mg | 24 | 78.99 | 100 | 23.985 042 | 24.3050 |
| | | 25 | 10.00 | 12.66 | 24.985 837 | |
| | | 26 | 11.01 | 13.94 | 25.982 593 | |
| 13 | Al | 27 | 100 | 100 | 26.981 539 | 26.9815 |
| 14 | Si | 28 | 92.21 | 100 | 27.976 927 | 28.0855 |
| | | 29 | 4.67 | 5.065 | 28.976 495 | |
| | | 30 | 3.10 | 3.336 | 29.973 770 | |
| 15 | P | 31 | 100 | 100 | 30.973 762 | 30.9738 |
| 16 | S | 32 | 95.03 | 100 | 31.972 070 | 32.066 |
| | | 33 | 0.75 | 0.789 | 32.971 456 | |
| | | 34 | 4.22 | 4.44 | 33.967 866 | |
| | | 36 | 0.02 | 0.021 | 35.967 080 | |
| 17 | Cl | 35 | 75.77 | 100 | 34.968 852 | 35.453 |
| | | 37 | 24.23 | 31.98 | 36.965 903 | |
| 18 | Ar | 36 | 0.337 | 0.338 | 35.967 545 | 39.948 |
| | | 38 | 0.063 | 0.0633 | 37.962 732 | |
| | | 40 | 99.600 | 100 | 39.962 384 | |
| 19 | K | 39 | 93.2581 | 100 | 38.963 707 | 39.0983 |
| | | 40 | 0.0117 | 0.0125 | 39.963 999 | |

| | | | | | | |
|---|---|---|---|---|---|---|
| | | 41 | 6.7302 | 7.22 | 40.961 825 | |
| 20 | Ca | 40 | 96.941 | 100 | 39.962 591 | 40.078 |
| | | 42 | 0.647 | 0.66742 | 41.958 618 | |
| | | 43 | 0.135 | 0.139 | 42.958 766 | |
| | | 44 | 2.086 | 2.152 | 43.955 480 | |
| | | 46 | 0.004 | 0.004 | 45.953 689 | |
| | | 48 | 0.187 | 0.193 | 47.952 533 | |
| 21 | Sc | 45 | 100 | 100 | 44.955 911 | 44.956 |
| 22 | Ti | 46 | 8.00 | 10.84 | 45.952 629 | 47.88 |
| | | 47 | 8.3 | 9.892 | 46.951 764 | |
| | | 48 | 73.8 | 100 | 47.947 947 | |
| | | 49 | 5.51 | 7.466 | 48.947 871 | |
| | | 50 | 5.4 | 7.317 | 49.944 792 | |
| 23 | V | 50 | 0.25 | 0.251 | 49.947 161 | 50.9415 |
| | | 51 | 99.75 | 100 | 50.943 962 | |
| 24 | Cr | 50 | 4.345 | 5.185 | 49.946 046 | 51.9961 |
| | | 52 | 83.79 | 100 | 51.940 509 | |
| | | 53 | 9.50 | 11.34 | 52.940 651 | |
| | | 54 | 2.365 | 2.82 | 53.938 882 | |
| 25 | Mn | 55 | 100 | 100 | 54.938 046 | 54.9380 |
| 26 | Fe | 54 | 5.9 | 6.43 | 53.939 612 | 55.847 |
| | | 56 | 91.72 | 100 | 55.934 939 | |
| | | 57 | 2.1 | 2.29 | 56.935 396 | |
| | | 58 | 0.28 | 0.305 | 57.933 277 | |
| 27 | Co | 59 | 100 | 100 | 58.933 198 | 58.9332 |
| 28 | Ni | 58 | 68.27 | 100 | 57.935 346 | 58.6934 |
| | | 60 | 26.10 | 38.23 | 59.930 788 | |
| | | 61 | 1.13 | 1.66 | 60.931 058 | |
| | | 62 | 3.59 | 5.26 | 61.928 346 | |
| | | 64 | 0.91 | 1.33 | 63.927 968 | |

| | | | | | | |
|---|---|---|---|---|---|---|
| 29 | Cu | 63 | 69.17 | 100 | 62.929 598 | 63.546 |
| | | 65 | 30.83 | 44.57 | 64.927 765 | |
| 30 | Zn | 64 | 48.6 | 100 | 63.929 145 | 65.39 |
| | | 66 | 27.9 | 57.41 | 65.926 034 | |
| | | 67 | 4.1 | 8.44 | 66.927 129 | |
| | | 68 | 18.8 | 38.68 | 67.924 846 | |
| | | 70 | 0.6 | 1.23 | 69.925 325 | |
| 31 | Ga | 69 | 60.108 | 100 | 68.925 580 | 69.723 |
| | | 71 | 39.892 | 66.37 | 70.924 700 | |
| 32 | Ge | 70 | 20.5 | 56.16 | 69.924 250 | 72.61 |
| | | 72 | 27.4 | 75.07 | 71.922 079 | |
| | | 73 | 7.8 | 21.37 | 72.923 463 | |
| | | 74 | 36.5 | 100 | 73.921 177 | |
| | | 76 | 7.8 | 21.37 | 75.921 401 | |
| 33 | As | 75 | 100 | 100 | 74.921 594 | 74.9216 |
| 34 | Se | 74 | 0.9 | 1.80 | 73.922 475 | 78.96 |
| | | 76 | 9.1 | 18.24 | 75.919 212 | |
| | | 77 | 7.6 | 15.23 | 76.919 912 | |
| | | 78 | 23.6 | 47.29 | 77.917 309 | |
| | | 80 | 49.9 | 100 | 79.916 520 | |
| | | 82 | 8.9 | 17.84 | 81.916 698 | |
| 35 | Br | 79 | 50.69 | 100 | 78.918 336 | 79.904 |
| | | 81 | 49.31 | 97.28 | 80.916 289 | |
| 36 | Kr | 78 | 0.35 | 0.614 | 77.920 401 | 83.80 |
| | | 80 | 2.25 | 3.947 | 79.916 380 | |
| | | 82 | 11.6 | 20.35 | 81.913 482 | |
| | | 83 | 11.5 | 20.175 | 82.914 135 | |
| | | 84 | 57.0 | 100 | 83.911 507 | |
| | | 86 | 17.3 | 30.35 | 85.910 610 | |
| 37 | Rb | 85 | 72.17 | 100 | 84.911 794 | 85.4678 |
| | | 87 | 27.83 | 38.562 | 86.909 187 | |

| | | | | | | |
|---|---|---|---|---|---|---|
| 38 | Sr | 84 | 0.56 | 0.68 | 83.913 431 | 87.62 |
| | | 86 | 9.86 | 11.94 | 85.909 267 | |
| | | 87 | 7.00 | 8.5 | 86.908 884 | |
| | | 88 | 82.58 | 100 | 87.905 619 | |
| 39 | Y | 89 | 100 | 100 | 88.905 849 | 88.906 |
| 40 | Zr | 90 | 51.45 | 100 | 89.904 703 | 91.224 |
| | | 91 | 11.22 | 21.73 | 90.905 643 | |
| | | 92 | 17.15 | 33.33 | 91.905 039 | |
| | | 94 | 17.38 | 33.78 | 93.906 314 | |
| | | 96 | 2.80 | 5.44 | 95.908 275 | |
| 41 | Nb | 93 | 100 | 100 | 92.906 377 | 92.906 |
| 42 | Mo | 92 | 14.84 | 61.50 | 91.906 808 | 95.94 |
| | | 94 | 9.25 | 38.33 | 93.905 085 | |
| | | 95 | 15.92 | 65.98 | 94.905 840 | |
| | | 96 | 16.68 | 69.13 | 95.904 678 | |
| | | 97 | 9.55 | 39.58 | 96.906 020 | |
| | | 98 | 24.13 | 100 | 97.905 406 | |
| | | 100 | 9.63 | 39.91 | 99.907 477 | |
| 43 | Tc | | | 100 | | |
| 44 | Ru | 96 | 5.54 | 17.53 | 95.907 599 | 101.07 |
| | | 98 | 1.86 | 5.89 | 97.905 267 | |
| | | 99 | 12.7 | 40.19 | 98.905 939 | |
| | | 100 | 12.6 | 38.87 | 99.904 219 | |
| | | 101 | 17.1 | 54.11 | 100.905 582 | |
| | | 102 | 31.6 | 100 | 101.904 348 | |
| | | 104 | 18.6 | 58.86 | 103.905 424 | |
| 45 | Rh | 103 | 100 | 100 | 102.905 500 | 102.905 |
| 46 | Pd | 102 | 1.02 | 3.73 | 101.905 634 | 106.42 |
| | | 104 | 11.14 | 40.76 | 103.904 029 | |
| | | 105 | 22.33 | 81.71 | 104.905 079 | |
| | | 106 | 27.33 | 100 | 105.903 478 | |
| | | 108 | 26.46 | 96.82 | 107.903 895 | |

| | | | | | | |
|---|---|---|---|---|---|---|
| | | 110 | 11.72 | 42.88 | 109.905 167 | |
| 47 | Ag | 107 | 51.839 | 100 | 106.905 092 | 107.868 |
| | | 109 | 48.161 | 94.90 | 108.904 757 | |
| 48 | Cd | 106 | 1.25 | 4.35 | 105.906 461 | 112.411 |
| | | 108 | 0.89 | 3.10 | 107.904 176 | |
| | | 110 | 12.49 | 43.47 | 109.903 005 | |
| | | 111 | 12.80 | 44.55 | 110.904 182 | |
| | | 112 | 24.13 | 83.99 | 111.902 758 | |
| | | 113 | 12.22 | 42.53 | 112.904 400 | |
| | | 114 | 28.73 | 100 | 113.903 357 | |
| | | 116 | 7.49 | 26.07 | 115.904 754 | |
| 49 | In | 113 | 4.3 | 4.49 | 112.904 061 | 114.82 |
| | | 115 | 95.7 | 100 | 114.903 880 | |
| 50 | Sn | 112 | 0.97 | 2.98 | 111.904 826 | 118.710 |
| | | 114 | 0.65 | 1.99 | 113.902 784 | |
| | | 115 | 0.36 | 1.10 | 114.903 348 | |
| | | 116 | 14.53 | 43.58 | 115.901 747 | |
| | | 117 | 7.68 | 23.57 | 116.902 956 | |
| | | 118 | 24.22 | 73.32 | 117.901 609 | |
| | | 119 | 8.58 | 26.33 | 118.903 310 | |
| | | 120 | 32.59 | 100 | 119.902 200 | |
| | | 122 | 4.63 | 14.21 | 121.903 440 | |
| | | 124 | 5.79 | 17.77 | 123.905 274 | |
| 51 | Sb | 121 | 57.4 | 100 | 120.903 821 | 121.752 |
| | | 123 | 42.6 | 74.22 | 122.904 216 | |
| 52 | Te | 120 | 0.095 | 0.28 | 119.904 048 | 127.60 |
| | | 122 | 2.59 | 7.65 | 121.903 054 | |
| | | 123 | 0.905 | 2.67 | 122.904 271 | |
| | | 124 | 4.79 | 14.14 | 123.902 823 | |
| | | 125 | 7.12 | 21.02 | 124.904 433 | |
| | | 126 | 18.93 | 55.89 | 125.903 314 | |
| | | 128 | 31.70 | 93.59 | 127.904 463 | |
| | | 130 | 33.87 | 100 | 129.906 229 | |

| | | | | | | |
|---|---|---|---|---|---|---|
| 53 | I | 127 | 100 | 100 | 126.904 476 | 126.9045 |
| 54 | Xe | 124 | 0.10 | 0.37 | 123.905 894 | 131.29 |
| | | 126 | 0.09 | 0.33 | 125.904 281 | |
| | | 128 | 1.91 | 7.10 | 127.903 531 | |
| | | 129 | 26.4 | 98.14 | 128.904 780 | |
| | | 130 | 4.1 | 15.24 | 129.903 509 | |
| | | 131 | 21.2 | 78.81 | 130.905 072 | |
| | | 132 | 26.9 | 100 | 131.904 144 | |
| | | 134 | 10.4 | 38.866 | 133.905 395 | |
| | | 136 | 8.9 | 33.09 | 135.907 214 | |
| 55 | Cs | 133 | 100 | 100 | 132.905 429 | 132.905 |
| 56 | Ba | 130 | 1.101 | 1.536 | 129.906 284 | 137.34 |
| | | 132 | 0.097 | 0.135 | 131.905 045 | |
| | | 134 | 2.42 | 3.77 | 133.904 493 | |
| | | 135 | 6.59 | 9.2 | 134.905 671 | |
| | | 136 | 7.81 | 10.9 | 135.904 559 | |
| | | 137 | 11.32 | 15.8 | 136.905 815 | |
| | | 138 | 71.66 | 100 | 137.905 235 | |
| 57 | La | 138 | 0.090 | 0.09 | 137.907 11 | 138.91 |
| | | 139 | 99.91 | 100 | 138.906 347 | |
| 72 | Hf | 174 | 0.162 | 0.46 | 173.940 044 | 178.49 |
| | | 176 | 5.206 | 14.83 | 175.941 406 | |
| | | 177 | 18.606 | 53.01 | 176.943 217 | |
| | | 178 | 27.297 | 77.77 | 177.943 696 | |
| | | 179 | 13.629 | 38.83 | 178.945 812 | |
| | | 180 | 35.100 | 100 | 179.946 545 | |
| 73 | Ta | 180 | 0.012 | 0.012 | 179.947 462 | 180.948 |
| | | 181 | 99.988 | 100 | 180.947 992 | |
| 74 | W | 180 | 0.12 | 0.39 | 179.946 701 | 183.85 |
| | | 182 | 26.3 | 85.67 | 181.948 202 | |
| | | 183 | 14.28 | 46.51 | 182.950 220 | |
| | | 184 | 30.7 | 100 | 183.950 928 | |
| | | 186 | 28.6 | 93.16 | 185.954 357 | |

| | | | | | | |
|---|---|---|---|---|---|---|
| 75 | Re | 185 | 37.40 | 59.74 | 184.952 951 | 186.207 |
| | | 187 | 62.40 | 100 | 186.955 744 | |
| 76 | Os | 184 | 0.02 | 0.05 | 183.952 488 | 190.2 |
| | | 186 | 1.58 | 3.85 | 185.953 830 | |
| | | 187 | 1.6 | 3.90 | 186.955 741 | |
| | | 188 | 13.3 | 32.44 | 187.955 860 | |
| | | 189 | 16.1 | 39.27 | 188.958 137 | |
| | | 190 | 26.4 | 64.39 | 189.958 436 | |
| | | 192 | 41.0 | 100 | 191.961 467 | |
| 77 | Ir | 191 | 37.3 | 59.49 | 190.960 584 | 192.22 |
| | | 193 | 62.7 | 100 | 192.962 917 | |
| 78 | Pt | 190 | 0.01 | 0.03 | 189.959 917 | 195.08 |
| | | 192 | 0.79 | 2.34 | 191.961 019 | |
| | | 194 | 32.9 | 97.34 | 193.962 655 | |
| | | 195 | 33.8 | 100 | 194.964 766 | |
| | | 196 | 25.3 | 74.85 | 195.964 926 | |
| | | 198 | 7.2 | 21.30 | 197.967 869 | |
| 79 | Au | 197 | 100 | 100 | 196.966 543 | 196.967 |
| 80 | Hg | 196 | 0.15 | 0.50 | 195.965 807 | 200.59 |
| | | 198 | 10.0 | 33.56 | 197.966 743 | |
| | | 199 | 16.9 | 56.71 | 198.968 254 | |
| | | 200 | 23.1 | 77.52 | 199.968 300 | |
| | | 201 | 13.2 | 44.30 | 200.970 277 | |
| | | 202 | 29.8 | 100 | 201.970 617 | |
| | | 204 | 6.85 | 22.99 | 203.973 467 | |
| 81 | Tl | 203 | 29.524 | 41.89 | 202.972 320 | 204.383 |
| | | 205 | 70.476 | 100 | 204.974 401 | |
| 82 | Pb | 204 | 1.4 | 2.67 | 203.973 020 | 207.2 |
| | | 206 | 24.1 | 45.99 | 205.974 440 | |
| | | 207 | 22.1 | 42.18 | 206.975 872 | |
| | | 208 | 52.4 | 100 | 207.976 627 | |

| | | | | | | |
|---|---|---|---|---|---|---|
| 83 | Bi | 209 | 100 | 100 | 208.980 374 | 208.980 |
| 90 | Th | 232 | 100 | 100 | 232.038 054 | 232.038 |
| 92 | U | 234 | 0.0055 | 0.0055 | 234.040 946 | 238.03 |
| | | 235 | 0.720 | 0.725 | 235.043 924 | |
| | | 238 | 99.2745 | 100 | 238.050 784 | |

**Appendix 2**

ISOTOPIC ABUNDANCES FOR VARIOUS ELEMENTAL COMPOSITIONS CHON[a]

| | $M+1$ | $M+2$ | Mass |
|---|---|---|---|
| 12 | | | |
| C | 1.11 | 0.00 | 12.0000 |
| 13 | | | |
| CH | 1.13 | 0.00 | 13.0078 |
| 14 | | | |
| N | 0.37 | 0.00 | 14.0031 |
| $CH_2$ | 1.14 | 0.00 | 14.0157 |
| 15 | | | |
| HN | 0.39 | 0.00 | 15.0109 |
| $CH_3$ | 1.16 | 0.00 | 15.0235 |
| 16 | | | |
| O | 0.04 | 0.20 | 15.9949 |
| $H_2N$ | 0.40 | 0.00 | 16.0187 |
| $CH_4$ | 1.17 | 0.00 | 16.0313 |
| 17 | | | |
| HO | 0.06 | 0.20 | 17.0027 |
| $H_3N$ | 0.42 | 0.00 | 17.0266 |
| 18 | | | |
| $H_2O$ | 0.07 | 0.20 | 18.0106 |
| 24 | | | |
| $C_2$ | 2.22 | 0.01 | 24.0000 |
| 25 | | | |
| $C_2H$ | 2.24 | 0.01 | 25.0078 |

| | $M+1$ | $M+2$ | Mass |
|---|---|---|---|
| CN | 1.48 | 0.00 | 26.0031 |
| $C_2H_2$ | 2.25 | 0.01 | 26.0157 |
| 27 | | | |
| CHN | 1.50 | 0.00 | 27.0109 |
| $C_2H_3$ | 2.27 | 0.01 | 27.0235 |
| 28 | | | |
| $N_2$ | 0.74 | 0.00 | 28.0062 |
| CO | 1.15 | 0.20 | 27.9949 |
| $CH_2N$ | 1.51 | 0.00 | 28.0187 |
| $C_2H_4$ | 2.28 | 0.01 | 28.0313 |
| 29 | | | |
| $HN_2$ | 0.76 | 0.00 | 29.0140 |
| CHO | 1.17 | 0.20 | 29.0027 |
| $CH_3N$ | 1.53 | 0.00 | 29.0266 |
| $C_2H_5$ | 2.30 | 0.01 | 29.0391 |
| 30 | | | |
| NO | 0.41 | 0.20 | 29.9980 |
| $H_2N_2$ | 0.77 | 0.00 | 30.0218 |
| $CH_2O$ | 1.18 | 0.20 | 30.0106 |
| $CH_4N$ | 1.54 | 0.01 | 30.0344 |
| $C_2H_6$ | 2.31 | 0.01 | 30.0470 |
| 31 | | | |
| HNO | 0.43 | 0.20 | 31.0058 |
| $H_3N_2$ | 0.79 | 0.00 | 31.0297 |
| $CH_3O$ | 1.20 | 0.02 | 31.0184 |
| $CH_5N$ | 1.56 | 0.01 | 31.0422 |

| | $M+1$ | $M+2$ | Mass |
|---|---|---|---|
| 32 | | | |
| $O_2$ | 0.08 | 0.40 | 31.9898 |
| $H_2NO$ | 0.44 | 0.20 | 32.0136 |
| $H_4N_2$ | 0.80 | 0.00 | 32.0375 |
| $CH_4O$ | 1.21 | 0.20 | 32.0262 |
| 33 | | | |
| $HO_2$ | 0.10 | 0.40 | 32.9976 |
| $H_3NO$ | 0.46 | 0.20 | 33.0215 |
| 34 | | | |
| $H_2O_2$ | 0.11 | 0.40 | 34.0054 |
| 36 | | | |
| $C_3$ | 3.33 | 0.04 | 36.0000 |
| 37 | | | |
| $C_3H$ | 3.35 | 0.04 | 37.0078 |
| 38 | | | |
| $C_2N$ | 2.59 | 0.02 | 38.0031 |
| $C_3H_2$ | 3.36 | 0.04 | 38.0157 |
| 39 | | | |
| $C_2HN$ | 2.61 | 0.02 | 39.0109 |
| $C_3H_3$ | 3.38 | 0.04 | 39.0235 |
| 40 | | | |
| $CN_2$ | 1.85 | 0.01 | 40.0062 |
| $C_2O$ | 2.26 | 0.21 | 39.9949 |
| $C_2H_2N$ | 2.62 | 0.02 | 40.0187 |
| $C_3H_4$ | 3.39 | 0.04 | 40.0313 |
| 41 | | | |
| $CHN_2$ | 1.87 | 0.01 | 41.0140 |
| $C_2HO$ | 2.28 | 0.21 | 41.0027 |
| $C_2H_3N$ | 2.64 | 0.02 | 41.0266 |
| $C_3H_5$ | 3.41 | 0.04 | 41.0391 |

| | $M+1$ | $M+2$ | Mass |
|---|---|---|---|
| 42 | | | |
| $N_3$ | 1.11 | 0.00 | 42.0093 |
| CNO | 1.52 | 0.21 | 41.9980 |
| $CH_2N_2$ | 1.88 | 0.01 | 42.0218 |
| $C_2H_2O$ | 2.29 | 0.21 | 42.0106 |
| $C_2H_4N$ | 2.65 | 0.02 | 42.0344 |
| $C_3H_6$ | 3.42 | 0.04 | 42.0470 |
| 43 | | | |
| $HN_3$ | 1.13 | 0.00 | 43.0171 |
| CHNO | 1.54 | 0.21 | 43.0058 |
| $CH_3N_2$ | 1.90 | 0.01 | 43.0297 |
| $C_2H_3O$ | 2.31 | 0.21 | 43.0184 |
| $C_2H_5N$ | 2.67 | 0.02 | 43.0422 |
| $C_3H_7$ | 3.44 | 0.04 | 43.0548 |
| 44 | | | |
| $N_2O$ | 0.78 | 0.20 | 44.0011 |
| $H_2N_3$ | 1.14 | 0.00 | 44.0249 |
| $CO_2$ | 1.19 | 0.40 | 43.9898 |
| $CH_2NO$ | 1.55 | 0.21 | 44.0136 |
| $CH_4N_2$ | 1.91 | 0.01 | 44.0375 |
| $C_2H_4O$ | 2.32 | 0.21 | 44.0262 |
| $C_2H_6N$ | 2.68 | 0.02 | 44.0501 |
| $C_3H_8$ | 3.45 | 0.04 | 44.0626 |
| 45 | | | |
| $HN_2O$ | 0.80 | 0.20 | 45.0089 |
| $H_3N_3$ | 1.16 | 0.00 | 45.0328 |
| $CHO_2$ | 1.21 | 0.40 | 44.9976 |
| $CH_3NO$ | 1.57 | 0.21 | 45.0215 |
| $CH_5N_2$ | 1.93 | 0.01 | 45.0453 |
| $C_2H_5O$ | 2.34 | 0.21 | 45.0340 |
| $C_2H_7N$ | 2.70 | 0.02 | 45.0579 |
| 46 | | | |
| $NO_2$ | 0.45 | 0.40 | 45.9929 |
| $H_2N_2O$ | 0.81 | 0.20 | 46.0167 |

| | $M+1$ | $M+2$ | Mass |
|---|---|---|---|
| $H_4N_3$ | 1.17 | 0.01 | 46.0406 |
| $CH_2O_2$ | 1.22 | 0.40 | 46.0054 |
| $CH_4NO$ | 1.58 | 0.21 | 46.0293 |
| $CH_6N_2$ | 1.94 | 0.01 | 46.0532 |
| $C_2H_6O$ | 2.35 | 0.22 | 46.0419 |
| 47 | | | |
| $HNO_2$ | 0.47 | 0.40 | 47.0007 |
| $H_3N_2O$ | 0.83 | 0.20 | 47.0248 |
| $H_5N_3$ | 1.19 | 0.01 | 47.0484 |
| $CH_3O_2$ | 1.24 | 0.40 | 47.0133 |
| $CH_5NO$ | 1.60 | 0.21 | 47.0371 |
| 48 | | | |
| $O_3$ | 0.12 | 0.60 | 47.9847 |
| $H_2NO_2$ | 0.48 | 0.40 | 48.0085 |
| $H_4N_2O$ | 0.84 | 0.20 | 48.0324 |
| $CH_4O_2$ | 1.25 | 0.40 | 48.0211 |
| $C_4$ | 4.44 | 0.07 | 48.0000 |
| 49 | | | |
| $HO_3$ | 0.14 | 0.60 | 48.9925 |
| $H_3NO_2$ | 0.50 | 0.40 | 49.0164 |
| $C_4H$ | 4.46 | 0.07 | 49.0078 |
| 50 | | | |
| $H_2O_3$ | 0.15 | 0.60 | 50.0003 |
| $C_3N$ | 3.70 | 0.05 | 50.0031 |
| $C_4H_2$ | 4.47 | 0.07 | 50.0157 |
| 51 | | | |
| $C_3HN$ | 3.72 | 0.05 | 51.0109 |
| $C_4H_3$ | 4.49 | 0.08 | 51.0235 |
| 52 | | | |
| $C_2N_2$ | 2.96 | 0.03 | 52.0062 |
| $C_3O$ | 3.37 | 0.24 | 51.9949 |
| $C_3H_2N$ | 3.73 | 0.05 | 52.0187 |
| $C_4H_4$ | 4.50 | 0.08 | 52.0313 |

| | $M+1$ | $M+2$ | Mass |
|---|---|---|---|
| 53 | | | |
| $C_2HN_2$ | 2.98 | 0.03 | 53.0140 |
| $C_3HO$ | 3.39 | 0.24 | 53.0027 |
| $C_3H_3N$ | 3.75 | 0.05 | 53.0266 |
| $C_4H_5$ | 4.52 | 0.08 | 53.0391 |
| 54 | | | |
| $CN_3$ | 2.22 | 0.02 | 54.0093 |
| $C_2NO$ | 2.63 | 0.22 | 53.9980 |
| $C_2H_2N_2$ | 2.98 | 0.03 | 54.0218 |
| $C_3H_2O$ | 3.40 | 0.24 | 54.0106 |
| $C_3H_4N$ | 3.76 | 0.05 | 54.0344 |
| $C_4H_6$ | 4.53 | 0.08 | 54.0470 |
| 55 | | | |
| $CHN_3$ | 2.24 | 0.02 | 55.0171 |
| $C_2HNO$ | 2.65 | 0.22 | 55.0058 |
| $C_2H_3N_2$ | 3.01 | 0.03 | 55.0297 |
| $C_3H_3O$ | 3.42 | 0.24 | 55.0184 |
| $C_3H_5N$ | 3.78 | 0.05 | 55.0422 |
| $C_4H_7$ | 4.55 | 0.08 | 55.0548 |
| 56 | | | |
| $N_4$ | 1.48 | 0.01 | 56.0124 |
| $CN_2O$ | 1.89 | 0.21 | 56.0011 |
| $CH_2N_3$ | 2.25 | 0.02 | 56.0249 |
| $C_2O_2$ | 2.30 | 0.41 | 55.9898 |
| $C_2H_2NO$ | 2.66 | 0.22 | 56.0136 |
| $C_2H_4N_2$ | 3.02 | 0.03 | 56.0375 |
| $C_3H_4O$ | 3.43 | 0.24 | 56.0262 |
| $C_3H_6N$ | 3.79 | 0.05 | 56.0501 |
| $C_4H_8$ | 4.56 | 0.08 | 56.0626 |
| 57 | | | |
| $HN_4$ | 1.50 | 0.01 | 57.0202 |
| $CHN_2O$ | 1.91 | 0.21 | 57.0089 |
| $CH_3N_3$ | 2.27 | 0.02 | 57.0328 |
| $C_2HO_2$ | 2.32 | 0.41 | 56.9976 |

| | $M+1$ | $M+2$ | Mass |
|---|---|---|---|
| $C_2H_3NO$ | 2.68 | 0.22 | 57.0215 |
| $C_2H_5N_2$ | 3.04 | 0.03 | 57.0453 |
| $C_3H_5O$ | 3.45 | 0.24 | 57.0340 |
| $C_3H_7N$ | 3.81 | 0.05 | 57.0579 |
| $C_4H_9$ | 4.58 | 0.08 | 57.0705 |
| **58** | | | |
| $N_3O$ | 1.15 | 0.20 | 58.0042 |
| $H_2N_4$ | 1.51 | 0.01 | 58.0280 |
| $CNO_2$ | 1.56 | 0.41 | 57.9929 |
| $CH_2N_2O$ | 1.92 | 0.21 | 58.0167 |
| $CH_4N_3$ | 2.28 | 0.02 | 58.0406 |
| $C_2H_2O_2$ | 2.33 | 0.42 | 58.0054 |
| $C_2H_4NO$ | 2.69 | 0.22 | 58.0293 |
| $C_2H_6N_2$ | 3.05 | 0.03 | 58.0532 |
| $C_3H_6O$ | 3.46 | 0.24 | 58.0419 |
| $C_3H_8N$ | 3.82 | 0.05 | 58.0657 |
| $C_4H_{10}$ | 4.59 | 0.08 | 58.0783 |
| **59** | | | |
| $HN_3O$ | 1.17 | 0.20 | 59.0120 |
| $H_3N_4$ | 1.53 | 0.01 | 59.0359 |
| $CHNO_2$ | 1.58 | 0.41 | 59.0007 |
| $CH_3N_2O$ | 1.94 | 0.21 | 59.0246 |
| $CH_5N_3$ | 2.30 | 0.02 | 59.0484 |
| $C_2H_3O_2$ | 2.35 | 0.42 | 59.0133 |
| $C_2H_5NO$ | 2.71 | 0.22 | 59.0371 |
| $C_2H_7N_2$ | 3.07 | 0.03 | 59.0610 |
| $C_3H_7O$ | 3.48 | 0.24 | 59.0497 |
| $C_3H_9N$ | 3.84 | 0.05 | 59.0736 |
| **60** | | | |
| $N_2O_2$ | 0.82 | 0.40 | 59.9960 |
| $H_2N_3O$ | 1.18 | 0.20 | 60.0198 |
| $H_4N_4$ | 1.54 | 0.01 | 60.0437 |
| $CO_3$ | 1.23 | 0.60 | 59.9847 |
| $CH_2NO_2$ | 1.59 | 0.41 | 60.0085 |
| $CH_4N_2O$ | 1.95 | 0.21 | 60.0324 |

| | $M+1$ | $M+2$ | Mass |
|---|---|---|---|
| $CH_6N_3$ | 2.31 | 0.02 | 60.0563 |
| $C_2H_4O_2$ | 2.36 | 0.42 | 60.0211 |
| $C_2H_6NO$ | 2.72 | 0.22 | 60.0449 |
| $C_2H_8N_2$ | 3.08 | 0.03 | 60.0688 |
| $C_3H_8O$ | 3.49 | 0.24 | 60.0575 |
| $C_5$ | 5.55 | 0.12 | 60.0000 |
| **61** | | | |
| $HN_2O_2$ | 0.84 | 0.40 | 61.0038 |
| $H_3N_3O$ | 1.20 | 0.21 | 61.0277 |
| $H_5N_4$ | 1.56 | 0.01 | 61.0515 |
| $CHO_3$ | 1.25 | 0.60 | 60.9925 |
| $CH_3NO_2$ | 1.61 | 0.41 | 61.0164 |
| $CH_5N_2O$ | 1.97 | 0.21 | 61.0402 |
| $CH_7N_3$ | 2.33 | 0.02 | 61.0641 |
| $C_2H_5O_2$ | 2.38 | 0.42 | 61.0289 |
| $C_2H_7NO$ | 2.74 | 0.22 | 61.0528 |
| $C_5H$ | 5.57 | 0.12 | 61.0078 |
| **62** | | | |
| $NO_3$ | 0.49 | 0.60 | 61.9878 |
| $H_2N_2O_2$ | 0.85 | 0.40 | 62.0116 |
| $H_4N_3O$ | 1.21 | 0.42 | 62.0368 |
| $H_6N_4$ | 1.57 | 0.01 | 62.0594 |
| $CH_2O_3$ | 1.26 | 0.60 | 62.0003 |
| $CH_4NO_2$ | 1.62 | 0.41 | 62.0242 |
| $CH_6N_2O$ | 1.98 | 0.21 | 62.0480 |
| $C_2H_6O_2$ | 2.39 | 0.42 | 62.0368 |
| $C_4N$ | 4.81 | 0.09 | 62.0031 |
| $C_5H_2$ | 5.58 | 0.12 | 62.0157 |
| **63** | | | |
| $HNO_3$ | 0.51 | 0.60 | 62.9956 |
| $H_3N_2O_2$ | 0.87 | 0.40 | 63.0195 |
| $H_5N_3O$ | 1.23 | 0.21 | 63.0433 |
| $CH_3O_3$ | 1.28 | 0.60 | 63.0082 |
| $CH_5NO_2$ | 1.64 | 0.41 | 63.0320 |
| $C_4HN$ | 4.83 | 0.09 | 63.0109 |

| | $M+1$ | $M+2$ | Mass |
|---|---|---|---|
| $C_5H_3$ | 5.60 | 0.12 | 63.0235 |
| 64 | | | |
| $O_4$ | 0.16 | 0.80 | 63.9796 |
| $H_2NO_3$ | 0.52 | 0.60 | 64.0034 |
| $H_4N_2O_2$ | 0.88 | 0.40 | 64.0273 |
| $CH_4O_3$ | 1.29 | 0.60 | 64.0160 |
| $C_3N_2$ | 4.07 | 0.06 | 64.0062 |
| $C_4O$ | 4.48 | 0.27 | 63.9949 |
| $C_4H_2N$ | 4.84 | 0.09 | 64.0187 |
| $C_5H_4$ | 5.61 | 0.12 | 64.0313 |
| 65 | | | |
| $HO_4$ | 0.18 | 0.80 | 64.9874 |
| $H_3NO_3$ | 0.54 | 0.60 | 65.0113 |
| $C_3HN_2$ | 4.09 | 0.06 | 65.0140 |
| $C_4HO$ | 4.50 | 0.27 | 65.0027 |
| $C_4H_3N$ | 4.86 | 0.09 | 65.0266 |
| $C_5H_5$ | 5.63 | 0.12 | 65.0391 |
| 66 | | | |
| $H_2O_4$ | 0.19 | 0.80 | 65.9953 |
| $C_2N_3$ | 3.33 | 0.04 | 66.0093 |
| $C_3NO$ | 3.74 | 0.25 | 65.9980 |
| $C_3H_2N_2$ | 4.10 | 0.06 | 66.0218 |
| $C_4H_2O$ | 4.51 | 0.27 | 66.0106 |
| $C_4H_4N$ | 4.87 | 0.09 | 66.0344 |
| $C_5H_6$ | 5.64 | 0.12 | 66.0470 |
| 67 | | | |
| $C_2HN_3$ | 3.35 | 0.04 | 67.0171 |
| $C_3HNO$ | 3.76 | 0.25 | 67.0058 |
| $C_3H_3N_2$ | 4.12 | 0.06 | 67.0297 |
| $C_4H_3O$ | 4.53 | 0.27 | 67.0184 |
| $C_4H_5N$ | 4.89 | 0.09 | 67.0422 |
| $C_5H_7$ | 5.66 | 0.12 | 67.0548 |
| 68 | | | |
| $CN_4$ | 2.59 | 0.02 | 68.0124 |
| $C_2N_2O$ | 3.00 | 0.23 | 68.0011 |
| $C_2H_2N_3$ | 3.36 | 0.04 | 68.0249 |
| $C_3O_2$ | 3.41 | 0.44 | 67.9898 |
| $C_3H_2NO$ | 3.77 | 0.25 | 68.0136 |
| $C_3H_4N_2$ | 4.13 | 0.06 | 68.0375 |
| $C_4H_4O$ | 4.54 | 0.28 | 68.0262 |
| $C_4H_6N$ | 4.90 | 0.09 | 68.0501 |
| $C_5H_8$ | 5.67 | 0.13 | 68.0626 |
| 69 | | | |
| $CHN_4$ | 2.61 | 0.03 | 69.0202 |
| $C_2HN_2O$ | 3.02 | 0.23 | 69.0089 |
| $C_2H_3N_3$ | 3.38 | 0.04 | 69.0328 |
| $C_3HO_2$ | 3.43 | 0.44 | 68.9976 |
| $C_3H_3NO$ | 3.79 | 0.25 | 69.0215 |
| $C_3H_5N_2$ | 4.15 | 0.06 | 69.0453 |
| $C_4H_5O$ | 4.56 | 0.28 | 69.0340 |
| $C_4H_7N$ | 4.92 | 0.09 | 69.0579 |
| $C_5H_9$ | 5.69 | 0.13 | 69.0705 |
| 70 | | | |
| $CHN_3O$ | 2.28 | 0.22 | 71.0120 |
| $CH_3N_4$ | 2.64 | 0.03 | 71.0359 |
| $C_2HNO_2$ | 2.69 | 0.42 | 71.0007 |
| $C_2H_3N_2O$ | 3.05 | 0.23 | 71.0246 |
| $C_2H_5N_3$ | 3.41 | 0.04 | 71.0484 |
| $C_3H_3O_2$ | 3.46 | 0.44 | 71.0133 |
| $C_3H_5NO$ | 3.82 | 0.25 | 71.0371 |
| $C_3H_7N_2$ | 4.18 | 0.07 | 71.0610 |
| $C_4H_7O$ | 4.59 | 0.28 | 71.0497 |
| $C_4H_9N$ | 4.95 | 0.10 | 71.0736 |
| $C_5H_{11}$ | 5.72 | 0.13 | 71.0861 |
| 72 | | | |
| $N_4O$ | 1.52 | 0.21 | 72.0073 |
| $CN_2O_2$ | 1.93 | 0.41 | 71.9960 |
| $CH_2N_3O$ | 2.29 | 0.22 | 72.0198 |

| | M + 1 | M + 2 | Mass |
|---|---|---|---|
| $CH_4N_4$ | 2.65 | 0.03 | 72.0437 |
| $C_2O_3$ | 2.34 | 0.62 | 71.9847 |
| $C_2H_2NO_2$ | 2.70 | 0.42 | 72.0085 |
| $C_2N_4N_2O$ | 3.06 | 0.23 | 72.0324 |
| $C_2H_6N_3$ | 3.42 | 0.04 | 72.0563 |
| $C_3H_4O_2$ | 3.47 | 0.44 | 72.0211 |
| $C_3H_6NO$ | 3.83 | 0.25 | 72.0449 |
| $C_3H_8N_2$ | 4.19 | 0.07 | 72.0688 |
| $C_4H_8O$ | 4.60 | 0.28 | 72.0575 |
| $C_4H_{10}N$ | 4.96 | 0.09 | 72.0814 |
| $C_5H_{12}$ | 5.73 | 0.13 | 72.0939 |
| $C_6$ | 6.66 | 0.18 | 72.0000 |
| 73 | | | |
| $HN_4O$ | 1.54 | 0.21 | 73.0151 |
| $CHN_2O_2$ | 1.95 | 0.41 | 73.0038 |
| $CH_3N_3O$ | 2.31 | 0.22 | 73.0277 |
| $CH_5N_4$ | 2.67 | 0.03 | 73.0515 |
| $C_2HO_3$ | 2.36 | 0.62 | 72.9925 |
| $C_2H_3NO_2$ | 2.72 | 0.42 | 73.0164 |
| $C_2H_5N_2O$ | 3.08 | 0.23 | 73.0402 |
| $C_2H_7N_3$ | 3.44 | 0.04 | 73.0641 |
| $C_3H_5O_2$ | 3.49 | 0.44 | 73.0289 |
| $C_3H_7NO$ | 3.85 | 0.25 | 73.0528 |
| $C_3H_9N_2$ | 4.21 | 0.07 | 73.0767 |
| $C_4H_9O$ | 4.62 | 0.28 | 73.0653 |
| $C_4H_{11}N$ | 4.98 | 0.09 | 73.0892 |
| $C_6H$ | 6.68 | 0.18 | 73.0078 |
| 74 | | | |
| $N_3O_2$ | 1.19 | 0.41 | 73.9991 |
| $H_2N_4O$ | 1.55 | 0.21 | 74.0229 |
| $CNO_3$ | 1.60 | 0.61 | 73.9878 |
| $CH_2N_2O_2$ | 1.96 | 0.41 | 74.0116 |
| $CH_4N_3O$ | 2.32 | 0.22 | 74.0355 |
| $CH_6N_4$ | 2.68 | 0.03 | 74.0594 |
| $C_2H_2O_3$ | 2.37 | 0.62 | 74.0003 |
| $C_2H_4NO_2$ | 2.73 | 0.42 | 74.0242 |
| $C_2H_6N_2O$ | 3.09 | 0.23 | 74.0480 |
| $C_2H_8N_3$ | 3.45 | 0.05 | 74.0719 |
| $C_3H_6O_2$ | 3.50 | 0.44 | 74.0368 |
| $C_3H_8NO$ | 3.86 | 0.25 | 74.0606 |
| $C_3H_{10}N_2$ | 4.22 | 0.07 | 74.0845 |
| $C_4H_{10}O$ | 4.63 | 0.28 | 74.0732 |
| $C_5N$ | 5.92 | 0.14 | 74.0031 |
| $C_6H_2$ | 6.69 | 0.18 | 74.0157 |
| 75 | | | |
| $HN_3O_2$ | 1.21 | 0.41 | 75.0069 |
| $H_3N_4O$ | 1.57 | 0.21 | 75.0308 |
| $CHNO_3$ | 1.62 | 0.61 | 74.9956 |
| $CH_3N_2O_2$ | 1.98 | 0.41 | 75.0195 |
| $CH_5N_3O$ | 2.34 | 0.22 | 75.0433 |
| $CH_7N_4$ | 2.70 | 0.03 | 75.0672 |
| $C_2H_3O_3$ | 2.39 | 0.62 | 75.0082 |
| $C_2H_5NO_2$ | 2.75 | 0.43 | 75.0320 |
| $C_2H_7N_2O$ | 3.11 | 0.23 | 75.0559 |
| $C_2H_9N_3$ | 3.47 | 0.05 | 75.0798 |
| $C_3H_7O_2$ | 3.52 | 0.44 | 75.0446 |
| $C_3H_9NO$ | 3.88 | 0.25 | 75.0684 |
| $C_5HN$ | 5.94 | 0.14 | 75.0109 |
| $C_6H_3$ | 6.71 | 0.18 | 75.0235 |
| 76 | | | |
| $N_2O_3$ | 0.86 | 0.60 | 75.9909 |
| $H_2N_3O_2$ | 1.22 | 0.41 | 76.0147 |
| $H_4N_4O$ | 1.58 | 0.21 | 76.0386 |
| $CO_4$ | 1.27 | 0.80 | 75.9796 |
| $CH_2NO_3$ | 1.63 | 0.61 | 76.0034 |
| $CH_4N_2O_2$ | 1.99 | 0.41 | 76.0273 |
| $CH_6N_3O$ | 2.35 | 0.22 | 76.0511 |
| $CH_8N_4$ | 2.71 | 0.03 | 76.0750 |
| $C_2H_4O_3$ | 2.40 | 0.62 | 76.0160 |
| $C_2H_6NO_2$ | 2.76 | 0.43 | 76.0399 |
| $C_2H_8N_2O$ | 3.12 | 0.24 | 76.0637 |
| $C_3H_8O_2$ | 3.53 | 0.44 | 76.0524 |

| | $M+1$ | $M+2$ | Mass |
|---|---|---|---|
| $C_4N_2$ | 5.18 | 0.10 | 76.0062 |
| $C_5O$ | 5.59 | 0.32 | 75.9949 |
| $C_5H_2N$ | 5.95 | 0.14 | 76.0187 |
| $C_6H_4$ | 6.72 | 0.19 | 76.0313 |
| 77 | | | |
| $HN_2O_3$ | 0.88 | 0.60 | 76.9987 |
| $H_3N_3O_2$ | 1.24 | 0.41 | 77.0226 |
| $H_5N_4O$ | 1.60 | 0.21 | 77.0464 |
| $CHO_4$ | 1.29 | 0.80 | 76.9874 |
| $CH_3NO_3$ | 1.65 | 0.61 | 77.0113 |
| $CH_5N_2O_2$ | 2.01 | 0.41 | 77.0351 |
| $CH_7N_3O$ | 2.37 | 0.22 | 77.0590 |
| $C_2H_5O_3$ | 2.42 | 0.62 | 77.0238 |
| $C_2H_7NO_2$ | 2.78 | 0.43 | 77.0477 |
| $C_4HN_2$ | 5.20 | 0.11 | 77.0140 |
| $C_5HO$ | 5.61 | 0.32 | 77.0027 |
| $C_5H_3N$ | 5.97 | 0.15 | 77.0266 |
| $C_6H_5$ | 6.74 | 0.19 | 77.0391 |
| 78 | | | |
| $NO_4$ | 0.53 | 0.80 | 77.9827 |
| $H_2N_2O_3$ | 0.89 | 0.60 | 78.0065 |
| $H_4N_3O_2$ | 1.25 | 0.41 | 78.0304 |
| $H_6N_4O$ | 1.61 | 0.21 | 78.0542 |
| $CH_2O_4$ | 1.30 | 0.80 | 77.9953 |
| $CH_4NO_3$ | 1.66 | 0.61 | 78.0191 |
| $CH_6N_2O_2$ | 2.02 | 0.41 | 78.0429 |
| $C_2H_6O_3$ | 2.43 | 0.62 | 78.0317 |
| $C_3N_3$ | 4.44 | 0.08 | 78.0093 |
| $C_4NO$ | 4.85 | 0.29 | 77.9980 |
| $C_4H_2N_2$ | 5.21 | 0.11 | 78.0218 |
| $C_5H_2O$ | 5.62 | 0.32 | 78.0106 |
| $C_5H_4N$ | 5.98 | 0.14 | 78.0344 |
| $C_6H_6$ | 6.75 | 0.19 | 78.0470 |
| 79 | | | |
| $HNO_4$ | 0.55 | 0.80 | 78.9905 |
| $H_3N_2O_3$ | 0.91 | 0.60 | 79.0144 |
| $H_5N_3O_2$ | 1.27 | 0.41 | 79.0382 |
| $CH_3O_4$ | 1.32 | 0.80 | 79.0031 |
| $CH_5NO_3$ | 1.68 | 0.61 | 79.0269 |
| $C_3HN_3$ | 4.46 | 0.08 | 79.0171 |
| $C_4HNO$ | 4.87 | 0.29 | 79.0058 |
| $C_4H_3N_2$ | 5.23 | 0.11 | 79.0297 |
| $C_5H_3O$ | 5.64 | 0.32 | 79.0184 |
| $C_5H_5N$ | 6.00 | 0.14 | 79.0422 |
| $C_6H_7$ | 6.77 | 0.19 | 79.0548 |
| 80 | | | |
| $H_2NO_4$ | 0.56 | 0.80 | 79.9983 |
| $H_4N_2O_3$ | 0.92 | 0.60 | 80.0222 |
| $CH_4O_4$ | 1.33 | 0.80 | 80.0109 |
| $C_2N_4$ | 3.70 | 0.05 | 80.0124 |
| $C_3N_2O$ | 4.11 | 0.26 | 80.0011 |
| $C_3H_2N_3$ | 4.47 | 0.08 | 80.0249 |
| $C_4O_2$ | 4.52 | 0.47 | 79.9898 |
| $C_4H_2NO$ | 4.88 | 0.29 | 80.0136 |
| $C_4H_4N_2$ | 5.24 | 0.11 | 80.0375 |
| $C_5H_4O$ | 5.65 | 0.32 | 80.0262 |
| $C_5H_6N$ | 6.01 | 0.14 | 80.0501 |
| $C_6H_8$ | 6.78 | 0.19 | 80.0626 |
| 81 | | | |
| $H_3NO_4$ | 0.58 | 0.08 | 81.0062 |
| $C_2HN_4$ | 3.72 | 0.50 | 81.0202 |
| $C_3HN_2O$ | 4.13 | 0.26 | 81.0089 |
| $C_3H_3N_3$ | 4.49 | 0.08 | 81.0328 |
| $C_4HO_2$ | 4.54 | 0.48 | 80.9976 |
| $C_4H_3NO$ | 4.90 | 0.29 | 81.0215 |
| $C_4H_5N_2$ | 5.26 | 0.11 | 81.0453 |
| $C_5H_5O$ | 5.67 | 0.32 | 81.0340 |
| $C_5H_7N$ | 6.03 | 0.14 | 81.0579 |
| $C_6H_9$ | 6.80 | 0.19 | 81.0705 |
| 82 | | | |

| | $M+1$ | $M+2$ | Mass |
|---|---|---|---|
| $C_2N_3O$ | 3.37 | 0.24 | 82.0042 |
| $C_2H_2N_4$ | 3.73 | 0.05 | 82.0280 |
| $C_3NO_2$ | 3.78 | 0.45 | 81.9929 |
| $C_3H_2N_2O$ | 4.14 | 0.26 | 82.0167 |
| $C_3H_4N_3$ | 4.50 | 0.08 | 82.0406 |
| $C_4H_2O_2$ | 4.55 | 0.48 | 82.0054 |
| $C_4H_4NO$ | 4.91 | 0.29 | 82.0293 |
| $C_4H_6N_2$ | 5.27 | 0.11 | 82.0532 |
| $C_5H_6O$ | 5.68 | 0.32 | 82.0419 |
| $C_5H_8N$ | 6.04 | 0.14 | 82.0657 |
| $C_6H_{10}$ | 6.81 | 0.19 | 82.0783 |
| 83 | | | |
| $C_2HN_3O$ | 3.39 | 0.24 | 83.0120 |
| $C_2H_3N_4$ | 3.75 | 0.06 | 83.0359 |
| $C_3HNO_2$ | 3.80 | 0.45 | 83.0007 |
| $C_3H_3N_2O$ | 4.16 | 0.27 | 83.0246 |
| $C_3H_5N_3$ | 4.52 | 0.08 | 83.0484 |
| $C_4H_3O_2$ | 4.57 | 0.48 | 83.0133 |
| $C_4H_5NO$ | 4.93 | 0.29 | 83.0371 |
| $C_4H_7N_2$ | 5.29 | 0.11 | 83.0610 |
| $C_5H_7O$ | 5.70 | 0.33 | 83.0497 |
| $C_5H_9N$ | 6.06 | 0.15 | 83.0736 |
| $C_6H_{11}$ | 6.83 | 0.19 | 83.0861 |
| 84 | | | |
| $CN_4O$ | 2.63 | 0.23 | 84.0073 |
| $C_2H_2N_3O$ | 3.40 | 0.24 | 84.0198 |
| $C_2N_2O_2$ | 3.04 | 0.43 | 83.9960 |
| $C_2H_4N_4$ | 3.76 | 0.06 | 84.0437 |
| $C_3O_3$ | 3.45 | 0.64 | 83.9847 |
| $C_3H_2NO_2$ | 3.81 | 0.45 | 84.0085 |
| $C_3H_4N_2O$ | 4.17 | 0.27 | 84.0324 |
| $C_3H_6N_3$ | 4.53 | 0.08 | 84.0563 |
| $C_4H_4O_2$ | 4.58 | 0.48 | 84.0211 |
| $C_4H_6NO$ | 4.94 | 0.29 | 84.0449 |
| $C_4H_8N_2$ | 5.30 | 0.11 | 84.0688 |

| | $M+1$ | $M+2$ | Mass |
|---|---|---|---|
| $C_5H_8O$ | 5.71 | 0.33 | 84.0575 |
| $C_5H_{10}N$ | 6.07 | 0.15 | 84.0814 |
| $C_6H_{12}$ | 6.84 | 0.19 | 84.0939 |
| $C_7$ | 7.77 | 0.26 | 84.0000 |
| 85 | | | |
| $CHN_4O$ | 2.65 | 0.23 | 85.0151 |
| $C_2HN_2O_2$ | 3.06 | 0.43 | 85.0038 |
| $C_2H_3N_3O$ | 3.42 | 0.24 | 85.0277 |
| $C_2H_5N_4$ | 3.78 | 0.06 | 85.0515 |
| $C_3HO_3$ | 3.47 | 0.64 | 84.9925 |
| $C_3H_3NO_2$ | 3.83 | 0.45 | 85.0164 |
| $C_3H_5N_2O$ | 4.19 | 0.27 | 85.0402 |
| $C_3H_7N_3$ | 4.55 | 0.08 | 85.0641 |
| $C_4H_5O_2$ | 4.60 | 0.48 | 85.0289 |
| $C_4H_7NO$ | 4.96 | 0.29 | 85.0528 |
| $C_4H_9N_2$ | 5.32 | 0.11 | 85.0767 |
| $C_5H_9O$ | 5.73 | 0.33 | 85.0653 |
| $C_5H_{11}N$ | 6.09 | 0.16 | 85.0892 |
| $C_6H_{13}$ | 6.86 | 0.20 | 85.1018 |
| $C_7H$ | 7.79 | 0.26 | 85.0078 |
| 86 | | | |
| $CN_3O_2$ | 2.30 | 0.41 | 85.9991 |
| $CH_2N_4O$ | 2.66 | 0.21 | 86.0229 |
| $C_2NO_3$ | 2.71 | 0.62 | 85.9878 |
| $C_2H_2N_2O_2$ | 3.07 | 0.43 | 86.0116 |
| $C_2H_4N_3O$ | 3.43 | 0.24 | 86.0355 |
| $C_2H_6N_4$ | 3.79 | 0.06 | 86.0594 |
| $C_3H_2O_3$ | 3.48 | 0.64 | 86.0003 |
| $C_3H_4NO_2$ | 3.84 | 0.45 | 86.0242 |
| $C_3H_6N_2O$ | 4.20 | 0.27 | 86.0480 |
| $C_3H_8N_3$ | 4.56 | 0.08 | 86.0719 |
| $C_4H_6O_2$ | 4.61 | 0.48 | 86.0368 |
| $C_4H_8NO$ | 4.97 | 0.30 | 86.0606 |
| $C_4H_{10}N_2$ | 5.33 | 0.11 | 86.0845 |
| $C_5H_{10}O$ | 5.74 | 0.33 | 86.0732 |
| $C_5H_{12}N$ | 6.10 | 0.16 | 86.0970 |

| | $M+1$ | $M+2$ | Mass |
|---|---|---|---|
| $C_6H_{14}$ | 6.87 | 0.21 | 86.1096 |
| $C_6N$ | 7.03 | 0.21 | 86.0031 |
| $C_7H_2$ | 7.80 | 0.26 | 86.0157 |
| 87 | | | |
| $CHN_3O_2$ | 2.32 | 0.42 | 87.0069 |
| $CH_3N_4O$ | 2.68 | 0.23 | 87.0308 |
| $C_2HNO_3$ | 2.73 | 0.62 | 86.9956 |
| $C_2H_3N_2O_2$ | 3.09 | 0.43 | 87.0195 |
| $C_2H_5N_3O$ | 3.45 | 0.25 | 87.0433 |
| $C_2H_7N_4$ | 3.81 | 0.06 | 87.0672 |
| $C_3H_3O_3$ | 3.50 | 0.64 | 87.0082 |
| $C_3H_5NO_2$ | 3.86 | 0.45 | 87.0320 |
| $C_3H_7N_2O$ | 4.22 | 0.27 | 87.0559 |
| $C_3H_9N_3$ | 4.58 | 0.08 | 87.0798 |
| $C_4H_7O_2$ | 4.63 | 0.48 | 87.0446 |
| $C_4H_9NO$ | 4.99 | 0.30 | 87.0684 |
| $C_4H_{11}N_2$ | 5.35 | 0.11 | 87.0923 |
| $C_5H_{11}O$ | 5.76 | 0.33 | 87.0810 |
| $C_5H_{13}N$ | 6.12 | 0.15 | 87.1049 |
| $C_6HN$ | 7.05 | 0.21 | 87.0109 |
| $C_7H_3$ | 7.82 | 0.26 | 87.0235 |
| 88 | | | |
| $N_4O_2$ | 1.56 | 0.41 | 88.0022 |
| $CN_2O_3$ | 1.97 | 0.61 | 87.9909 |
| $CH_2N_3O_2$ | 2.33 | 0.42 | 88.0147 |
| $CH_4N_4O$ | 2.69 | 0.23 | 88.0386 |
| $C_2O_4$ | 2.38 | 0.82 | 87.9796 |
| $C_2H_2NO_3$ | 2.74 | 0.63 | 88.0034 |
| $C_2H_4N_2O_2$ | 3.10 | 0.43 | 88.0273 |
| $C_2H_6N_3O$ | 3.46 | 0.25 | 88.0511 |
| $C_2H_8N_4$ | 3.82 | 0.06 | 88.0750 |
| $C_3H_4O_3$ | 3.51 | 0.64 | 88.0160 |
| $C_3H_6NO_2$ | 3.87 | 0.45 | 88.0399 |
| $C_3H_8N_2O$ | 4.23 | 0.27 | 88.0637 |
| $C_3H_{10}N_3$ | 4.59 | 0.08 | 88.0876 |
| $C_4H_8O_2$ | 4.64 | 0.48 | 88.0524 |
| $C_4H_{10}NO$ | 5.00 | 0.30 | 88.0763 |
| $C_4H_{12}N_2$ | 5.36 | 0.11 | 88.1001 |
| $C_5H_{12}O$ | 5.77 | 0.33 | 88.0888 |
| $C_5N_2$ | 6.29 | 0.16 | 88.0062 |
| $C_6O$ | 6.70 | 0.38 | 87.9949 |
| $C_6H_2N$ | 7.06 | 0.21 | 88.0187 |
| $C_7H_4$ | 7.83 | 0.26 | 88.0313 |
| 89 | | | |
| $HN_4O_2$ | 1.58 | 0.41 | 89.0100 |
| $CHN_{2O}3$ | 1.99 | 0.61 | 88.9987 |
| $CH_3N_3O_2$ | 2.35 | 0.42 | 89.0226 |
| $CH_5N_4O$ | 2.71 | 0.23 | 89.0464 |
| $C_2HO_4$ | 2.40 | 0.82 | 88.9874 |
| $C_2H_3NO_3$ | 2.76 | 0.63 | 89.0113 |
| $C_2H_5N_2O_2$ | 3.12 | 0.44 | 89.0351 |
| $C_2H_7N_3O$ | 3.48 | 0.25 | 89.0590 |
| $C_2H_9N_4$ | 3.84 | 0.06 | 89.0829 |
| $C_3H_5O_3$ | 3.53 | 0.64 | 89.0238 |
| $C_3H_7NO_2$ | 3.89 | 0.46 | 89.0477 |
| $C_3H_9N_2O$ | 4.25 | 0.27 | 89.0715 |
| $C_3H_{11}N_3$ | 4.61 | 0.08 | 89.0954 |
| $C_4H_9O_2$ | 4.66 | 0.48 | 89.0603 |
| $C_4H_{11}NO$ | 5.02 | 0.30 | 89.0841 |
| $C_5HN_2$ | 6.31 | 0.16 | 89.0140 |
| $C_6HO$ | 6.72 | 0.38 | 89.0027 |
| $C_6H_3N$ | 7.08 | 0.21 | 89.0266 |
| $C_7H_5$ | 7.85 | 0.26 | 89.0391 |
| 90 | | | |
| $N_3O_3$ | 1.23 | 0.60 | 89.9940 |
| $H_2N_4O_2$ | 1.59 | 0.40 | 90.0178 |
| $CNO_4$ | 1.64 | 0.80 | 89.9827 |
| $CH_2N_2O_3$ | 2.00 | 0.61 | 90.0065 |
| $CH_4N_3O_2$ | 2.36 | 0.42 | 90.0304 |
| $CH_6N_4O$ | 2.72 | 0.23 | 90.0542 |
| $C_2H_2O_4$ | 2.41 | 0.82 | 89.9953 |
| $C_2H_4NO_3$ | 2.77 | 0.63 | 90.0191 |

| | $M+1$ | $M+2$ | Mass |
|---|---|---|---|
| $C_2H_6N_2O_2$ | 3.13 | 0.44 | 90.0429 |
| $C_2H_8N_3O$ | 3.49 | 0.25 | 90.0668 |
| $C_2H_{10}N_4$ | 3.85 | 0.06 | 90.0907 |
| $C_3H_6O_3$ | 3.54 | 0.64 | 90.0317 |
| $C_3H_8NO_2$ | 3.90 | 0.46 | 90.0555 |
| $C_3H_{10}N_2O$ | 4.26 | 0.27 | 90.0794 |
| $C_4H_{10}O_2$ | 4.67 | 0.48 | 90.0681 |
| $C_4N_3$ | 5.55 | 0.13 | 90.0093 |
| $C_5NO$ | 5.96 | 0.34 | 89.9980 |
| $C_5H_2N_2$ | 6.32 | 0.17 | 90.0218 |
| $C_6H_2O$ | 6.73 | 0.38 | 90.0106 |
| $C_6H_4N$ | 7.09 | 0.21 | 90.0344 |
| $C_7H_6$ | 7.86 | 0.26 | 90.0470 |
| 91 | | | |
| $HN_3O_3$ | 1.25 | 0.60 | 91.0018 |
| $H_3N_4O_2$ | 1.61 | 0.41 | 91.0257 |
| $CHNO_4$ | 1.66 | 0.81 | 90.9905 |
| $CH_3N_2O_3$ | 2.02 | 0.61 | 91.0144 |
| $CH_5N_3O_2$ | 2.38 | 0.42 | 91.0382 |
| $CH_7N_4O$ | 2.74 | 0.23 | 91.0621 |
| $C_2H_3O_4$ | 2.43 | 0.82 | 91.0031 |
| $C_2H_5NO_3$ | 2.79 | 0.63 | 91.0269 |
| $C_2H_7N_2O_2$ | 3.15 | 0.44 | 91.0508 |
| $C_2H_9N_3O$ | 3.51 | 0.25 | 91.0746 |
| $C_3H_7O_3$ | 3.56 | 0.64 | 91.0395 |
| $C_3H_9NO_2$ | 3.92 | 0.46 | 91.0634 |
| $C_4HN_3$ | 5.57 | 0.13 | 91.0171 |
| $C_5HNO$ | 5.98 | 0.34 | 91.0058 |
| $C_5H_3N_2$ | 6.34 | 0.17 | 91.0297 |
| $C_6H_3O$ | 6.75 | 0.38 | 91.0184 |
| $C_6H_5N$ | 7.11 | 0.21 | 91.0422 |
| $C_7H_7$ | 7.88 | 0.26 | 91.0548 |
| 92 | | | |
| $N_2O_4$ | 0.90 | 0.80 | 91.9858 |
| $H_2N_3O_3$ | 1.26 | 0.60 | 92.0096 |
| $H_4N_4O_2$ | 1.62 | 0.41 | 92.0335 |

| | $M+1$ | $M+2$ | Mass |
|---|---|---|---|
| $CH_2NO_4$ | 1.67 | 0.81 | 91.9983 |
| $CH_4N_2O_3$ | 2.03 | 0.61 | 92.0222 |
| $CH_6N_3O_2$ | 2.39 | 0.42 | 92.0460 |
| $CH_8N_4O$ | 2.75 | 0.23 | 92.0699 |
| $C_2H_4O_4$ | 2.44 | 0.82 | 92.0109 |
| $C_2H_6NO_3$ | 2.80 | 0.63 | 92.0348 |
| $C_2H_8N_2O_2$ | 3.16 | 0.44 | 92.0586 |
| $C_3H_8O_3$ | 3.57 | 0.64 | 92.0473 |
| $C_3N_4$ | 4.81 | 0.09 | 92.0124 |
| $C_4N_2O$ | 5.22 | 0.31 | 92.0011 |
| $C_4H_2N_3$ | 5.58 | 0.13 | 92.0249 |
| $C_5O_2$ | 5.63 | 0.52 | 91.9898 |
| $C_5H_2NO$ | 5.99 | 0.34 | 92.0136 |
| $C_5H_4N_2$ | 6.35 | 0.17 | 92.0375 |
| $C_6H_4O$ | 5.76 | 0.38 | 92.0262 |
| $C_6H_6N$ | 7.12 | 0.21 | 92.0501 |
| $C_7H_8$ | 7.89 | 0.27 | 92.0626 |
| 93 | | | |
| $HN_2O_4$ | 0.92 | 0.80 | 92.9936 |
| $H_3N_3O_3$ | 1.28 | 0.60 | 93.0175 |
| $H_5N_4O_2$ | 1.64 | 0.41 | 93.0413 |
| $CH_3NO_4$ | 1.69 | 0.81 | 93.0062 |
| $CH_5N_2O_3$ | 2.05 | 0.61 | 93.0300 |
| $CH_7N_3O_2$ | 2.41 | 0.42 | 93.0539 |
| $C_2H_5O_4$ | 2.46 | 0.82 | 93.0187 |
| $C_2H_7NO_3$ | 2.82 | 0.63 | 93.0426 |
| $C_3HN_4$ | 4.83 | 0.09 | 93.0202 |
| $C_4HN_2O$ | 5.24 | 0.31 | 93.0089 |
| $C_4H_3N_3$ | 5.60 | 0.13 | 93.0328 |
| $C_5HO_2$ | 5.65 | 0.52 | 92.9976 |
| $C_5H_3NO$ | 6.01 | 0.35 | 93.0215 |
| $C_5H_5N_2$ | 6.37 | 0.17 | 93.0453 |
| $C_6H_5O$ | 6.78 | 0.38 | 93.0340 |
| $C_6H_7N$ | 7.14 | 0.22 | 93.0579 |
| $C_7H_9$ | 7.91 | 0.27 | 93.0705 |
| 94 | | | |

| | $M+1$ | $M+2$ | Mass |
|---|---|---|---|
| $H_2N_2O_4$ | 0.93 | 0.80 | 94.0014 |
| $H_4N_3O_3$ | 1.29 | 0.61 | 94.0253 |
| $H_6N_4O_2$ | 1.65 | 0.41 | 94.0491 |
| $CH_4NO_4$ | 1.70 | 0.81 | 94.0140 |
| $CH_6N_2O_3$ | 2.06 | 0.62 | 94.0379 |
| $C_2H_6O_4$ | 2.47 | 0.82 | 94.0266 |
| $C_3N_3O$ | 4.48 | 0.28 | 94.0042 |
| $C_3H_2N_4$ | 4.84 | 0.09 | 94.0280 |
| $C_4NO_2$ | 4.89 | 0.49 | 93.9929 |
| $C_4H_2N_2O$ | 5.25 | 0.31 | 94.0167 |
| $C_4H_4N_3$ | 5.61 | 0.13 | 94.0406 |
| $C_5H_2O_2$ | 5.66 | 0.52 | 94.0054 |
| $C_5H_4NO$ | 6.02 | 0.35 | 94.0293 |
| $C_5H_6N_2$ | 6.38 | 0.17 | 94.0532 |
| $C_6H_6O$ | 6.79 | 0.38 | 94.0419 |
| $C_6H_8N$ | 7.15 | 0.22 | 94.0657 |
| $C_7H_{10}$ | 7.92 | 0.27 | 94.0783 |
| 95 | | | |
| $H_3N_2O_4$ | 0.95 | 0.80 | 95.0093 |
| $H_5N_3O_3$ | 1.31 | 0.60 | 95.0331 |
| $CH_5NO_4$ | 1.72 | 0.81 | 95.0218 |
| $C_3HN_3O$ | 4.50 | 0.28 | 95.0120 |
| $C_3H_3N_4$ | 4.86 | 0.10 | 95.0359 |
| $C_4HNO_2$ | 4.91 | 0.49 | 95.0007 |
| $C_4H_3N_2O$ | 5.27 | 0.31 | 95.0246 |
| $C_4H_5N_3$ | 5.63 | 0.13 | 95.0484 |
| $C_5H_3O_2$ | 5.68 | 0.52 | 95.0133 |
| $C_5H_5NO$ | 6.04 | 0.35 | 95.0371 |
| $C_5H_7N_2$ | 6.40 | 0.17 | 95.0610 |
| $C_6H_7O$ | 6.81 | 0.39 | 95.0497 |
| $C_6H_9N$ | 7.17 | 0.22 | 95.0736 |
| $C_7H_{11}$ | 7.94 | 0.27 | 95.0861 |
| 96 | | | |
| $H_4N_2O_4$ | 0.96 | 0.80 | 96.0171 |
| $C_2N_4O$ | 3.74 | 0.26 | 96.0073 |
| $C_3N_2O_2$ | 4.15 | 0.47 | 95.9960 |
| $C_3H_2N_3O$ | 4.51 | 0.28 | 96.0198 |
| $C_3H_4N_4$ | 4.87 | 0.10 | 96.0437 |
| $C_4O_3$ | 4.56 | 0.67 | 95.9847 |
| $C_4H_2NO_2$ | 4.92 | 0.49 | 96.0085 |
| $C_4H_4N_2O$ | 5.28 | 0.31 | 96.0324 |
| $C_4H_6N_3$ | 5.64 | 0.13 | 96.0563 |
| $C_5H_4O_2$ | 5.69 | 0.53 | 96.0211 |
| $C_5H_6NO$ | 6.05 | 0.35 | 96.0449 |
| $C_5H_8N_2$ | 6.41 | 0.17 | 96.0688 |
| $C_6H_8O$ | 6.82 | 0.39 | 96.0575 |
| $C_6H_{10}N$ | 7.18 | 0.22 | 96.0814 |
| $C_7H_{12}$ | 7.95 | 0.27 | 96.0939 |
| $C_8$ | 8.88 | 0.34 | 96.0000 |
| 97 | | | |
| $C_2HN_4O$ | 3.76 | 0.26 | 97.0151 |
| $C_3HN_2O_2$ | 4.17 | 0.47 | 97.0038 |
| $C_3H_3N_3O$ | 4.53 | 0.28 | 97.0277 |
| $C_3H_5N_4$ | 4.89 | 0.10 | 97.0515 |
| $C_4HO_4$ | 4.58 | 0.68 | 96.9925 |
| $C_4H_3NO_2$ | 4.94 | 0.49 | 97.0164 |
| $C_4H_5N_2O$ | 5.30 | 0.31 | 97.0402 |
| $C_4H_7N_3$ | 5.66 | 0.13 | 97.0641 |
| $C_5H_5O_2$ | 5.71 | 0.53 | 97.0289 |
| $C_5H_7NO$ | 6.07 | 0.35 | 97.0528 |
| $C_5H_9N_2$ | 6.43 | 0.17 | 97.0767 |
| $C_6H_9O$ | 6.84 | 0.39 | 97.0653 |
| $C_6H_{11}N$ | 7.20 | 0.22 | 97.0892 |
| $C_7H_{13}$ | 7.97 | 0.27 | 97.1018 |
| $C_8H$ | 8.90 | 0.34 | 97.0078 |
| 98 | | | |
| $C_2N_3O_2$ | 3.41 | 0.44 | 97.9991 |
| $C_2H_2N_4O$ | 3.77 | 0.26 | 98.0229 |
| $C_3NO_3$ | 3.82 | 0.65 | 97.9878 |
| $C_3H_2N_2O_2$ | 4.18 | 0.47 | 98.0116 |
| $C_3H_4N_3O$ | 4.54 | 0.28 | 98.0355 |
| $C_3H_6N_4$ | 4.90 | 0.10 | 98.0594 |

| | $M+1$ | $M+2$ | Mass |
|---|---|---|---|
| $C_4H_2O_3$ | 4.59 | 0.68 | 98.0003 |
| $C_4H_4NO_2$ | 4.95 | 0.49 | 98.0242 |
| $C_4H_6N_2O$ | 5.31 | 0.31 | 98.0480 |
| $C_4H_8N_3$ | 5.67 | 0.13 | 98.0719 |
| $C_5H_6O_2$ | 5.72 | 0.53 | 98.0368 |
| $C_5H_8NO$ | 6.08 | 0.35 | 98.0606 |
| $C_5H_{10}N_2$ | 6.44 | 0.17 | 98.0645 |
| $C_6H_{10}O$ | 6.85 | 0.39 | 98.0732 |
| $C_6H_{12}N$ | 7.21 | 0.21 | 98.0970 |
| $C_7H_{14}$ | 7.98 | 0.26 | 98.1096 |
| $C_7N$ | 8.14 | 0.27 | 98.0031 |
| $C_8H_2$ | 8.91 | 0.33 | 98.0157 |
| 99 | | | |
| $C_2HN_3O_2$ | 3.43 | 0.44 | 99.0069 |
| $C_2H_3N_4O$ | 3.79 | 0.25 | 99.0308 |
| $C_3HNO_3$ | 3.84 | 0.65 | 98.9956 |
| $C_3H_3N_2O_2$ | 4.20 | 0.47 | 99.0195 |
| $C_3H_5N_3O$ | 4.56 | 0.28 | 99.0433 |
| $C_3H_7N_4$ | 4.92 | 0.10 | 99.0672 |
| $C_4H_3O_3$ | 4.61 | 0.68 | 99.0082 |
| $C_4H_5NO_2$ | 4.97 | 0.49 | 99.0320 |
| $C_4H_7N_2O$ | 5.33 | 0.31 | 99.0559 |
| $C_4H_9N_3$ | 5.69 | 0.13 | 99.0798 |
| $C_5H_7O_2$ | 5.74 | 0.53 | 99.0446 |
| $C_5H_9NO$ | 6.11 | 0.35 | 99.0684 |
| $C_5H_{11}N_2$ | 6.46 | 0.17 | 99.0923 |
| $C_6H_{11}O$ | 6.86 | 0.39 | 99.0810 |
| $C_6H_{13}N$ | 7.23 | 0.22 | 99.1049 |
| $C_7H_{15}$ | 8.00 | 0.27 | 99.1174 |
| $C_7HN$ | 8.16 | 0.29 | 99.0109 |
| $C_8H_3$ | 8.93 | 0.35 | 99.0235 |
| 100 | | | |
| $CN_4O_2$ | 2.67 | 0.43 | 100.0022 |
| $C_2N_2O_3$ | 3.08 | 0.63 | 99.9909 |
| $C_2H_2N_3O_2$ | 3.44 | 0.45 | 100.0147 |
| $C_2H_4N_4O$ | 3.80 | 0.26 | 100.0386 |
| $C_3O_4$ | 3.45 | 0.84 | 99.9796 |
| $C_3H_2NO_3$ | 3.85 | 0.65 | 100.0034 |
| $C_3H_4N_2O_2$ | 4.21 | 0.47 | 100.0273 |
| $C_3H_6N_3O$ | 4.57 | 0.28 | 100.0511 |
| $C_3H_8N_4$ | 4.94 | 0.10 | 100.0750 |
| $C_4H_4O_3$ | 4.62 | 0.68 | 100.0160 |
| $C_4H_6NO_2$ | 4.98 | 0.49 | 100.0399 |
| $C_4H_8N_2O$ | 5.34 | 0.31 | 100.0637 |
| $C_4H_{10}N_3$ | 5.70 | 0.13 | 100.0876 |
| $C_5H_8O_2$ | 5.76 | 0.53 | 100.0524 |
| $C_5H_{10}NO$ | 6.11 | 0.35 | 100.0763 |
| $C_5H_{12}N_2$ | 6.47 | 0.18 | 100.1001 |
| $C_6H_{12}O$ | 6.88 | 0.39 | 100.0888 |
| $C_6H_{14}N$ | 7.24 | 0.22 | 100.1127 |
| $C_6N_2$ | 7.40 | 0.23 | 100.0062 |
| $C_7H_{16}$ | 8.01 | 0.28 | 100.1253 |
| $C_7O$ | 7.81 | 0.46 | 99.9949 |
| $C_7H_2N$ | 8.17 | 0.29 | 100.0187 |
| $C_8H_4$ | 8.94 | 0.35 | 100.0313 |

[a]Values in %; $M$ = 100%

# Self-Assessment Questions and Responses

**SAQ 1.2a**

Calculate *m/z* values (to the nearest whole numbers) for the following molecular ions:

(a) $C_6H_6^{+\bullet}$

(b) $C_2H_5NH_2^{+\bullet}$

(c) $CH_3OH^{+\bullet}$

(d) $C_6H_5COOH^{+\bullet}$

(e) $C_3H_6^{+\bullet}$

(f) $C_2H_5CN^{+\bullet}$

(g) $C_3H_7SN^{+\bullet}$

(h) $CH_3CHONH_2^{+\bullet}$

(i) $(CH_3O)_3P^{+\bullet}$

(j) $C_6H_5NH_2^{2+}$

**Response**

(a) 78

(b) 45

(c) 32

(d) 122

(e) 44

(f) 55

(g) 76

(h) 60

(i) 124

(j) 93

**SAQ 1.2.b**

Express the mass of an electron in atomic mass units and use this result to calculate the mass difference (in amu) between the following pairs, given that the mass of an electron is $9.110 \times 10^{-31}$g and 1 amu is $1.661 \times 10^{-27}$g. Take the relative molecular masses to the nearest whole numbers.

(a) $C_2H_6$ and $C_2H_6^{+\cdot}$

(b) $C_{60}$ and $C_{60}^{+\cdot}$

**Response**

The mass of an electron is $(9.110 \times 10^{-27})/(1.661 \times 10^{-27}) = 5.49 \times 10^{-4}$.

(a) The relative molecular mass of $C_2H_6$ is 30 (to the nearest whole number, i.e. $(12 \times 2) + (1 \times 6)$).

The mass of $C_2H_6^{+}$ is $30 - (5.49 \times 10^{-4})$, i.e. 29.99(95) amu.

The mass difference is therefore $(1.83 \times 10^{-3})$ %.

(b) The relative molecular mass of $C_{60}$ is 720 (to the nearest whole number, i.e. $12 \times 60$).

The mass of $C_{60}^{+}$ is $720 - (5.49 \times 10^{-4})$, i.e. 719.99(95) amu.

The mass difference is therefore $(7.62 \times 10^{-5})$ %.

**SAQ 1.2c**

> Let us take a simple molecule, e.g. $C_2H_4$. Draw three fragmentation pathways for the molecular ion, each involving only C—H bond cleavage but which lead to the production of a neutral atom, a neutral radical and a neutral molecule, respectively.

**Response**

We can, in fact, write four processes, where one produces an atom, another produces a radical and the others both produce molecules. These can be represented as follows:

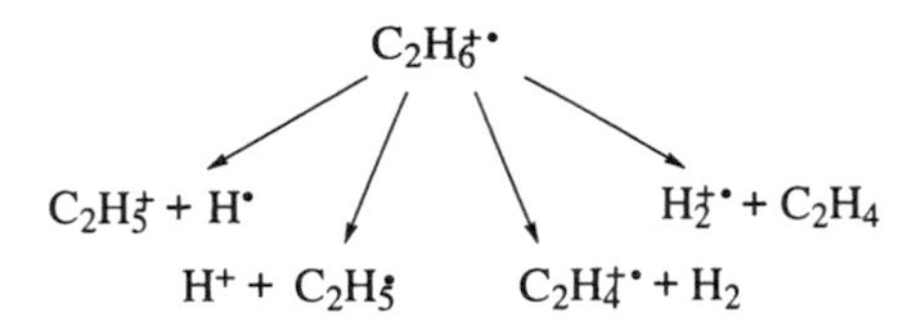

**SAQ 1.2d**

Decide whether ions of the following compounds are odd- or even-electron in nature:

(a) HI

(b) $C_7H_{15}$

(c) $C_6H_5NH_2$

(d) $CH_3CO$

(e) $C_6H_5COOC_2H_5$

(f) $C_6H_5CH_2CN$

(g) $C_6H_4(NH_2)_2$

(h) $C_6H_4$

(i) $C_7H_7$

(j) $(CH_3O)_3P$

**Response**

(a) $HI^{+\bullet}$

(b) $C_7H_{15}^+$

(c) $C_6H_5NH_2^{+\bullet}$

(d) $CH_3CO^+$

(e) $C_6H_5COOC_2H_5^{+\bullet}$

(f) $C_6H_5CH_2CN^{+\bullet}$

(g) $C_6H_4(NH_2)_2^{+\bullet}$

(h) $C_6H_4^{+\bullet}$

(i) $C_7H_7^+$

(j) $(CH_3O)_3P^{+\bullet}$

**SAQ 1.3a**

> Draw up a fragmentation pattern for methanol based on the above assignments and underline the most intense ion in the spectrum.

**Response**

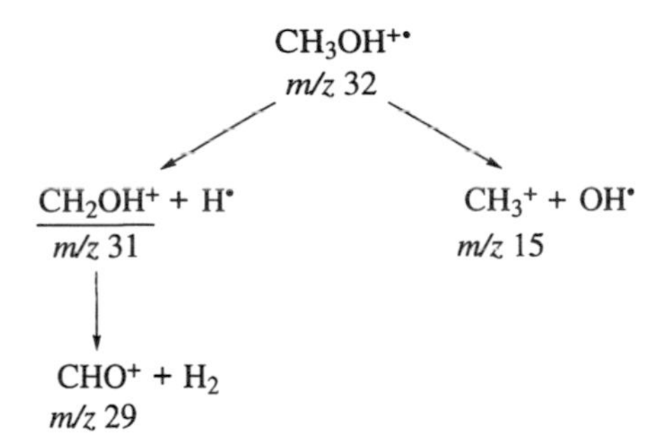


**SAQ 1.3b**

> Which ion(s) could be responsible for the *m/z* values of 33 and 15.5 in the mass spectrum of methanol?

**Response**

The first of these, *m/z* 33, is an isotope peak arising from the small proportion of methanol molecules in the sample which contain either a $^{13}C$ rather than a $^{12}C$ atom, or a single $^{2}H$ and three $^{1}H$ atoms rather than four $^{1}H$ atoms.

The second, *m/z* 15.5, is the doubly charged ion of mass 31.

Notice how small this peak is when compared to *m/z* 31.

**SAQ 1.3c**

> The ion $CH_3OH^{2+}$ has an *m/z* value of 16. Which singly charged ions with this value might you expect to observe in the methanol spectrum?

**Response**

The two possibilities are $CH_4^{+\bullet}$, which is formed by eliminating an oxygen atom from $CH_3OH^{+\bullet}$, and the $O^+$ ion.

**SAQ 1.3d**

> Why is the *m/z* axis non-linear in the scan trace?

**Response**

The *m/z* axis is non-linear in the scan trace because it is based on a magnetic-field scan, which follows the general form:

$$B_1/B_2 = \phi(m_2/m_1)$$

where $B_1$ and $B_2$ are the magnetic field strengths of two selected ions and $m_1$ and $m_2$ are the masses of these ions.

**SAQ 1.3e**

The mass spectrum of ethanol, $C_2H_5OH$, is shown in Figure 1.3b. Use this to carry out the following:

(a) write down the *m/z* value of the molecular ion;

(b) write down the *m/z* value of the base peak;

(c) the base peak is formed from the molecular ion via a single step—write down a pathway for this process;

(d) write down two fragmentation pathways for the production of ions of *m/z* 29 which may have different formulae;

(e) write down fragmentation pathways for the following stepwise decompositions:

$$m/z\ 46 \longrightarrow m/z\ 45 \longrightarrow m/z\ 43$$

(f) assign a formula to *m/z* 18 and give a fragmentation pathway for its direct formation from the molecular ion;

(g) combine the fragmentation pathways mentioned above to give the fragmentation pattern for ethanol.

→

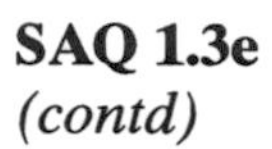

**SAQ 1.3e**
*(contd)*

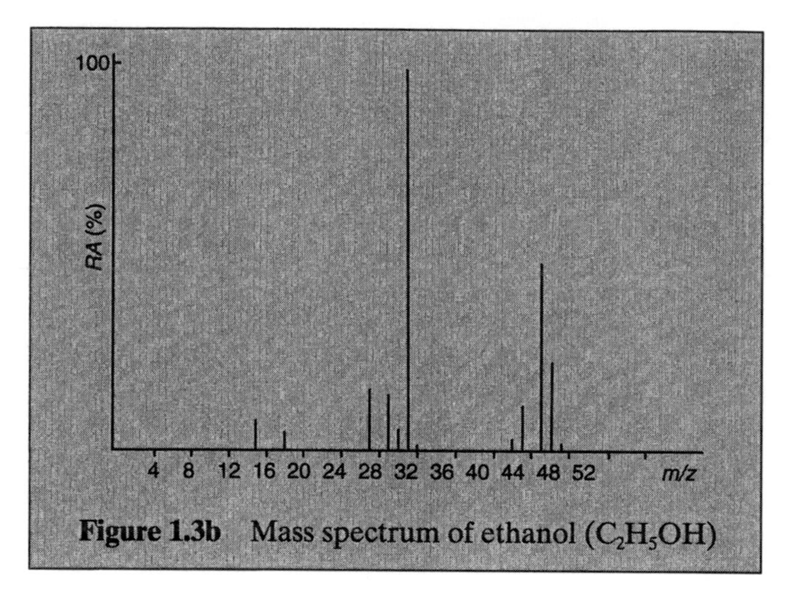


**Figure 1.3b** Mass spectrum of ethanol ($C_2H_5OH$)

**Response**

(a) *m/z* 46 (molecular ion)

(b) *m/z* 31 (base peak)

(c) $CH_3CH_2OH^{+\bullet} \longrightarrow CH_2OH^+ + CH_3^\bullet$

The base peak has the same formula as that in the methanol spectrum. It is formed by C—C bond cleavage in the case of ethanol.

(d) In methanol, *m/z* 29 was formulated as $CHO^+$, which arose from the fragmentation:

$$CH_2OH^+ \longrightarrow CHO^+ + H_2$$

This may also happen with ethanol, but the *m/z* 29 ion may also be formulated as $C_2H_5^+$, which may be formed from the fragmentation:

$$C_2H_5OH^{+\bullet} \longrightarrow C_2H_5^+ + OH^\bullet$$

This second process is analogous to:

$$CH_3OH^{+\bullet} \longrightarrow CH_3^+ + OH^\bullet$$

in methanol.

(e) The steps are the loss of a hydrogen atom from the molecular ion (*m/z* 46) to give *m/z* 45, followed by further loss of a hydrogen molecule from *m/z* 45 to give *m/z* 43.

This can be represented as

$$CH_3CH_2OH^{+\bullet} \longrightarrow CH_3CHOH^+ + H^\bullet$$
$$\downarrow$$
$$CH_3CO^+ + H_2$$

The fragmentations exactly parallel those of methanol.

(f) *m/z* 18 is $H_2O^{+\bullet}$, i.e.

$$C_2H_5OH^{+\bullet} \longrightarrow H_2O^{+\bullet} + C_2H_4$$

(g) The fragmentation pattern is as follows:

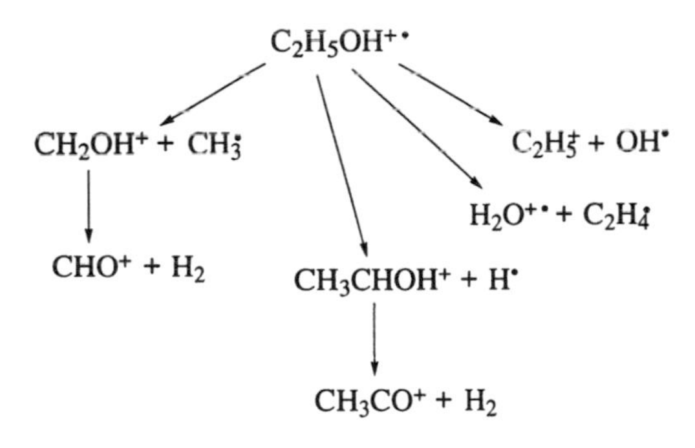

**SAQ 2.1**

Can you think of any negative ions or species that are often encountered in routine analysis and which you might expect to dominate a negative-ion mass spectrum of the molecules of which they are a part?

**Response**

If the molecule contains a nitro group, the spectrum is dominated by the ion $NO_2^-$ and there will be few other ions present which will give any structural information. Similarly, the halogens and OH are relatively electronegative and very prone to electron attachment.

**SAQ 2.2a**

Which of the above species is acting as an acid and which as a base?

**Response**

$CH_4^{+\bullet}$ is acting as a proton donator, i.e. an acid, while $CH_4$ is acting as a proton acceptor, i.e. a base.

**SAQ 2.2b**

Suggest how you might expect $NH_4^+$ and $(CH_3)_3C^+$ to react in ion–molecule reactions.

**Response**

$NH_4^+$ and $(CH_3)_3C^+$ both protonate samples to give $(M+H)^+$ ions. In the latter case, butene is formed:

$$(CH_3)_3C^+ + M \longrightarrow (MH)^+ + CH_2{=}C(CH_3)_2$$

The use of these alternative reactant gases can be important in CI mass spectrometry as they may lead to different amounts of fragmentation of the same quasi-molecular ion.

**SAQ 2.2c**

If you look at Figure 2.2b, you will see a major fragment ion at *m/z* 70. This represents the loss of 45 mass units from $M^{+\bullet}$ in the EI mass spectrum and 46 mass units from $(MH)^+$ in the CI spectrum. Suggest a formula for the neutral species produced and two possible structures for the quasi-molecular ion.

**Response**

45 mass units corresponds to —COOH, and 46 mass units corresponds to $C(OH)_2$ or an isomer of it. The quasi-molecular ion could be either structure A or B (see below). In drawing these structures, we have indicated the masses of both the fragment ion and the neutral species. As the formation of the quasi-molecular ion is an acid–base reaction, it is reasonable to write the product as involving electron-pair donation by a lone pair on the hetero-atom.

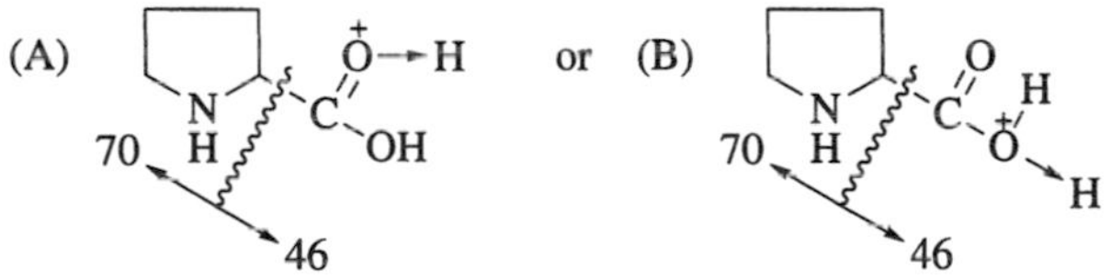

(A) 70 46 or (B) 70 46

**SAQ 2.2d**

Suggest mechanisms for loss of water from the quasi-molecular ions you have drawn in SAQ 2.2c.

**Response**

The small fragment-ion peak at *m/z* 98 corresponds to loss of $H_2O$ from the quasi-molecular ion. This can easily be explained in terms of structure B:

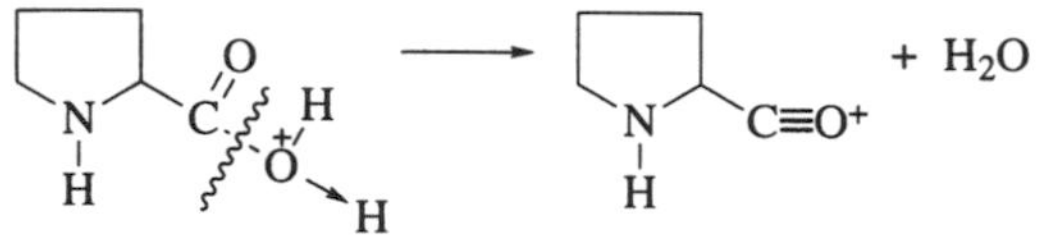

This does not necessarily exclude structure A, as a hydrogen transfer may occur, but this reaction is, however, less likely:

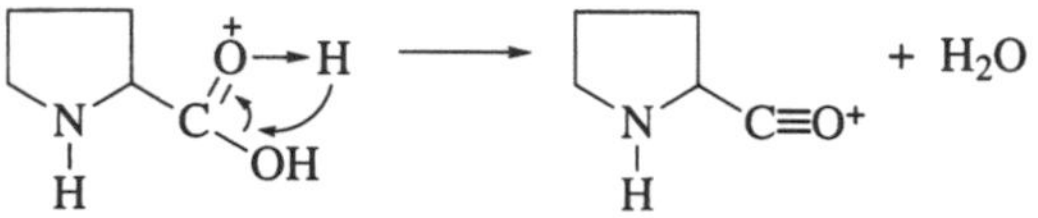

Fragmentations involving loss of *m/z* 46 ($CO_2H_2$) and $H_2O$ do seem to suggest that protonation does not occur at the nitrogen atom with an available lone pair of electrons.

**SAQ 2.2e**

Explain why the fragmentation of methy phenyl ketone, $C_6H_5COCH_3$, with isobutane reactant gas gives no fragmentation products, but with methane gives fragments, including *m/z* 43, 77, and 105.

**Response**

This happens because the amount of energy transferred to a sample molecule on protonation depends on both the proton affinity of the sample and the acidity (the willingness to release a proton) of the reactant ion.

Acidity increases in the order $NH_4^+ < C_4H_9^+ < CH_5^+$, and thus the internal energies of the quasi-molecular ion formed by these reactant ions follow the same order. As the amount of fragmentation depends on the internal energy of the ion, this helps to explain the observations made for methyl phenyl ketone.

**SAQ 2.2f**

List the advantages and disadvantages of using EI and CI as ion sources in mass spectrometry.

**Response**

The problems outlined for EI mass spectrometry were that:

(*i*) in some cases molecular ions were very weak or not observed because of fragmentation;

(*ii*) isomers could not be distinguished;

(*iii*) thermally unstable molecules do not give useful mass spectra;

(*iv*) mass spectra cannot be obtained for involatile molecules

In a large number of cases CI mass spectrometry can help with problem (*a*), in that quasi-molecular ions usually have a high relative intensity. There is a problem with some molecules in knowing what type of quasi-molecular ion they are going to form:

$(MH)^+$, $(M - H)^+$ or $(M + C_2H_5)^+$ etc and this may introduce some uncertainty in interpreting the results.

Because fragmentation is reduced, the structural information given by the fragments is reduced or completely lost. However, for some molecules fragmentation can reflect fine structural differences allowing differentiation between isomers.

Chemical ionisation still requires volatilisation of the sample since the ion-molecule reactions on which it depends are gas phase reactions. Thus, if the sample is thermally unstable or involatile, CI is not the answer.

**SAQ 2.3a**

List the advantages and disadvantages of FI over EI or CI ion sources.

**Response**

As with CI, because FI is a soft-ionisation technique, $M^{+\bullet}$ or $(MH)^+$ ions of considerable relative abundance are usually formed. Therefore, the method is useful for relative-molecular-mass determinations. Its value is, however, lowered somewhat by the uncertainty about whether $M^{+\bullet}$ or $(MH)^+$ ions will be formed. Unlike CI, no higher adduct ions such as $(M + C_2H_5)^+$ are observed. The amount of fragmentation is less than for EI mass spectrometry, but some structurally informative ions may be formed.

The sample requires vaporisation prior to ionisation and the method is, therefore, not suitable for thermally unstable or involatile samples.

**SAQ 2.3b**

List the advantages and disadvantages of FD over the other techniques mentioned.

**Response**

The advantages are:

(a) there is no need to vaporise the sample;

(b) intense $M^{+\bullet}$ or $(MH)^+$ ions aid relative-molecular-mass determination.

The disadvantages are:

(a) no structural information is produced, as very little, if any, fragmentation occurs;

(b) the compound must be soluble in a suitable volatile solvent in order to be applied to the emitter;

(c) preparation of the emitter is a specialised technique;

(d) ion beams may not be persistent, and therefore may not give sufficient time for tuning the spectrometer.

**SAQ 2.4a**

What problems could we have when introducing a matrix material into the ion source?

**Response**

Problems can arise in that the matrix also forms ions on bombardment, in addition to those formed by the sample. This obviously complicates the spectrum, although allowances, however, can be made for this.

**SAQ 2.4b**

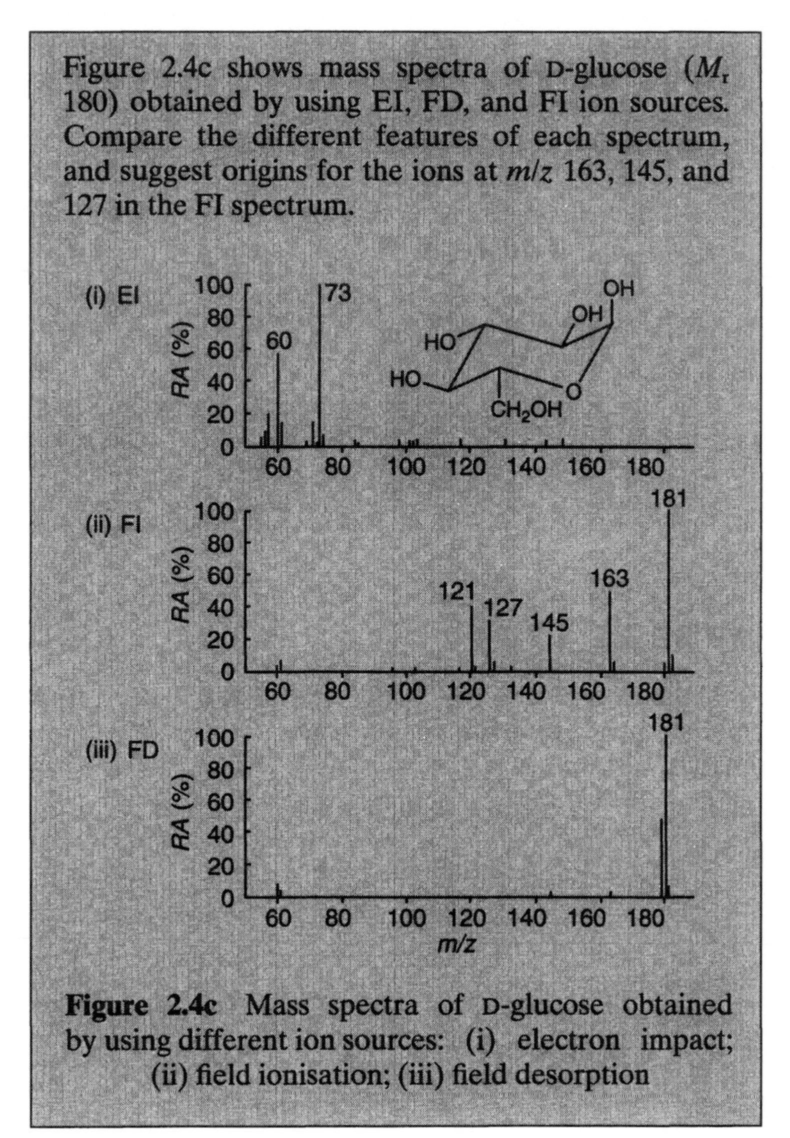

Figure 2.4c shows mass spectra of D-glucose ($M_r$ 180) obtained by using EI, FD, and FI ion sources. Compare the different features of each spectrum, and suggest origins for the ions at *m/z* 163, 145, and 127 in the FI spectrum.



**Figure 2.4c** Mass spectra of D-glucose obtained by using different ion sources: (i) electron impact; (ii) field ionisation; (iii) field desorption

**Response**

The EI spectrum shows no molecular ion and no fragment ions above *m/z* 73. It would be almost impossible to identify the molecule from this spectrum. Glucose has a relative molecular mass of 180 and thus both the FI and FD spectra give intense $(MH)^+$ ions. In the FD spectrum, $M^{+\bullet}$ ions are also formed. In the FI spectrum, there are a number of fragment ions, but none are observed in the FD spectrum which have a relative abundance of greater than 10%.

The fragment ions at *m/z* 163, 145, and 127 represent successive losses of $H_2O$ from $(MH)^+$,

**SAQ 2.8a**

When handling hot samples, why is it preferable to construct the whole inlet system out of glass rather than metal? (An 'all-glass heated inlet system' is often called an *AGHIS* system.)

**Response**

The problem with passing hot organic vapours over metal components is that, in some cases, the sample may suffer a metal-catalysed decomposition, which must obviously be avoided.

**SAQ 2.8b**

Which inlet system best meets the requirements needed for analysis of the following:

(a) a mixture of insect sex pheromones which can be separated by liquid chromatography;

(b) a relatively non-volatile organometallic compound or polymeric solid;

(c) a high-vapour-pressure gas?

**Response**

(a) septum inlet;

(b) direct-insertion probe;

(c) cold inlet.

**SAQ 3.1a**

Which gaseous ion produced from the atmosphere might you expect to be observed in a mass spectrometer operating at a pressure that was too high?

**Response**

You might expect to find $N_2$ (*m/z* 28), $O_2$ (*m/z* 32), and $CO_2$ (*m/z* 44).

**SAQ 3.1b**

Sometimes, EI mass spectrometers are used to produce negative ions. What alterations to the polarity of the ion-source components would have to be made in order for negative-ion mass spectrometry to take place?

**Response**

None of the actual components would have to be changed, but the polarity of the electrical connection between the ion repeller and the first accelerating plate would have to be reversed (repeller held at a negative potential with respect to the first plate). This would allow the negatively charged repeller to cause the negatively charged ions to move out of the source.

**SAQ 3.1c**

Figure 3.1c shows the trajectories of ions of two different masses, but with similar charges, that do not follow the correct flight path to reach the detector. One of these follows a flight path (i) that has a radius larger than that required, while the other takes a flight path (ii) with a radius smaller than that required. Which of the two ions is the heavier?

**Response**

Ions of type (i) are heavier.

Recall equation (3.11):

$$m/z = B^2r^2/2V \qquad (3.11)$$

If $B$, $V$, and $z$ are constant, then ions following a path of larger radius must be heavier, and those following a path of smaller radius must be lighter than the ions reaching the detector.

Another way of looking at this is to consider the centrifugal force associated with the ions, which is given by $mv^2/r$.

The heavier the ion, then the larger the centrifugal force it develops and the more likely it is to follow a flight path tangential to the circle, unless the magnitude of the magnetic force is increased. This is the case for ions of type (i).

For lighter ions, the magnetic force is greater than the centrifugal force and thus these ions are directed in towards the centre of the circle, as shown by ions of type (ii).

**SAQ 3.1d**

What accelerating potential ($V$) will be required to direct a singly charged methyl ion through the exit slit of a mass spectrometer if the magnet has a field strength of 0.250 T and a radius of curvature of the ion through the magnetic field of 10.5 cm?

**Response**

If we use the following equation:

$$m/z = B^2r^2/2V \tag{3.11}$$

and rearrange, we obtain:

$$V = B^2r^2z/2m$$

Here, $m = 15 \times 1.660566 \times 10^{-27}\ \text{g} = 3.736274 \times 10^{-29}\ \text{kg}$

$B = 0.250$ T

$r = 0.150$ m

$z = 1.60219 \times 10^{-19}$ C

Substituting the above values, gives $V = 2.22$ kV

**SAQ 3.1e**

As a revision exercise, can you list the main components of a mass spectrometer, labelling them in the order in which the ion beam would see them?

**Response**

The main components are ion source, mass analyser, detector and recorder.

**SAQ 3.2a**

Let us suppose that we are using an instrument with a resolving power of 4000, but find that the mass region we are interested in is 400. How accurately could we measure such masses?

**Response**

Resolving power = $m/\Delta m$, so:

$$4000 = 400/\Delta m$$

and therefore:

$$\Delta m = 0.1$$

Hence, with an instrument of resolving power 4000, we can measure masses of around 400 to an accuracy of 0.1 amu. Similarly, we can measure masses of around 40 to an accuracy of 0.01 amu.

**SAQ 3.2b**

> The nominal mass of the ethanol molecular ion, $CH_3OH^{+\bullet}$, is 32. Calculate its accurate mass, assuming, as before, that we are dealing with the predominant isotopes for each, i.e. $^{12}C$, $^{1}H$, and $^{16}O$ ($^{16}O$, $A_r$ = 15.994 92).

**Response**

The answer is 32.02624 amu.

**SAQ 3.2c**

> The molecular ion of a colourless gas is examined by mass spectrometry and gives a nominal mass of 32, which is similar to the nominal mass of methanol. However, its accurately measured mass is found to be 31.9898 ± 0.0001. Use the above list of accurately measured $A_r$ values to identify the molecule.

**Response**

The relative atomic mass ($A_r$) values of the four most common elements found in organic chemistry are:

$^{12}C$, 12.000 00;

$^{1}H$, 1.007 83;

$^{16}O$, 15.994 92;

$^{14}N$, 14.003 07.

The molecular ion of mass 31.989 8 must contain oxygen as this is the only element in the above list which is mass-deficient. In fact, 31.989 8 is the accurate mass of $O_2^{+\bullet}$.

**SAQ 3.2d**

> If we had a mixture of methanol and oxygen in the mass spectrometer, what resolving power would be needed to distinguish between the molecular ions?

**Response**

The mass difference between $CH_3OH^{+\bullet}$ and $O_2^{+\bullet}$ is 32.026 24 − 31.989 84 = 0.036 40. The resolving power is therefore:

$$32/0.036\,4 = 879.12$$

Thus, it should be a relatively simple task to differentiate between these two ions by using a magnetic analyser of the type discussed later in this chapter (see Sections 3.3 and 3.4).

**SAQ 3.2e**

> Calculate the resolving power needed to distinguish between the following pairs of ions. If the resolution of the magnetic analyser was 5000, which of these pairs could be differentiated?
>
> (a) $C_7H_{14}^{+\bullet}$ and $C_6H_{10}O^{+\bullet}$
>
> (b) $C_{23}H_{46}^{+\bullet}$ and $C_{22}H_{42}O^{+\bullet}$
>
> (c) $C_{39}H_{78}^{+\bullet}$ and $C_{38}H_{74}O^{+\bullet}$

**Response**

(a) The mass of $C_8H_{16}^{+\bullet}$ is 112.12528 and that of $C_7H_{12}O^{+\bullet}$ is 112.08888. The resolving power required is therefore:

$$112/0.03640 = 3077$$

(b) The mass of $C_{24}H_{50}^{+\bullet}$ is 338.39150 and that of $C_{23}H_{46}O^{+\bullet}$ is 338.35510. The resolving power required is therefore:

$$338/0.03640 = 9286$$

(c) The mass of $C_{40}H_{82}^{+\bullet}$ is 562.64206 and that of $C_{39}H_{78}O^{+\bullet}$ is 562.60566. The resolving power required is therefore:

$$562/0.03640 = 15440$$

Therefore, only pair (a) could be easily differentiated with this magnetic analyser.

**SAQ 3.3**

In Figure 3.3e, which of the flight paths, a, b, or c, will the ions of highest kinetic energy follow?

**Response**

The ion following flight path (a) has the highest kinetic energy. This ion has the highest centrifugal force, which will override the electric force and therefore the ion follows the circular path with the largest radius.

**SAQ 3.4**

What is the frequency of a normal household power supply?

**Response**

In the UK, this is 50 Hz.

**SAQ 3.5a**

If a spectrometer is scanned at a rate of 10 s $decade^{-1}$, how long would it take to scan from *m/z* 1 to 10 000?

**Response**

It would take 30 s, i.e. 10 s each for *m/z* 1 to 10, *m/z* 10 to 100, and *m/z* 100 to 1000. This appears to be a most peculiar way of measuring scan time, but it originates from magnetic-scanning instruments where the magnet follows an exponential scan law.

**SAQ 3.5b**

Suppose that you are the head of the experimental division at NASA and you have been asked to recommend the design of a mass spectrometer to be included in an unmanned space probe to Mars (take the 1997 Pathfinder mission, for example). What type of ion source and analyser would you suggest for the instrument, bearing in mind the following important points?

(a) The spectrometer will be analysing relatively simple organic molecules of low $M_r$ values ($\leqslant 150$), although absolute identification is required.

(b) The instrument must be small and not too heavy to fit into the probe, but it must also be robust in order to withstand take-off and landing operations and it needs to be reliable.

(c) It also needs to be computer-controlled, i.e. it is an unmanned flight.

**Response**

Most organic molecules with low $M_r$ values are volatile, so as long as you can get the sample into the instrument, one should not need an ion source designed for involatile materials. In order to obtain positive identification of compounds by mass spectrometry, one ideally requires a high-resolution measurement of the mass of the molecular ion to yield a molecular formula and fragmentation data to confirm the structure. All of this suggests an electron-impact ion source would be best. Possibly, a chemical-ionisation ion source would be more suitable for some molecules, but the increased complexity of this type of source may rule it out for this particular application.

High-resolution mass measurements require a double-focusing magnetic-sector instrument, but these are often large and complex.

All of the other requirements of size, weight, robustness, reliability and the need for computer control are available in the quadrupole analyser, but this is of limited resolution. Here one must weigh up the positive and negative features of a double-focussing magnetic-sector instrument against one with a quadrupole analyser.

**SAQ 3.7a**

How will the time resolution be changed by increasing the drift-tube length or decreasing the accelerating potential?

**Response**

The time resolution will increase with increased drift-tube length and will decrease with an increased accelerating potential. It can be shown that the following approximate relationship applies:

$$m/\Delta m = t/(2\Delta t)$$

**SAQ 3.7b**

What can we say about the speed of the data-acquisition system that is required?

**Response**

Needless to say, it must be fast.

**SAQ 3.8a**

What does the magnitude of this AC current depend upon?

**Response**

The number of ions produced in the cell.

**SAQ 3.8b**

What causes the decay of the image current?

**Response**

The ions revert back to their original (inner) circular path defined only by the magnetic field. This does not induce electron flow in either the upper or lower plate.

**SAQ 3.8c**

By use of which equation can we convert the frequency spectrum into a mass spectrum?

**Response**

Frequency and mass are related via equation 3.19

$$\omega_c = {}^{v}/_{r} = Bz/m$$

**SAQ 4.1a**

Where else in spectroscopy might you have seen a photomultiplier tube in operation?

**Response**

Gamma spectroscopy in radiochemistry, or elemental analysis by inductively coupled plasma mass spectrometry (ICP–MS) (see Chapter 9).

**SAQ 4.1b**

Why is the response time important in ion-beam measurements?

**Response**

A fast response time is required because mass spectra are usually scanned at relatively rapid rates.

**SAQ 4.1c**

Where else might you have seen scintillation counting used in analysis?

**Response**

Liquid scintillation counting in radiochemistry is one example.

**SAQ 4.2a**

Why do you think a photographic chart recorder is used in preference to a pen recorder? Remember that a typical spectrometer scan speed, when using a chart recorder, is 10 s $decade^{-1}$.

**Response**

A photographic chart recorder has a much faster response than a pen recorder. In the latter there is a considerable time lag between the pen reaching the maximum of a peak and its return to the baseline to begin recording another.

**SAQ 4.2b**

Can you remember why we often need more than one trace?

**Response**

This is so that we can have traces recorded with differing sensitivities, thus enabling both weak and strong peaks to be recorded.

**SAQ 4.2c**

Can you think of any other ways of representing a mass scan?

**Response**

The data could be presented in tabulated from ($m/z$ values versus relative abundances), or in the form of bar diagrams.

**SAQ 4.2d**

For what general class(es) of measurement would the time-based, signal-average technique be useful?

**Response**

This technique is generally useful for acquiring weak or transient mass spectra associated with rapid scanning across eluting gas chromatography peaks.

**SAQ 4.2e**

What can we say about the time value obtained and the true time value from peak (vi) in Figure 4.2h?

**Response**

It can be easily seen that the computer has found an 'average' centroid, based on incomplete resolution of the doublet at low sampling rates. Therefore, for the major peak, the measured time value will be too high, while for the minor peak it will be too low.

**SAQ 4.2f**

List some of the advantages of normalising peaks on the base peak.

**Response**

You can gain some quantitative idea of the abundance of fragment ions, and therefore on their stability. You can also compare homologues easily and directly. Library reference data often index values in this way.

**SAQ 4.2g**

Draw a bar diagram corresponding to the data recorded in Table 4.2a.

**Response**

The bar diagram that you produce (using the data for benzoic acid) should look like the following example (in this particular case, methanol):

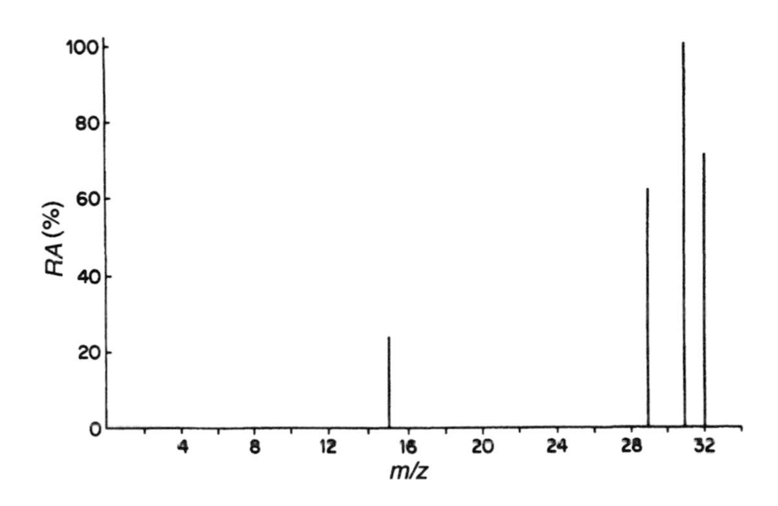


**SAQ 4.2h**

> What are the two most important pieces of information that the computer obtains from an initial scan of an unknown sample?

**Response**

For each peak, the computer obtains both a time value for the centroid of that peak and the peak area. (It converts the time value to a mass value with reference to the spectrum of a known compound.)

**SAQ 4.2i**

> What factors do you think determine the sensitivity of measurement for a particular ion in a mass spectrometer?

**Response**

The operating conditions of the ion source, ionisation probability of the ion and its stability, gain and response of the detector, and overall transmission efficiency.

**SAQ 5.2a**

> You have obtained a soil sample from a contaminated land site previously used by local industry. Name some elements that you *would not* expect to be present in the sample, by considering their abundances in nature.

**Response**

Technetium would not be present as it is an artificially made element. In addition, those elements heavier than uranium (the transuranics) would not be present, unless the industry was involved in nuclear reprocessing, and then you would not expect to see them at measurable levels. It perhaps goes without saying that volatile elements, i.e. gases, would be present in only minute concentrations in the 'soil air'.

**SAQ 5.2b**

> Can you think of some examples of monoisotopic elements? (If you have access to a 'chart of the nuclides' or a suitable reference book, then this will help you out, but don't worry if you can't think of many.)

**Response**

If you thought of one or more of the elements in the table below, well done. If not, don't worry too much, as a knowledge of the distribution of isotopes is rarely required outside mass spectrometry.

| Type | Element | $A_r$ |
|---|---|---|
| Metal | Beryllium | 9 |
| | Sodium | 23 |
| | Aluminium | 27 |
| | Scandium | 45 |
| | Cobalt | 59 |
| | Niobium | 93 |
| | Rhodium | 103 |
| | Caesium | 133 |
| | Gold | 197 |
| | Bismuth | 209 |
| Non-metal | Fluorine | 19 |
| | Phosphorus | 31 |
| | Arsenic | 75 |
| | Iodine | 127 |

**SAQ 5.2c**

Using the data given in Table 5.2b, calculate and plot the normalised spectrum (with reference to the base peak) for the $Pb^{+\bullet}$ ion cluster.

**Response**

Step 1:

For the most intense isotope (*m/z* 208) determine the factor by which its relative abundance would be converted to 100%, i.e. 100/51.7 = 1.93.

Step 2:

Now all of the other relative abundance values for *m/z* 204, 206, and

207 can be converted to the appropriate value on the same 0 to 100 scale by multiplying by 1.93.

For plotting purposes, it is sufficiently accurate to round all figures off to the nearest whole percentage value. You should now have obtained the following data:

| Atomic number | *m/z* | Relative abundance | % |
|---|---|---|---|
| 82 | 204 | 1.4 | 3 |
| | 206 | 25.2 | 49 |
| | 207 | 21.7 | 42 |
| | 208 | 51.7 | 100 |

and plotted a graph that looks like this:

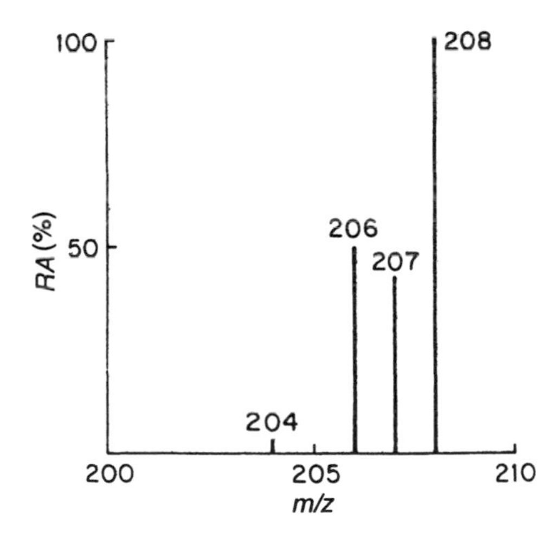

**SAQ 5.2d**

A normalised spectrum of a molecular-ion cluster is shown in Figure 5.2c. Calculate the isotopic abundance of the element concerned. Given that the species has the formula $(CH_3)_2X$, what is this element? Use mass tables is you are unsure.

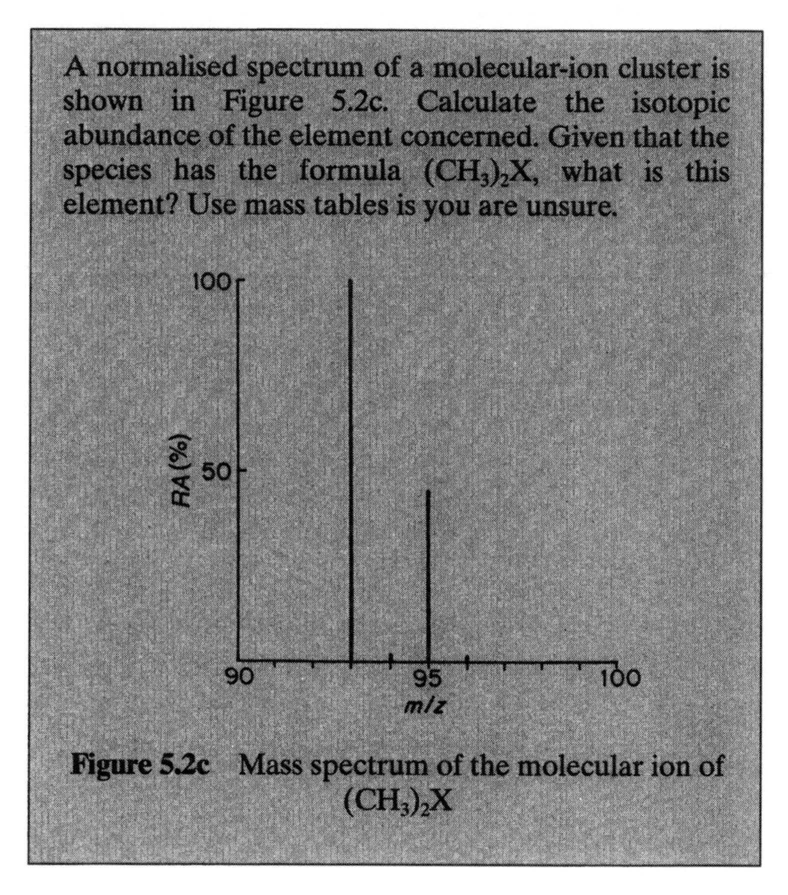


**Figure 5.2c** Mass spectrum of the molecular ion of $(CH_3)_2X$

## Response

The base peak here is *m/z* 93, and *m/z* 95 is 44.5% (or thereabouts, as near as the graph can be read). Therefore, of the total positive-ion intensity of 144.5%, the most abundant isotope is (100/144.5) or 69.2%, while the less abundant species is (44.5/144.5) or 30.8%. Since the mass of two $CH_3$ groups is 30 amu, the atomic weights of the isotopes must be 63 (i.e. 93–30) and 65 (i.e. 95–30), respectively. Reference to mass and abundance tables of the isotopes of the elements shows that the element in this case is *copper* ($^{63}Cu$, 69.1%; $^{65}Cu$, 30.9%). You probably guessed this from your residual memory of common atomic weights, even if you didn't know that copper had these two isotopes.

**SAQ 5.2e**

(a) Calculate the abundances of the (M+1) peaks of CO, $N_2$ and$CH_2{=}CH_2$ ($M^{+\bullet}$ = 28).

(b) Calculate the abundance of the (M+1) peak of naphthalene, $C_{10}H_8$ ($M^{+\bullet}$ = 128). Would you expect to be able to distinguish naphthalene from cyclohexanecarboxylic acid, $C_7H_{12}O_2$ (also $M_r$ 128), from the height of its (M+1) peak alone?

(c) Calculate the abundances of the (M+1) and (M+2) peaks in the mass spectrum of cholesterol, $C_{27}H_{45}OH$.

(d) Calculate the abundance of the (M+2) peak in the mass spectrum of trimethyl phosphate $(CH_3O)_3PO$.

(e) At what carbon number (approximatcly), would you expect the (M+2) peak of an organic compound, containing C, H, N, F, P, and/or I, and four oxygen and/atoms, to appear above a general background level of 2% (relative to $M^{+\bullet}$ = 100%)?

**Response**

(a) For this question, you need to use equation (5.2) and the data in Table 5.2c:

$$(M + 1)/M = (1.1 \times 1)\% = 1.1\%, \text{ for CO}$$

which has one carbon atom and no hydrogens or nitrogen. The abundance of the (M + 1) isotope of O, $^{17}O$, is 0.04% (quite negligible). For $N_2$, there is no carbon or hydrogen, so only the third term in equation (5.2) matters, i.e.

$$(M + 1)/M = (0.36 \times 2)\% = 0.72\%.$$

For ethene ($CH_2{=}CH_2$), there is no nitrogen, so we need only calculate the first two terms, i.e.

$$(M + 1)/M = (1.1 \times 2)\% + (0.016 \times 4)\% = 2.68\%.$$

These calculations illustrate an interesting point. If you could measure the height of the (M + 1) peak (*m/z* 29) accurately, you could distinguish the isomeric compounds CO, $N_2$ and $CH_2{=}CH_2$ by the relative abundances of their (M + 1) peaks alone. In practice, the background in the mass spectrometer would have to be very low in order to achieve this, say 10% of the intensity of the smallest *m/z*, namely 29 (that of $N_2$), i.e. 0.07% relative to the $N_2$ molecular ion at *m/z* 28.

(b) Taking naphthalene first, for $C_{10}H_8$ equation (5.2) gives:

$$(M + 1)/M = (1.1 \times 10)\% + (0.016 \times 8)\% = (11.0 + 0.13)\% = 11.13\%.$$

Notice that the bulk of the (M + 1) peak consists of $^{12}C_9{}^{13}C^1H_8$, with $^{12}C_{10}{}^1H_7{}^2H$ making a very minor contribution (0.13%). In practice, bearing in mind the inherent inaccuracy in determining relative abundances in ordinary mass spectrometers, you can neglect the deuterium in organic compounds until the number of hydrogen atoms reaches 20 or so. At this point, (M + 1) for $^2H$ is $(0.016 \times 20)\% = 0.32\%$.

Now considering cyclohexanecarboxylic acid, $C_7H_{12}O_2$:

$$(M + 1)/M = (1.1 \times 7)\% + (0.016 \times 12)\% = (7.7 + 0.19)\% = 7.9\%.$$

You can now see that the difference in the relative abundances of the (M + 1) peaks for these two isomeric compounds is 3.24%, so it should be possible to distinguish them (but bear in mind the point made about background levels in (a)). You can also see from this example that molecules containing a large proportion of carbon, i.e. hydrocarbons in general, or molecules having large hydrocarbon residues within them, *have relatively intense (M + 1) peaks.*

(c) For this answer, both equations (5.2) and (5.3) are used:

$$(M + 1)/M = (1.1 \times 27)\% + (0.016 \times 46)\%$$

$$= (29.7 + 0.72)\%$$

$$= 30.42\%.$$

$$(M + 2)/M = (1.1 \times 27)^2/200\% + (0.2 \times 1)\%$$

$$= (4.4 + 0.2)\ \%$$

$$= 4.6\%.$$

The (M + 2) peak consists of contributions from $^{12}C_{25}{}^{13}C_2{}^1H_{45}OH$ and $^{12}C_{27}{}^1H_{45}{}^{18}OH$. Note that these larger (C, H and O)-containing organic molecules have substantial (M + 1) peaks which are mainly due to $^{13}C$, and that their (M + 2) peaks are readily observable due to the finite chance that *two* $^{13}C$ atoms are incorporated in the *same* molecule.

(d) Here we need to check first in Table 5.2c whether P has a significant (M + 2) isotope. It does not — it is one of the few monoisotopic elements. Check also SAQ 5.2b for this. Therefore, we can apply equation (5.3) with no special modifications:

$$(M + 2)/M = (1.1 \times 3)^2/200\ \% + (0.2 \times 4)\%$$

$$= 0.05\% + 0.8\%$$

$$= 0.85\%.$$

In this example, note that the 2 × $^{13}C$ isotope contribution given by the first term is really negligible, but that the $^{18}O$ isotope term starts to become appreciable as the number of O atoms increases to four.

(e) None of the atoms mentioned, except C and O, have isotopes which contribute significantly to (M + 2). The answer to (d) showed that the contribution to this peak from four $^{18}O$ atoms is 0.8%. This problem, then, boils down to finding out how large $n$ will be for the term $(1.1 \times n)^2/200$ to reach (2 − 0.8)% i.e. 1.2%.

When this happens, the (M + 2) peak will exceed 2% and appears above the background. Therefore:

$$(1.1 \times n)^2/200 = 1.2$$

$$1.21n^2=240$$

$$n^2=198$$

$$n=14.08$$

Therefore, a compound would have to contain *14* carbon atoms to satisfy this criterion. A couple of benzene rings, plus a small side-chain, would do the trick.

**SAQ 5.2f**

> What are the $m/z$ values of the molecular ions of the following compounds:
>
> (a) 1-chloronaphthalene, $C_{10}H_7Cl$;
>
> (b) diethyl mercury, $(C_2H_5)_2Hg$;
>
> (c) tetraethyl lead, $(C_2H_5)_4Pb$;
>
> (d) ferrocene (dicyclopentadienyliron), $(C_5H_5)_2Fe$?

**Response**

In all of these calculations, the $A_r$ of the most abundant isotope is used.

(a) $C_{10}H_7Cl$: $M_r = (12 \times 10) + (1 \times 7) + (35 \times 1) = 162$

$M^{+\bullet} = m/z$ 162

(b) $(C_2H_5)_2Hg$: $M_r = (12 \times 4) + (1 \times 10) + (202 \times 1) = 260$

$M^{+\bullet} = m/z$ 260

If you obtained $M^{+\bullet} = m/z$ 258 here, have another look at Table 5.2b where you will see values of 200 (= 23.1%), and 202 (= 29.8%) for Hg; therefore, 202 is the correct $A_r$ to use for calculating the $M_r$ of Hg compounds.

(c) $(C_2H_5)_4$ Pb: $M_r = (12 \times 8) + (1 \times 20) + (208 \times 1) = 324$

$M^{+\bullet} = m/z$ 324

You should not have used $A_r$ values of 206 or 207 here for Pb, because these are not the most abundant Pb isotopes (see Table 5.2b).

(d) $(C_5H_5)_2$ Fe: $M_r = (12 \times 10) + (1 \times 10) + (56 \times 1) = 186$

$M^{+\bullet} = m/z$ 186

The principal isotope of Fe is $^{56}Fe$ (91.52%).

**SAQ 5.3a**

> Use equation (5.5) to calculate the ratios of the isotope peaks expected for a $Br_2$– containing ion. Compare your predictions with Figure 5.3d (ii).

**Response**

The ratios of the $Br_2$ isotope peaks are obtained from $a^2 + 2ab + b_2$, where $a = b = 1$.

In this way, three peaks are predicted, i.e. $(1^2) + (2 \times 1 \times 1) + 1^2$, which becomes 1/2/1. If you examine Figure 5.3d(ii), you will see that M/(M + 2)/(M + 4) is very nearly 1/2/1.

**SAQ 5.3b**

> Use equation (5.6) to calculate the ratios of the isotope peaks expected for a $Br_3$-containing ion. Compare your predictions with Figure 5.3d (iii).

**Response**

The ratios of the $Br_3$ isotope peaks are 1/3/3/1, which are calculated by using equation (5.6):

$$a^3 + 3a^2b + 3ab^2 + b_3, \text{ which becomes:}$$

$$(1^3) + (3 \times 1^2 \times 1) + (3 \times 1 \times 1^2) + 1^3, \text{ or } 1/3/3/1.$$

Inspection of Figure 5.3d(iii) shows that

$$M/(M + 2)/(M + 4)/(M + 6) \text{ is very nearly } 1/3/3/1.$$

**SAQ 5.3c**

Figure 5.3e shows the unnormalised mass spectrum of 4-bromochlorobenzene. Are the molecular ions present in the predicted 3/4/1 ratio? What other ions containing halogens are present in the spectrum?

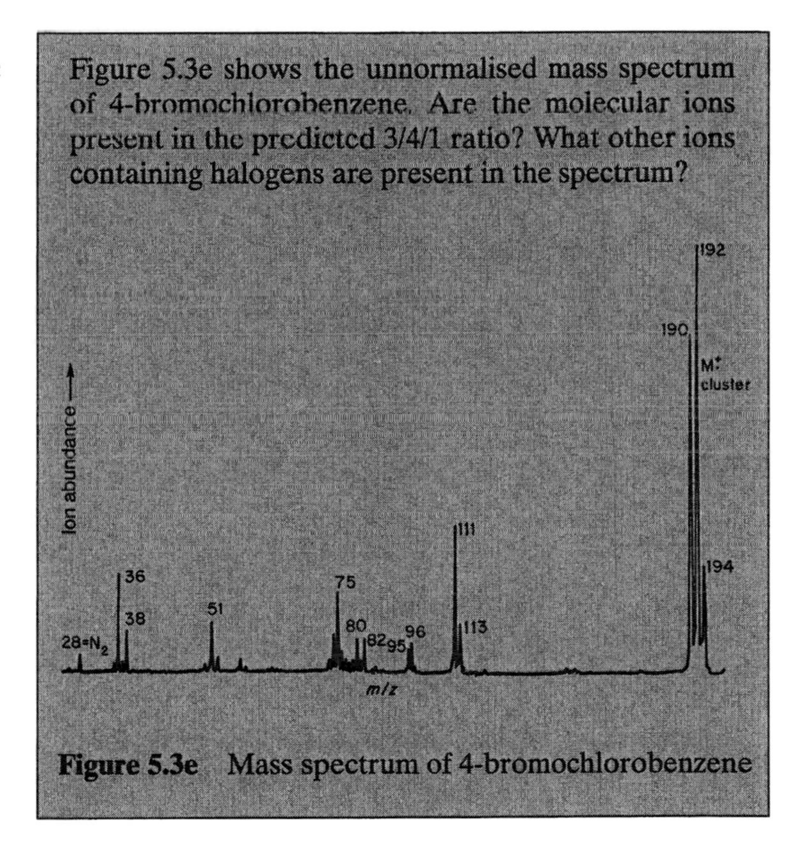


**Figure 5.3e** Mass spectrum of 4-bromochlorobenzene

**Response**

First, draw a baseline across the molecular-ion region under the *m/z*

190–194 peaks, and then measure the peak heights. In mm, the ratios are *m/z* 190/192/194 = 56/71/17, or 3.3/4.2/1. This is not too good an agreement with the expected 3/4/1 ratio, but ions do occur at *m/z* 111 and 113, and 36 and 38, in the 3/1 ratio expected of chlorine, and 80 and 81 in the 1/1 ratio expected of bromine. Now (190–79) is 111, and (192–79) is 113, showing that $M^{+\bullet}$ decomposes by loss of bromine to give an ion containing *one* chlorine. Therefore, the sample is confirmed as bromochlorobenzene. In addition, *m/z* 111/113 is $C_6H_4Cl^+$, *m/z* 80/82 is $HBr^{+\bullet}$, and *m/z* 36/38 is $HCl^{+\bullet}$.

**SAQ 5.3d**

Calculate the relative abundances, using the simplified method, of the major isotope peaks in clusters containing:

(a) $CuBr_2$;

(b) SClBr.

**Response**

(a) $CuB_r2_2$

For copper, $^{63}Cu/^{65}Cu$=69.1/30.9, or 2.24/1 (SAQ 5.2d).

For $Br_2$, $(a + b)_2 = (1 + 1)^2 = 1/2/1$.

Combining these relative abundances gives:

$$
\begin{array}{llrcl}
(2.24/1) & (1/2/1) & (2.24/1) \times 1 & = & 2.24/1 \\
 & & (2.24/1) \times 2 & = & \quad 4.48/2 \\
 & & (2.24/1) \times 1 & = & \qquad\quad 2.24/1 \\
\hline
 & & \text{Total} & = & 2.24/5.48/4.24/1
\end{array}
$$

Therefore, $M^{+\bullet}/(M+2)^{+\bullet}/(M+4)^{+\bullet}/(M+6)^{+\bullet}$ is 2.24/5.48/4.24/1.

(b) SClBr

Combining the three relative abundances, we have (95/4)(3/1)(1/1) or (95/4)(3/4/1) when Cl and Br are combined as in Section 5.3.2 (Table 5.3b). Therefore:

| | | | | | |
|---|---|---|---|---|---|
| (95/4) × 3 = | 285 / | 12 | | | |
| (95/4) × 4 = | | 380 / | 16 | | |
| (95/4) × 1 = | | | 95 / | 4 | |
| Total = | 285 / | 392 / | 111 / | 4 | |

Hence, $M^{+\bullet}/(M+2)^{+\bullet}/(M+4)^{+\bullet}/(M+6)^{+\bullet}$ is 71/98/28/1.

Notice how close the ratio is to the 3/4/1 expected for an ion containing ClBr, if we neglect the $(M+6)^{+\bullet}$ ion, i.e. 2.5/3.5/1.

**SAQ 5.4a**

Reference to Table 5.2c shows that $^{33}S$ has a natural abundance of 0.78% relative to $^{32}S$ (100%).

(a) What is the relative abundance of $(M+1)^{+\bullet}$ in the $S_8^{+\bullet}$ cluster? (Hint—use the nmemonic.)

(b) What is the contribution made by species containing two $^{33}S$ to the $(M+2)^{+\bullet}$ ion?

(c) Would $(M+3)^{+\bullet}$ be observable in the $S_8^{+\bullet}$ cluster?

In attempting parts (b) and (c) you may find the binomial approximation more helpful (see equation (5.9)).

**Response**

(a) The simple formula in equation (5.1) applies here:

$$\% \, (M+1)^{+\bullet} = (0.78 \times 8) = 6.24\%.$$

(b) It is best to evaluate equation (5.9) here to the fourth term, as this will give the answer to (c) also. The ratio of $^{32}S/^{33}S$ is 100/0.78 or 128/1. Therefore, $M^{+\bullet}/(M+1)^{+\bullet}/(M+2)^{+\bullet}/(M+3)^{+\bullet}$ for $^{33}S$ is:

$$(128)^3/(8 \times 128^2 \times 1)/\tfrac{8}{2}(7)(128 \times 1^2)/\tfrac{8}{6}(7)(6)1^3$$

$$= (2.097 \times 10^6)/(1.31 \times 10^5)/3584/56$$

$$\simeq 100/6.24/0.17/0.003.$$

The contribution to $(M+2)^{+\bullet}$ due to two $^{33}S$ atoms is only 0.17%.

(c) The answer to (b) shows that $(M+3)^{+\bullet}$ due to $^{33}S$ is only 0.003%. You would not expect to be able to observe such a small peak by using an ordinary mass spectrometer.

You might like to note that both the simple formula (equation 5.1)) and the more complicated equation (5.9) give the same answer for the intensity of $(M+1)^{+\bullet}$ to two decimal places. We hope that you find this reassuring. The moral is — use equation (5.1) whenever possible.

**SAQ 5.4b**

A common method for analysing polyhydroxy compounds, such as glucose, is to convert each —OH group into a $-OSi(CH_3)_3$ (trimethylsilyl, TMS) group; this is then used in GC/MS analysis. How intense would you expect the $(M+1)^{+\bullet}$ ions to be for the penta-TMS derivative of glucose ($C_6H_{12}O_6$), which has the formula, $C_{21}H_{47}O_6Si_5$?

**Response**

If five —OH groups are converted to five $-OSi(CH_3)_3$ groups, the formula of the derivative is $C_{21}H_{47}O_6Si_5$. The molecular ion contains mostly $^{13}C$, $^2H$, and $^{29}Si$, and so equation (5.2) applies:

$$(M + 1)^{+\bullet} = [1.1 \times 21)\% + (0.016 \times 47)\% + (5.07 \times 5)\%]$$

where natural abundances are taken from Table 5.2c.

Therefore, we have:

$$(M + 1)^{+\bullet} = (23.1 + 0.75 + 25.35)\%$$

$$= 49.2\%.$$

$(M + 2)^{+\bullet}$ will contain $^{18}O$, $^{30}Si$, and species containing *two* $^{13}C$ or $^{29}Si$ atoms. Here a combination of equations (5.1) and (5.3) can be used:

$$(M + 2)^{+\bullet} = (0.2 \times 6)\% + (3.31 \times 5)\% + \frac{(1.1 \times 21)^2}{200}\% + \frac{(5.07 \times 5)^2}{200}\%$$

$$= (1.2 + 2.43 + 16.55 + 3.21)\%$$

$$= 23.39\%.$$

$(M + 3)^{+\bullet}$ will consist of three possible species, namely ions containing $^{13}C_2{}^{29}Si$, $^{13}C^{30}Si$, and $^{29}Si_3$ (neglecting any $^2H$ contributors which will be very small). Therefore:

$$\%\ ^{13}C_2{}^{29}Si = (5 \times 2.43 \times 0.0507)\% = 0.62\%.$$

$$\%\ ^{13}C^{30}Si = (21 \times 0.011 \times 3.31 \times 5)\% = 3.82\%.$$

$\%\ ^{29}Si_3$ can be obtained from equation (5.9), where $^{28}Si/^{29}Si = 20/1$ and $n = 5$. Therefore:

$$M^{+\bullet}/(M + 1)^{+\bullet}/(M + 2)^{+\bullet}/(M + 3)^{+\bullet}$$

$$= 20^3/\left(5 \times 20^2 \times 1\right)/\left(\frac{5}{2} \times 4 \times 20 \times 1^2\right)/\left(\frac{5}{2} \times 4 \times 3 \times 1^3\right)$$

$$= 8000/2000/200/10$$

$$= 100/25/2.5/0.125.$$

Hence $(M + 3)^{+\bullet}$ due to $^{29}Si_3$ is 0.125%.

Therefore, the total intensity of $(M + 3)^{+\bullet}$ is 4.57%.

**SAQ 5.4c**

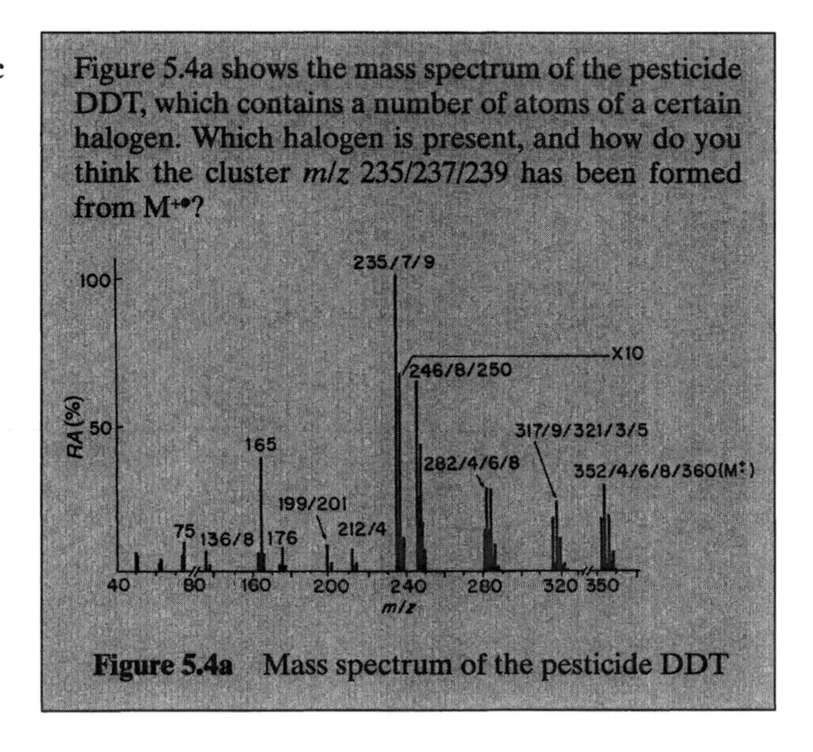

Figure 5.4a shows the mass spectrum of the pesticide DDT, which contains a number of atoms of a certain halogen. Which halogen is present, and how do you think the cluster *m/z* 235/237/239 has been formed from $M^{+\bullet}$?



**Figure 5.4a** Mass spectrum of the pesticide DDT

**Response**

The most intense peaks in the clusters in Figure 5.4a from *m/z* 235 onwards are all separated by 2 amu, thus indicating chlorine or bromine rather than fluorine or iodine. Since you are told a halogen is present, the general lack of symmetry of the clusters tends to rule out bromine. In addition, there are mass differences of 35 between some of the clusters. It appears that DDT contains $Cl_n$. You should then work out the normalised intensities of the $Cl_n$ clusters for $n = 2$–6 from the data given in Table 5.3b, for comparison with the $M^{+\bullet}$ cluster at $m/z$ 352, etc. We produced the following table when we did this:

| $Cl_n$ | $M^{+\bullet}$ | $(M+2)^{+\bullet}$ | $(M+4)^{+\bullet}$ | $(M+6)^{+\bullet}$ | $(M+8)^{+\bullet}$ | $(M+10)^{+\bullet}$ | $(M+12)^{+\bullet}$ |
|---|---|---|---|---|---|---|---|
| 2 | 100 | 65 | 10 | | | | |
| 3 | 100 | 100 | 33 | 4 | | | |
| 4 | 81 | 100 | 56 | 13 | 0.6 | | |
| 5 | 59 | 100 | 67 | 22 | 4 | 0.4 | |
| 6 | 49 | 100 | 84 | 37 | 10 | 1 | 0.1 |
| DDT | 61 | 100 | 64 | 22 | 4 | – | – |

The best agreement is with $n = 5$. DDT contains 5 chlorine atoms and has the structure:

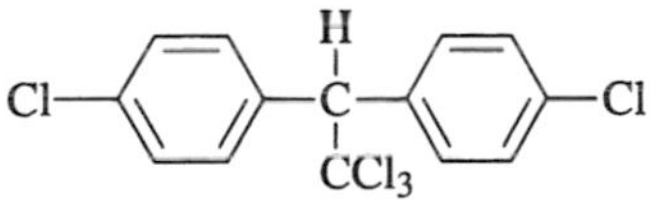

H
Cl C Cl
$CCl_3$

Its full name is bis-2,2-(4-chlorophenyl)-1,1,1-trichloroethane.

Note that the clusters at $m/z$ 317–325, 282–288, and 246–250 show the intensity distributions expected for $Cl_4$, $Cl_3$, and $Cl_2$, respectively. These ions arise by successive losses of chlorine from $M^{+\bullet}$. The $m/z$ 235/237/239 clearly contains two chlorines because it has the characteristic 100/65/10 ratio expected for $Cl_2$. However, it cannot have been formed by loss of three chlorines because this fragment would lead to the $m/z$ 246/248/250 cluster (which includes the loss of an additional hydrogen atom). The loss from $M^{+\bullet}$ to give $m/z$ 235/237/239 is 117 amu. This corresponds to a loss of $Cl_3C$, leading to:

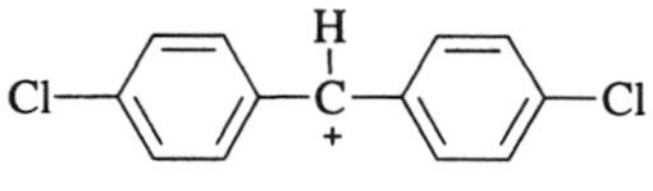

Such an ion would be specially stabilised (see Chapter 6). If you managed to deduce that DDT must contain a $CCl_3$ group, then well done.

**SAQ 6.1a**

Expand and rearrange equation (6.3) to show that $v_1 = v_2$.

**Response**

$$m_1v_1 = m_2v_2 + (m_1 - m_2)v_2 \tag{6.3}$$

$$m_1v_1 = m_2v_2 + m_1v_2 - m_2v_2$$

$$m_1v_1 = m_1v_2$$

which gives:

$$v_1 = v_2.$$

**SAQ 6.1b**

Suppose that an ion of $m/z$ 60 decomposes by loss of a hydrogen atom to $m/z$ 59. What would be the corresponding $m^*$ value?

**Response**

$$m^* = m_2^2/m_1$$

which gives:

$$59^2/60 = 58.02.$$

**SAQ 6.2a**

All of the metastable ions observed in the mass spectrum of benzene arise from the molecular ion ($m/z$ 78). Using the formula, $m^* = m_2^2/m_1$, assign the fragment ions ($M_2^+$). As an example, the fragmentation:

$$C_6H_6^{+\bullet} \longrightarrow C_6H_5^+ + H^\bullet$$

$m/z$ 78 $m/z$ 77

shows a metastable ion at an $m/z$ value of $77^2/78$, i.e. 76.

**Response**

$$\underset{78}{C_6H_6^{+\bullet}} \longrightarrow \underset{76}{C_6H_4^{+\bullet}} + H_2$$

$m^* = 76^2/78 = 74.1.$

$$\underset{78}{C_6H_6^{+\bullet}} \longrightarrow \underset{52}{C_4H_4^{+\bullet}} + C_2H_2$$

$m^* = 52^2/78 = 34.7.$

$$\underset{78}{C_6H_6^{+\bullet}} \longrightarrow \underset{39}{C_3H_3^+} + C_3H_3^\bullet$$

$m^* = 39^2/78 = 19.5.$

**SAQ 6.2b**

Other metastable ions are observed in the benzene spectrum:

| $M^*$ | $M_1^+$ | $\longrightarrow$ | $M_2^+$ |
|---|---|---|---|
| 50.0 | 52 | $\longrightarrow$ | 51 |
| 48.1 | 52 | $\longrightarrow$ | 50 |
| 33.8 | 77 | $\longrightarrow$ | 51 |
| 32.9 | 76 | $\longrightarrow$ | 50 |

Use these metastable ions, and those used in SAQ 6.2a for the $C_6H_6^{+\bullet}$ ion, to draw up a complete fragmentation pattern for benzene.

**Response**

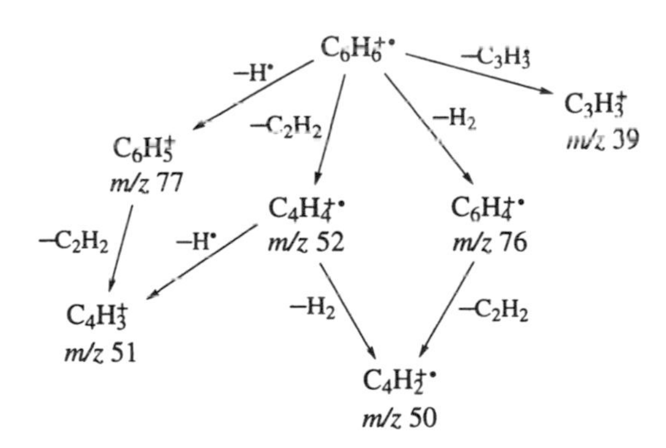

**SAQ 6.2c**

Figure 6.1a is actually the mass spectrum of aniline, $C_6H_5NH_2$. The two metastable ions shown at *m/z* 46.8 and 45.9 correspond to fragmentations of the molecular ion ($M^{+\bullet}$) and the fragment ion $(M-H)^+$, respectively. What are the daughter ions formed in these two fragmentation processes?

Can you suggest a formula for the neutral species that is lost?

**Response**

The metastable ion at *m/z* 46.8 arises from the molecular ion of aniline (*m/z* 93). Therefore, as the mass of the daughter ion from this fragmentation is given by:

$$m_2^2 = m^* m_1$$

we calculate that the daughter ion has an *m/z* value of 66.

The fragmentation is therefore as follows:

$$C_6H_5NH_2^{+\bullet}(m/z\ 93) \longrightarrow m/z\ 66 + 27\ \text{amu.}$$

Similarly, *m/z* 45.9 arises from $(M-H)^{+\bullet}$(*m/z* 92), and here the mass of the daughter ion is *m/z* 65.

The fragmentation here is therefore:

$$(M-H)^+(m/z\ 92) \longrightarrow m/z\ 65 + 27\ \text{amu.}$$

At first sight, fragmentations resulting in the loss of a neutral of mass 27 seem unlikely, since we have just seen that benzene, a closely related molecule, loses $C_2H_2$ (26 amu) or $C_3H_3$ (39 amu), and therefore aniline might be expected to behave similarly and not lose $C_2H_3$. Alternatively, one might expect aniline to lose $^{\bullet}NH_2$ (16 amu).

The fact that the neutral species has an *odd* mass is a vital clue. This means that it may well contain nitrogen and thus a likely formulation is HCN. The formulae of the fragment ions at *m/z* 66 and 65 must therefore be $C_5H_6^{+\bullet}$ and $C_5H_5^{+}$, respectively.

**SAQ 6.3a**

Figure 6.3b shows the mass spectra of (i) benzene, (ii) naphthalene, and (iii) anthracene. In each case, which ions are formed from the loss of HC≡CH?

**Response**

(i) There are 26 amu in HC≡CH, so we are looking in each spectrum for mass differences of 26. In the benzene spectrum, there are two such losses, *m/z* 78 ⟶ 52, and *m/z* ⟶ 51. You probably spotted the first one easily enough, but don't forget that the molecular ions of organic compounds often lose a single hydrogen, as here, i.e. *m/z* 78 ⟶ 77. This daughter ion then also undergoes the typical loss of HC≡CH.

(ii) There is only one clear loss of HC≡CH here, i.e. *m/z* 128 ⟶ 102.

(iii) In the case of anthracene, both *m/z* 178 and 177 lose 26 amu each to give *m/z* 152 and 151, respectively.

**SAQ 6.3b**

Compare the resonance hybrids, A to H, given above. Which hybrids are equivalent?

**Response**

G is equivalent to B or E.

H is equivalent to C or A.

**SAQ 6.3c**

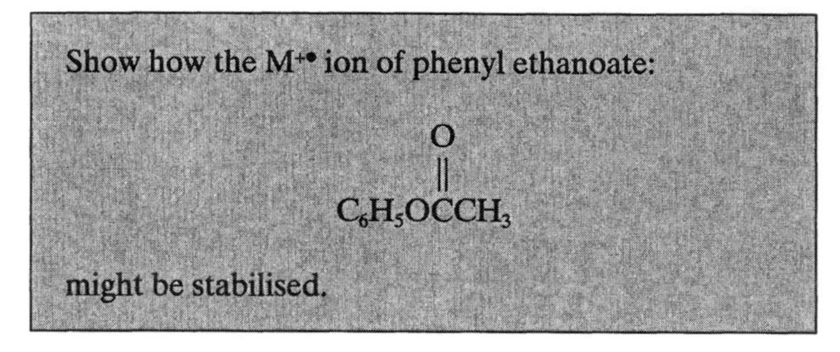

Show how the $M^{+\bullet}$ ion of phenyl ethanoate:

$$C_6H_5OC(=O)CH_3$$

might be stabilised.

## Response

If ionisation occurs on the oxygen next to the benzene ring, or on the benzene ring itself (which was shown to be equivalent in Section 6.3.2(ii)) we can stabilise $M^{+\bullet}$ as follows:

$CH_3C(=O)-\ddot{O}^{+}$ H ⟷ $CH_3C(=O)-\ddot{O}^{+}$ I ⟷ $CH_3C(=O)-\ddot{O}^{+}$ J

↕

etc ⟷ $CH_3C(=O)-\ddot{O}$ L ⟷ $CH_3C(=O)-\ddot{O}^{+}$ K

However, if ionisation occurs on the C=O group, resonance is limited to the following:

$-\ddot{O}-C(CH_3)=\dot{O}^{+}$ ⟷ $-\ddot{O}^{+}=C(CH_3)-\ddot{O}^{\bullet}$

M N

so presumably, most of the molecular ions would be of the H–K type, although we cannot prove this from the mass spectrum alone.

**SAQ 6.3d**

Would you expect the $M^{+\bullet}$ ion of benzyl ethanoate:

$$C_6H_5CH_2OC(=O)CH_3$$

to be as stable as the $M^{+\bullet}$ ion of phenyl ethanoate?

**Response**

$$C_6H_5-CH_2-O-C(=O)-CH_3$$

could have three $M^{+\bullet}$ structures:

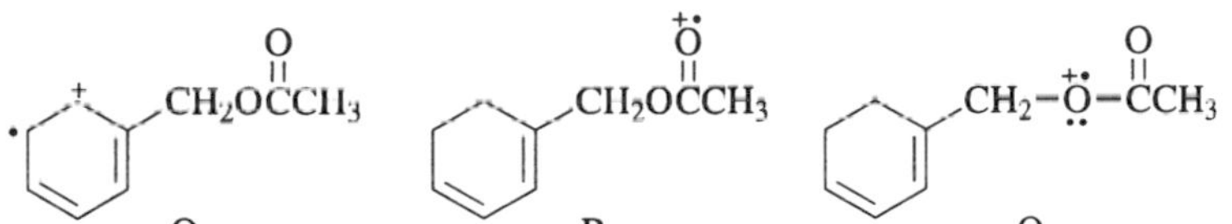

O (radical cation on the benzene ring, $CH_2OC(=O)CH_3$ substituent), P ($C_6H_5CH_2OC(=\overset{+\bullet}{O})CH_3$, charge on carbonyl oxygen), Q ($C_6H_5CH_2-\overset{+\bullet}{O}-C(=O)CH_3$, charge on ether oxygen)

While O and P have resonance forms as already described (Sections 6.3.2 ((i) and (ii)), and SAQ 6.3c) and Q has the possible resonance forms:

$$-CH_2-\overset{+\bullet}{\ddot{O}}-C(=O)-CH_3 \rightleftharpoons -CH_2-\overset{+}{\ddot{O}}=C(-O^{\bullet})-CH_3$$

the presence of the $-CH_2-$ group means O, P, and Q are isolated from one another. The $M^{+\bullet}$ of benzyl ethanoate would *not* be as stable as that of phenyl ethanoate, $C_6H_5-OCOCH_3$, because when the positive charge is on the $\overset{\bullet+}{O}-C_6H_5$ group is can be delocalised around the benzene ring, thus giving more resonance forms of $M^{+\bullet}$ for this compound. Remember, the more the number of resonance forms, then the greater the stability of the ion.

**SAQ 6.3e**

Show how the nitro group:

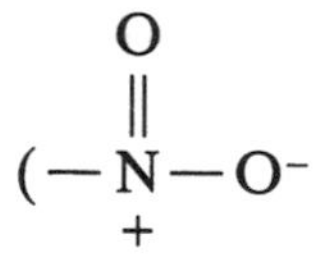

$$(-\overset{+}{N}(=O)-O^-$$

acts to destabilise the molecular ion of 4-nitrophenol:

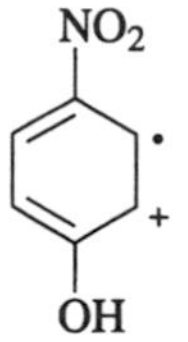

$NO_2$

OH

**Response**

The key ion is R. In order to try to stabilise it, electrons are drawn out of the ring to give S:

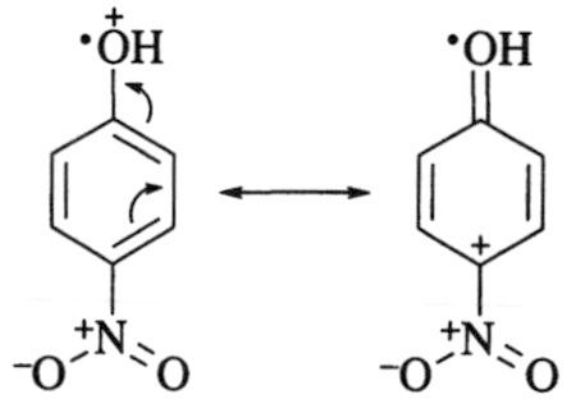

R ⟷ S

S shows two positive charges on adjacent atoms, clearly a highly unstable arrangement. If you started from one of the resonance forms which have a positive charge on the ring, you should have obtained a form such as T, which also has two adjacent positive charges:

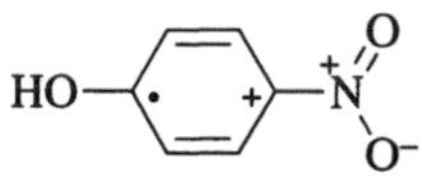

T

**SAQ 6.3f**

In compounds of the type $RC_6H_4COX$, would the following substituents (i.e. R) stabilise, destabilise, or have little effect on the benzoyl ion that is produced:

(a) 3-methoxy;
(b) 4-methoxy;
(c) 3-cyano;
(d) 4-nitro?

Try to explain your answers by means of resonance forms of the ions concerned.

**Response**

(a) Little effect

The crucial ion would be:

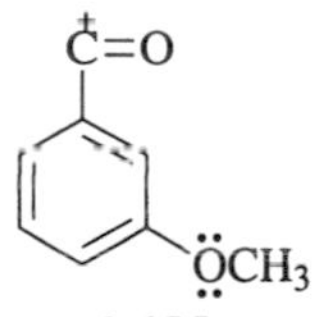

*m/z* 135

Can this be stabilised by the 3-$OCH_3$ substituent? Use of curly arrows to show the **+M**-effect of the oxygen in this group shows that it could *not* stabilise the benzoyl ion:

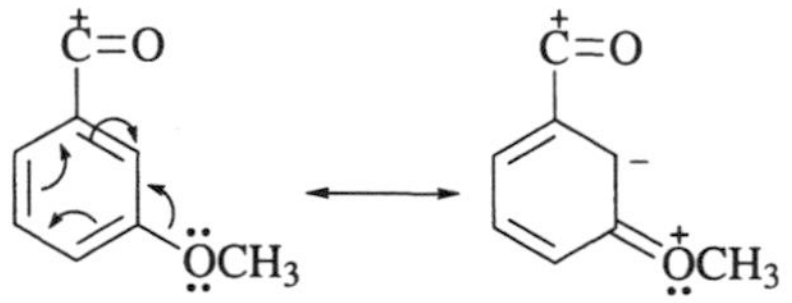

*m/z* 135

Therefore, the effect on the relative abundances of *m/z* 135 would not be great, because the 1-position where the $-\overset{+}{C}{=}O$ is located is bypassed in the resonance.

*This is a general result for 3-substituents, whether* **+*M*** *or* **−*M***.

(b) Stabilising effect

When the methoxy group is at the 4-position, the benzoyl ion is stabilised:

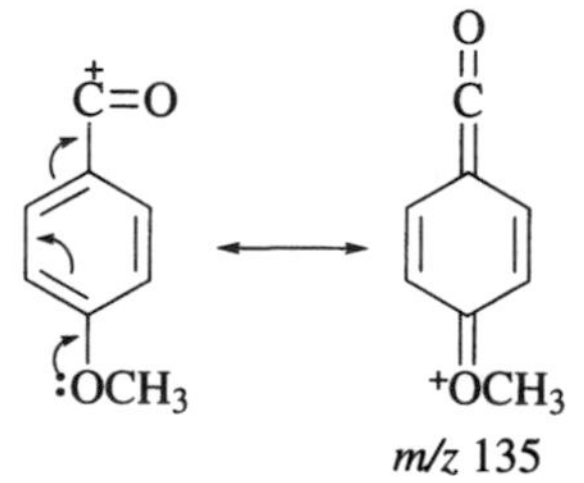

$\overset{+}{C}{=}O$ ... $:OCH_3$ ⟷ $O{=}C$ ... $^{+}OCH_3$

*m/z* 135

Therefore, this ion, or fragments derived from it, would be expected to be enhanced in the spectrum.

(c) Little effect

The $-C{\equiv}N$ group is −**M** (see Table 6.3b), so it would destabilise the benzoyl ion if it were conjugated with the $-\overset{+}{C}{=}O$ group. However, in the 3-position it is not conjugated:

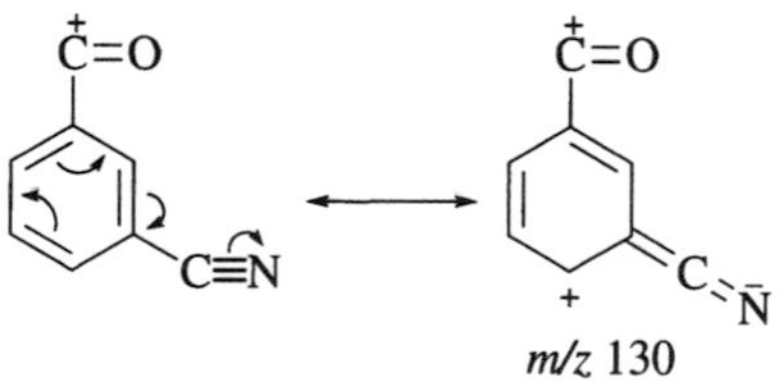

$\overset{+}{C}{=}O$ ... $C{\equiv}N$ ⟷ $\overset{+}{C}{=}O$ ... $C{=}\bar{N}$ +

*m/z* 130

There might be a general destabilising effect from the withdrawal of electron density from the benzene ring, due to the −**I**-effect of the −CN group at the 3-position, thus leading to a small decrease in *m/z* 130.

(d) Destabilising effect

The $-NO_2$ group is strongly $-\mathbf{M}$, so in the 4-position it would destabilise the benzoyl ion, *m/z* 150:

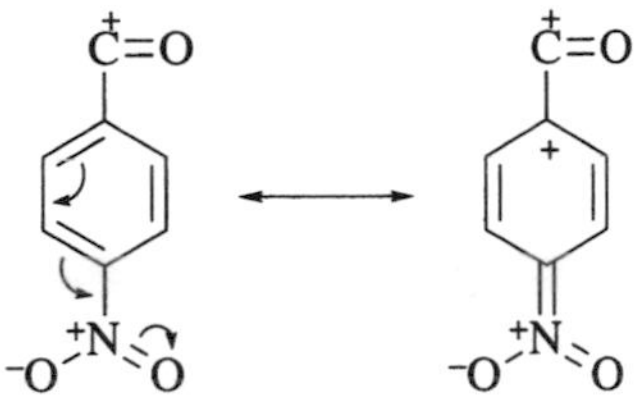

*m/z* 150

A less prominent *m/z* 150 would be expected. In practice, the ratio of $(M-CH_3)/M-Ar)$ in the spectrum of 4-nitrophenylethanone is 4.3/1, compared to 5.6/1 in the case of 1-phenylethanone, i.e. there is relatively less of the 4-nitrobenzoyl ion, *m/z* 150.

**SAQ 6.3g**

(This following optional question is intended for those who feel they need practice in predicting ion stabilisation.)

Show how the following ions are stabilised:

(a) $C_6H_5\overset{+}{C}H_2$; (b) $CH_3CH{=}CH{-}CH{-}\overset{+}{C}O$; (c) $CH_3\ddot{S}{-}CH{=}CH{-}\overset{+}{C}H_2$;

(d) ; (e) ; (f) $(Cl{-}C_6H_4)_2\overset{+}{C}H$

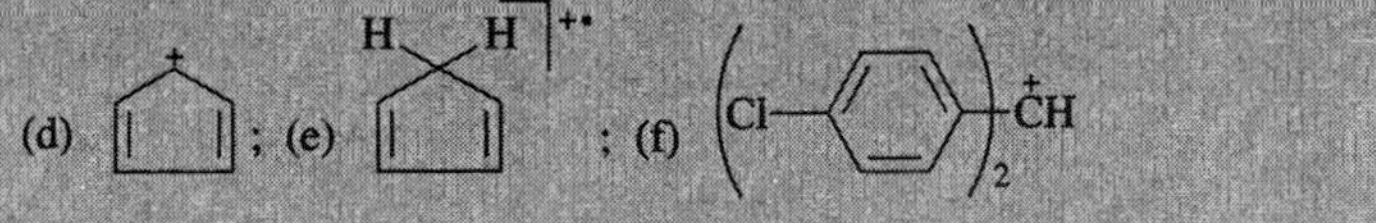

**Response**

(a) *m/z* 91

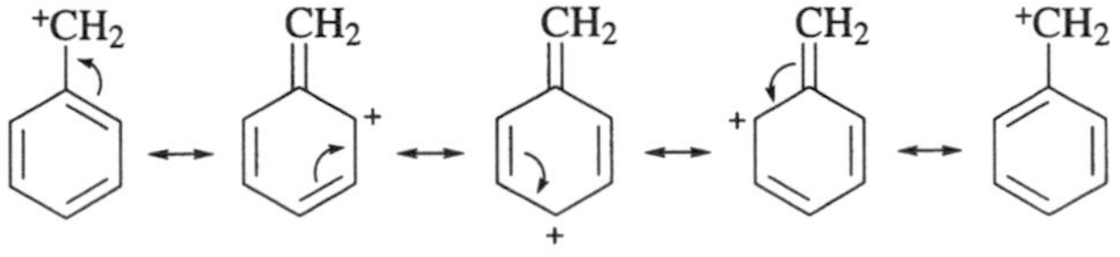

(b) *m/z* 69 $CH_3CH{=}CH{-}\overset{+}{C}{=}O \rightleftharpoons CH_3\overset{+}{C}H{-}CH{=}C{=}O$

(c) *m/z* 87 $CH_3\ddot{S}{-}CH{=}CH{-}\overset{+}{C}H_2 \rightleftharpoons CH_3{-}\overset{+}{S}{=}CH{-}CH{=}CH_2$

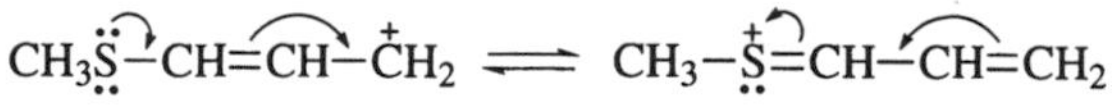

(d) *m/z* 65

Note that all five structures are the same in (d). Therefore, this must be a particularly stable ion.

(e) *m/z* 66 H H H H or H H H H

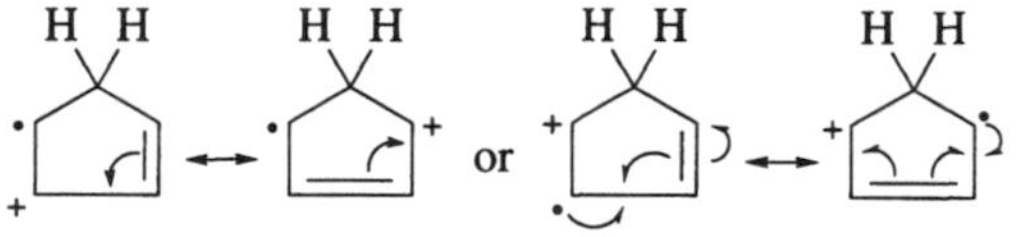

(f) *m/z* 235/237/239

Cl— =CH— —Cl

Cl— $\overset{+}{C}H$— —Cl

Cl— =CH— —Cl

Cl— $\overset{+}{C}H$— —Cl

$\overset{+}{Cl}$= =CH— —Cl

Cl— =CH— —Cl

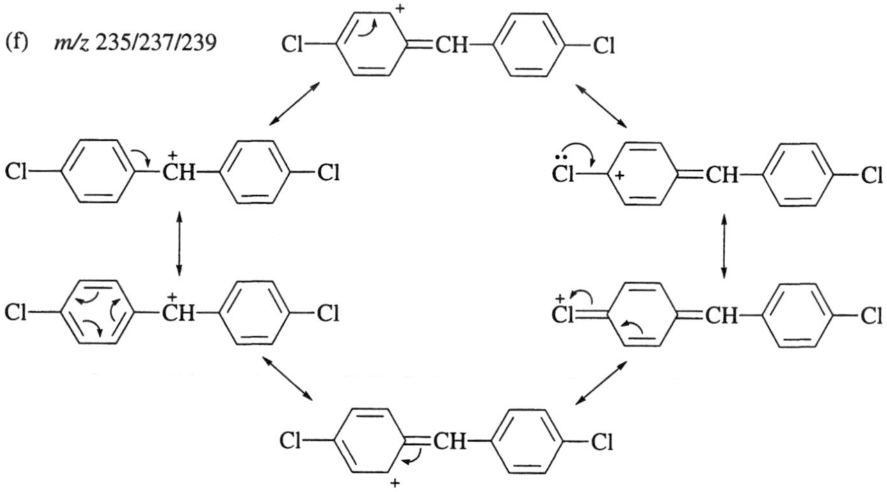

There are, of course, a further four possible structures which involve the right-hand benzene ring, thus making nine in all. This explains why this ion is the base peak of the DDT spectrum (see Figure 5.4a).

**SAQ 6.3h**

How would you expect the molecular ion of 1-bromo-1-phenylethane, $C_6H_5CHBrCH_3$, to fragment? Which of the ions formed by heterolytic and homolytic cleavages (α- and β-) would be stabilised by **+M**- and/or **+I**-effects?

**Response**

The four types of cleavage occur with the following results:

(a) Heterolytic α-cleavage:

$$C_6H_5-\underset{CH_3}{\overset{H}{C}}-Br^{+\bullet} \longrightarrow C_6H_5-\underset{+}{\overset{H}{C}}-CH_3 + Br^{\bullet}$$

*m/z* 105

This should be an intense ion in the spectrum since the positive charge on the carbon which is α- to the benzene ring is stabilised both by delocalisation, and the **+I**, $CH_3$— group.

(b) Heterolytic β-cleavage:

$$\text{SAQ 40} \quad C_6H_5-\underset{CH_3}{\overset{H}{C}}-Br^{+\bullet} \longrightarrow C_6H_5^{+} + CH_3CH{=}Br^{\bullet}$$

*m/z* 77

This is much less likely, as neither $C_6H_5^{+}$ nor $CH_3CH{=}Br$ are particularly stabilised.

(c) Homolytic α-cleavage:

$$C_6H_5-\underset{CH_3}{\overset{H}{C}}-Br^{+\bullet} \longrightarrow C_6H_5\dot{C}HCH_3 + Br^+$$

*m/z* 79/81

$Br^+$ is not stabilised, so this cleavage is not likely to be favoured.

(d) Homolytic β-cleavage:

$$C_6H_5-\underset{CH_3}{\overset{H}{C}}-Br^{+\bullet} \longrightarrow C_6H_5-\overset{H}{C}{=}Br^+ + CH_3^{\bullet}$$

*m/z* 169/171

This cleavage might be favoured to some extent as the $(M\text{–}CH_3)^+$ ion is conjugated through to the **+M** phenyl group:

$$C_6H_5-CH{=}\overset{+}{Br} \longleftrightarrow {}^{+}C_6H_5{=}CH-Br \longleftrightarrow \text{etc.}$$

In practice, *m/z* 105 is the base peak of the spectrum with no other ion being greater than 16% relative abundance, so (a) is the favoured process.

If you correctly obtained the ions described, well done, and if you decided *m/z* 105 and 169/171 were likely to be the most intense, then you have grasped the ideas of resonance stabilisation too!

**SAQ 6.3i**

How would you expect the molecular ions of propan-2-ol, $CH_3CHOHCH_3$, to fragment? Which of the ions formed by heterolytic and homolytic cleavages (α- and β-) would be stabilised by **+M**- and/or **+I**-effects?

The four types of cleavage occur with the following results:

(a) Heterolytic α-cleavage:

$$(CH_3)_2CH{-}\overset{+\bullet}{O}H \longrightarrow (CH_3)_2\overset{+}{C}H + \dot{O}H$$

*m/z* 43

*m/z* 43 is a secondary carbocation and would be stabilised to some extent by the **+I**-effects of the $CH_3$ groups. It would be reasonable to expect this ion to be observed in the spectrum.

(b) Heterolytic β-cleavage:

Did you spot that this could occur in two ways? The following are possible:

$$(CH_3)_2CH{-}\overset{+\bullet}{O}H \longrightarrow (CH_3)_2C{=}\dot{O}H + H^+$$

*m/z* 1

and

$$(CH_3)_2CH{-}\overset{+\bullet}{O}H \longrightarrow CH_3CH{=}\dot{O}H + CH_3^+$$

*m/z* 15

Neither $H^+$ nor $CH_3^+$ are stabilised ions, so these cleavages would not be expected to occur to any great extent.

(c) Homolytic α-cleavage:

$$(CH_3)_2CH\text{—}\dot{\overset{+}{O}}H \longrightarrow (CH_3)_2\dot{C}H + \overset{+}{O}H$$

$m/z$ 17

This is not favoured because $^{+}OH$ is unstable.

(d) Homolytic β-cleavage:

Here again, this can occur in two ways — did you realise this? The two possibilities are:

$$(CH_3)_2CH\text{—}\dot{\overset{+}{O}}H \longrightarrow (CH_3)_2C{=}\overset{+}{O}H + H^{\bullet}$$

$m/z$ 59

and

$$CH_3(H)CH(CH_3)\text{—}\dot{\overset{+}{O}}H \longrightarrow CH_3CH{=}\overset{+}{O}H + CH_3^{\bullet}$$

$m/z$ 45

The ions *m/z* 59 and 45 are both resonance stabilised, e.g.

$$(CH_3)_2C{=}\overset{+}{O}H \longleftrightarrow (CH_3)_2\overset{+}{C}OH$$

so these would be expected to be favoured over the only other fairly stable ion, i.e. $(CH_3)_2\overset{+}{C}H$. If you correctly obtained the ions described, well done, and if you decided that *m/z* 43, 45, and 59 were the most likely to be formed, very well done indeed. You have certainly grasped the fundamental principles of Chapter 6!

**SAQ 7.1a**

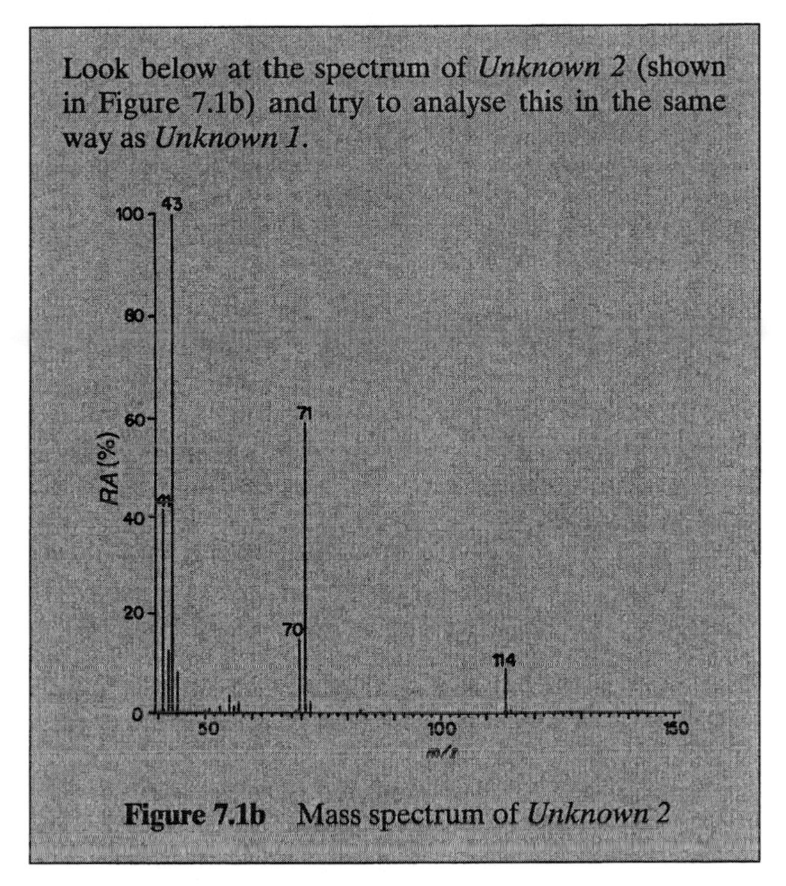


**Figure 7.1b** Mass spectrum of *Unknown 2*

## Response

You should have obtained an analysis table similar to Table 7.1d overleaf.

**Table 7.1d** Analysis table for *Unknown 2*

| *m/z* | Possible structure | Associated X-loss | Inference |
|---|---|---|---|
| 114 | – | – | $M^{+\bullet}$ |
| 71 | $C_5H_{11}$, $C_3H_7CO$ | 43 | propyl, methyl ketone, acetyl |
| 70 | $C_5H_{10}$ | 1 | stable alkenic structure(?) |
| 114 | – | – | $M^{+\bullet}$ |
| 70 | $C_5H_{10}$ | 44 | aldehyde with γ-hydrogen, ester, acid anhydride |
| 114 | – | – | $M^{+\bullet}$ |
| 43 | $CH_3CO$, $C_3H_7$ | 71 | Complementary ion to *m/z* 71 |
| 71 | $C_5H_{11}$, $C_3H_7CO$ | – | Parent for *m/z* 42 or 41(?) |
| 43 | $CH_3CO$, $C_3H_7$ | 28 | CO or $CH_2{=}CH_2$ present(?) |
| 41 | $C_3H_5$ | 30 | Not sensible from *m/z* 71 |
| 41 | $C_3H_5$ | 2 | Loss of $H_2$ from *m/z* 43, implies *m/z* 43 is $C_3H_7^+$ |

Deduction: $(CH_3CH_2CH_2)_2CO$ or $((CH_3)_2CH)_2CO$

The analysis suggests the presence of propyl (Pr) or isopropyl (*i*-Pr) groups, CO, or $CH_3CO$. The relatively few ions in the spectrum show that there are only two or three easily fragmented bonds, which suggests cleavage α- to the C=O group. One possible process, 114 → 70, suggests the presence of $-CH_2CHO$ or $-CO_2-$, but since there is no other support for an aldehyde, ester, acid or anhydride, and *m/z* 70 is only 17% relative abundance, this can be dismissed. The loss of one or two hydrogens from an aliphatic ion to give an unsaturated species is quite common, so *m/z* 70 is probably formed from *m/z* 71.

If *Unknown 2* contained both Pr and $CH_3CO$, it would have to be $PrCOCH_3$ but this has a $M^{+\bullet}$ of only 86. Any other structure involving $COCH_3$ and the correct number of carbon atoms (now seen to be seven) to give an $M_r$ of 114, would have a hydrocarbon chain of five carbons, i.e. $CH_3-CH_2-CH_2-CH_2-CH_2^+$, *m/z* 71. Although this could lose $CH_2{=}CH_2$ and give $CH_3CH_2CH_2^+$, *m/z* 43, by the process shown, it would also be expected to fragment in other ways, e.g. by the losses of $CH_3^{\bullet}$ and $CH_3CH_2^{\bullet}$ from $M^{+\bullet}$ and *m/z* 71, and these are not observed. In particular, *m/z* 99 is *not* found. This *should* form in a methyl ketone because it is an acylium ion and fairly stable:

$$CH_3\text{—}\overset{\overset{+\bullet}{O}}{\overset{\|}{C}}\text{—}(CH_2)_4CH_3 \longrightarrow \underset{m/z\ 99}{CH_3(CH_2)_4\overset{+}{C}{=}O} + CH_3^{\bullet}$$

Therefore, the structures $(i\text{-Pr})_2CO$ or $(Pr)_2CO$ are most likely. In fact, *Unknown 2* is $(i\text{-Pr})_2CO$. Its main fragmentations are:

$$i\text{-Pr}\text{—}\overset{\overset{+\bullet}{O}}{\overset{\|}{C}}\text{—}i\text{-Pr} \longrightarrow i\text{-Pr}^{\bullet} + \underset{m/z\ 71}{i\text{-Pr}\text{—}C{\equiv}\overset{+}{O}} \longrightarrow \underset{m/z\ 43}{i\text{-Pr}^{+}} + :C{=}O$$

You could not be certain from the mass spectrum alone which structural isomer you had. Indeed, it could have been PrCO*i*-Pr! To decide which of the three you had you could (a) consult reference lists of mass spectra, (b) use a computer library database to get a match, or (c) obtain the NMR spectrum.

**SAQ 7.1b**

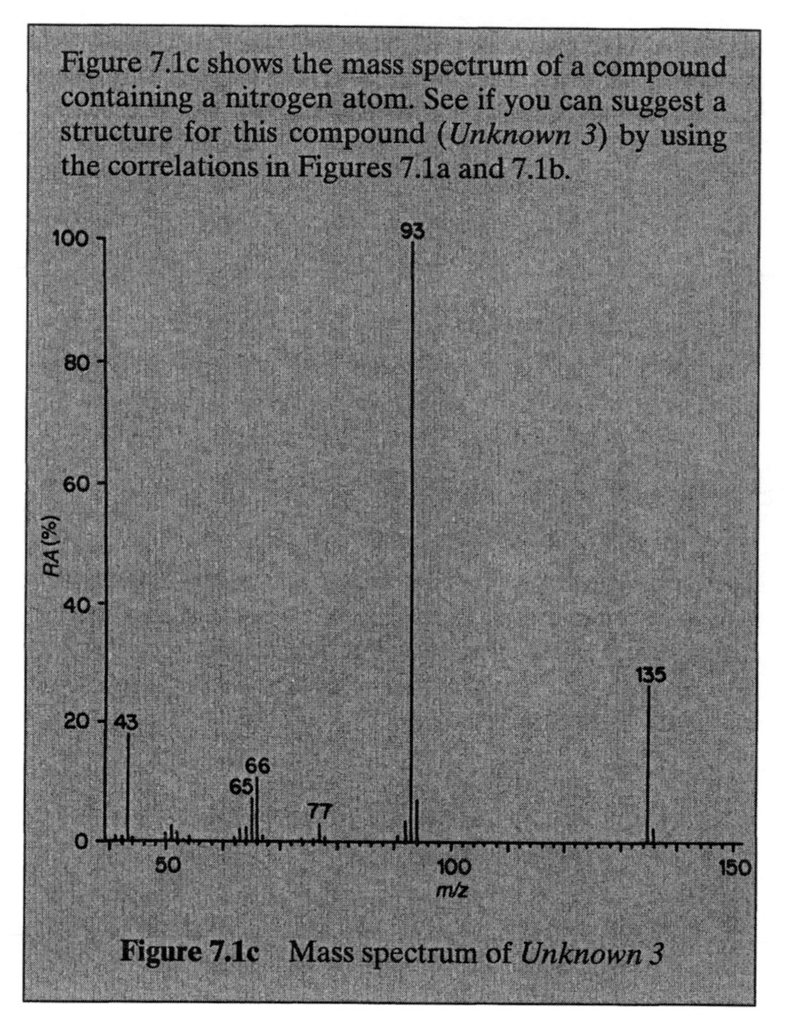

Figure 7.1c shows the mass spectrum of a compound containing a nitrogen atom. See if you can suggest a structure for this compound (*Unknown 3*) by using the correlations in Figures 7.1a and 7.1b.



**Figure 7.1c** Mass spectrum of *Unknown 3*

## Response

The sort of analysis table that you should have obtained is shown in Table 7.1e below.

**Table 7.1e** Analysis table for *Unknown 3*

| *m/z* | Possible structure | Associated X-loss | Inference |
|---|---|---|---|
| 135 | – | – | $M^{+\bullet}$ odd, so 1,3,5, etc. N atoms (in fact, 1 N is given) |
| 93 | $C_6H_7N$ | 42 | $C_6H_5NHX$, *N*-acetyl compound |
| | $C_7H_9$ | 42 | Terpene derivative(?) |
| 77 | $C_6H_5$ | 58 | RNCO, RCNO(?) |
| 93 | $C_6H_7N$ | – | Parent for *m/z* 66, 65(?) |
| 66 | $C_5H_6$ | 27 | HCN from $C_6H_7N$ |
| 65 | $C_5H_5$ | 1 | H from $C_5H_6$ |
| 43 | $CH_3CO$, $C_3H_7$ | 50 | Not very likely(!) *m/z* 43 not from *m/z* 93 |
| 135 | – | – | $M^{+\bullet}$ |
| 43 | $CH_3CO$, $C_3H_7$ | 92 | Complementary ion to *m/z* 93, with hydrogen transfer(?) |
| | | | Deduction: $C_6H_5NHCOCH_3$ |

From the general appearance of this spectrum, the compound appears to be aromatic. The appearance of an *m/z* 77 ion, although small, indicates a $C_6H_5$ unit. This leaves 68 amu to account for. The loss of 42 amu from $M^{+\bullet}$($CH_2{=}C{=}O$) is very characteristic of *N*-acetyl compounds. Therefore, *Unknown 3* is $C_6H_5NHCOCH_3$. This is confirmed by the remainder of the spectrum looking very like that of aniline, since *m/z* 93 *is* $C_6H_5NH_2^{+\bullet}$, with the confirming presence of $CH_3CO^+$, *m/z* 43. If this were a propyl compound, it could not fragment by loss of ketene, so this possibility can be dismissed immediately. The mechanisms of the processes involved are complex, apart from the formation of $CH_3CO^+$:

$$C_6H_5{-}NH{-}\overset{\overset{+\bullet}{O}}{\overset{\|}{C}}{-}CH_3 \longrightarrow C_6H_5\dot{N}H + CH_3C{\equiv}\overset{+}{O}$$

**SAQ 7.2a**

Examine the mass spectra of the three isomers of butanol shown in Figure 7.2b and see if Stevenson's rule applies.

**Response**

Butan-2-ol (Figure 7.2b(ii)) cleaves to give 100% $(M–CH_3CH_2)^+$, *m/z* 45, and 20% $(M–CH_3)$, *m/z* 59, a ratio of 5/1 in favour of the loss of $CH_3CH_2^\bullet$. In fact, the ion $CH_3CH{=}\overset{+}{O}H$ is typical of '2-ols', because in the structure $CH_3CHOHR$ (if $R{=}CH_3$), $CH_3^\bullet$ will be lost because $CH_3^\bullet$ > $H^\bullet$. For any other R group, it will be larger than $CH_3$, so $R^\bullet$ is lost according to Stevenson's rule. 2-Methylpropan-2-ol (*t*-butyl alcohol) (Figure 7.2b (iii)), is really analogous to methanol because it has three identical groups around the α-carbon and can only lose $CH_3^\bullet$ to form $(CH_3)_2C{=}\overset{+}{O}H$ as the base peak. These cleavages are summarised as follows:

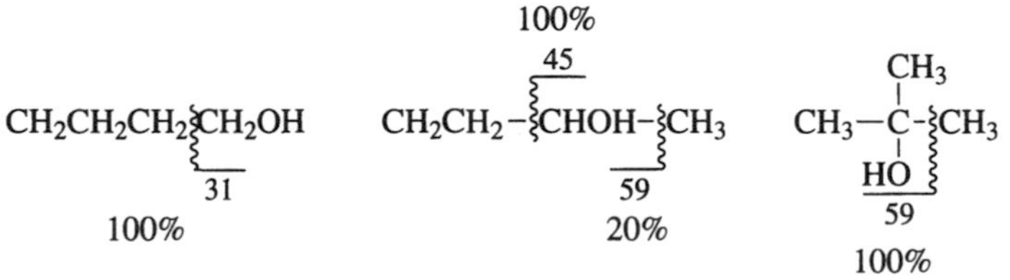

$CH_2CH_2CH_2 \wr CH_2OH$ — 31, 100%

$CH_2CH_2 – \wr CHOH – \wr CH_3$ — 45 (100%); 59 (20%)

$CH_3–C(CH_3)(HO)–\wr CH_3$ — 59, 100%

In this shorthand presentation, the bond cleavage is shown broken by a wavy line, with the ion formed and its intensity being shown above or below the horizontal straight line. This is a very useful way of summarising the main features of a spectrum.

**SAQ 7.2b**

Why do you think the *m/z* 57 peak is so abundant in most cyclanol mass spectra?

**Response**

The answer is shown in Figure 7.2d. This ion is conjugated, so the positive charge is stabilised by resonance. This is yet another example of the importance of **+M** groups in determining the directions and mechanisms by which molecular ions fragment.

**SAQ 7.2c**

Show how the $M^{+\bullet}$ of 3-methylcyclohexanol can give rise to both *m/z* 57 and *m/z* 71 peaks in its mass spectrum. Why is *m/z* 71 more intense than *m/z* 57 in this spectrum?

**Response**

The molecular ion can undergo α-cleavage in two ways, namely path (a) or path (b), with each being equally likely. When the ring-opened $M^{+\bullet}$ ions fragment, they each follow the mechanism of Figure 7.2d, leading to *m/z* 57 or 71, respectively:

$M^{+\bullet}$, *m/z* 114 —(b)→ ring-opened ion → $^{+}OH{=}CH{-}CH{=}CH{-}CH_3$ (H, $CH_3$) + $CH_3^{\bullet}$ + $H_2C{=}CH_2$, *m/z* 71 (64%)

$M^{+\bullet}$, *m/z* 114 —(a)→ ring-opened ion → $^{+}OH{=}CH{-}C(H){=}CH_2$ + $CH_3^{\bullet}$ + $CH_3CH{=}CH_2$, *m/z* 57 (36%)

*m/z* 71 is more intense than *m/z* 57 because the double-bond system in *m/z* 71 is further stabilised by the **+I** $CH_3$ group.

**SAQ 7.2d**

Tropylium ions have a great number of resonance forms. Can you draw resonance forms for the hydroxytropylium ion (A in Figure 7.2f)?

**Response**

Tropylium ions have a larger number of resonance forms than other aromatic species of isomeric structure, and are stabilised as follows:

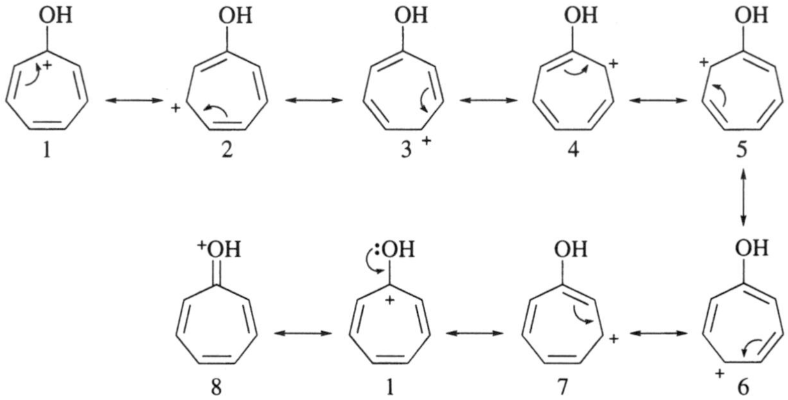

OH OH OH OH OH

1 2 3 4 5

+OH :OH OH OH

8 1 7 6

We hope that you found most of these. Did you remember to use the lone-pair electrons on the —OH in order to get structure (8) and explain why a **+M** group would further stabilise such an ion? This ion has eight resonance forms, more than you have seen in any previous example, thus explaining its surprising formation and stability.

**SAQ 7.2e**

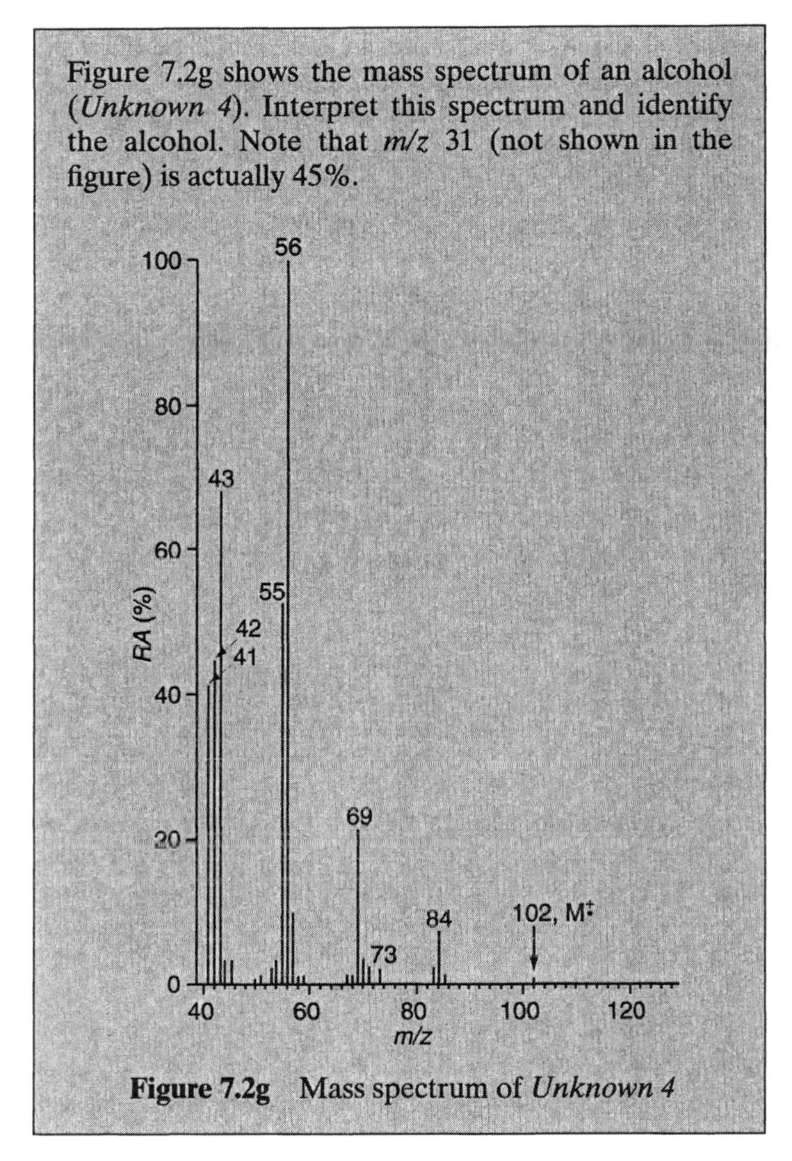

Figure 7.2g shows the mass spectrum of an alcohol (*Unknown 4*). Interpret this spectrum and identify the alcohol. Note that *m/z* 31 (not shown in the figure) is actually 45%.



**Figure 7.2g** Mass spectrum of *Unknown 4*

**Response**

*Unknown 4* is hexanol, $CH_3(CH_2)OH$. The $M_r$ of 102 indicates a molecular formula of $C_6H_{14}O$, i.e. a saturated alcohol, and its low intensity is typical of primary alcohols. The base peak in this spectrum is *m/z* 56, i.e. $(M - 46)^+$. This is typical of the concerted loss of $H_2O$ and $H_2C{=}CH_2$ shown by longer-chain alcohols.

$$\underset{m/z\ 102}{\text{(cyclic } H_2C{-}O^{+\bullet}{-}H \cdots H{-}CH(CH_2CH_3){-}CH_2{-}H_2C\text{)}} \longrightarrow H_2O + H_2C{=}CH_2 + \underset{m/z\ 56}{[H_2C{=}CHCH_2CH_3]^{+\bullet}}$$

The peak *m/z* 31, $CH_2\overset{+}{O}H$, is not shown in the spectrum but in fact is *ca.* 45% relative abundance. Other peaks are $(M{-}H_2O)$, *m/z* 84, $(M{-}(H_2O{-}CH_3))^+$, *m/z* 69, $(M{-}(H_2O{-}CH_2CH_3))^+$, *m/z* 55, and $CH_3CH_2CH_2^+$, *m/z* 43. Apart from *m/z* 43, hydrocarbon ions are of low abundance (see, for example, *m/z* 57, 71, and 85).

**SAQ 7.2f**

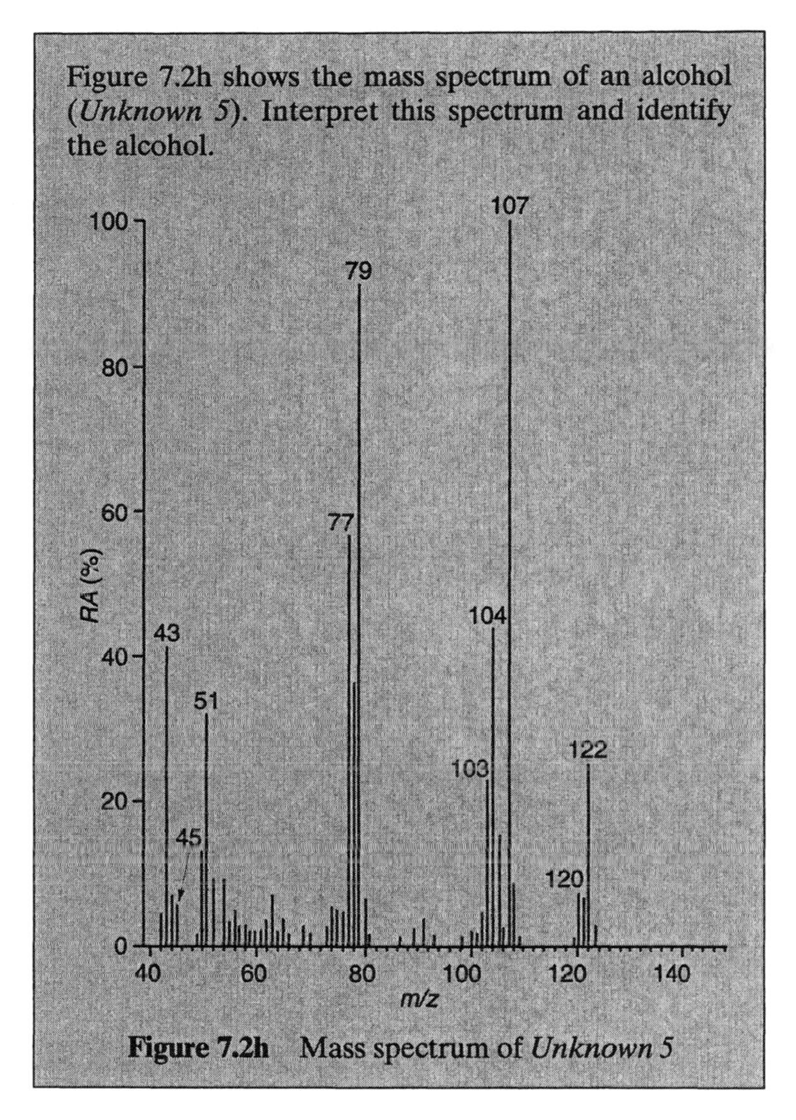

Figure 7.2h shows the mass spectrum of an alcohol (*Unknown 5*). Interpret this spectrum and identify the alcohol.



**Figure 7.2h** Mass spectrum of *Unknown 5*

**Response**

*Unknown 5* is 1-phenylethanol, $C_6H_5CH(OH)CH_3$. The $M^{+\bullet}$ indicates that this is an isomer of 2-phenylethanol but the lack of *m/z* 91 shows that the $C_6H_5CH_2$ grouping is absent. The presence of *m/z* 107, combined with *m/z* 79, 77 and 51, is very typical of the hydroxytropylium ion and its fragments, so *Unknown 5* must contain a structural unit capable of forming this ion, but not containing $C_6H_5CH_2O$. The answer has to be a secondary alcohol, which loses $CH_3^{\bullet}$ preferentially against the predictions of Stevenson's rule, because of the high stability of the hydroxytropylium ion:

$$C_6H_5-\underset{CH_3}{\overset{H}{C}}-\overset{+\bullet}{O}H \longrightarrow C_6H_5CH=\overset{+}{O}H \longrightarrow \text{(hydroxytropylium ion, } m/z \text{ 107)} \longleftrightarrow \text{etc}$$

The peak $m/z$ 104 results from $(M-H_2O)^+$ and $m/z$ 103 from $((M-H)-H_2O)^+$. In this spectrum, $m/z$ 45 ($CH_3CH=\overset{+}{O}H$) is replaced by $m/z$ 43 ($CH_3CO^+$). It appears that $M^{+\bullet}$ loses two hydrogen atoms to give $C_6H_5COCH_3^{+\bullet}$ (common in alcohol mass spectra since the ketone has a more stable $M^{+\bullet}$, with $CH_3CO^+$ being formed by α-cleavage of this ion.)

**SAQ 7.3**

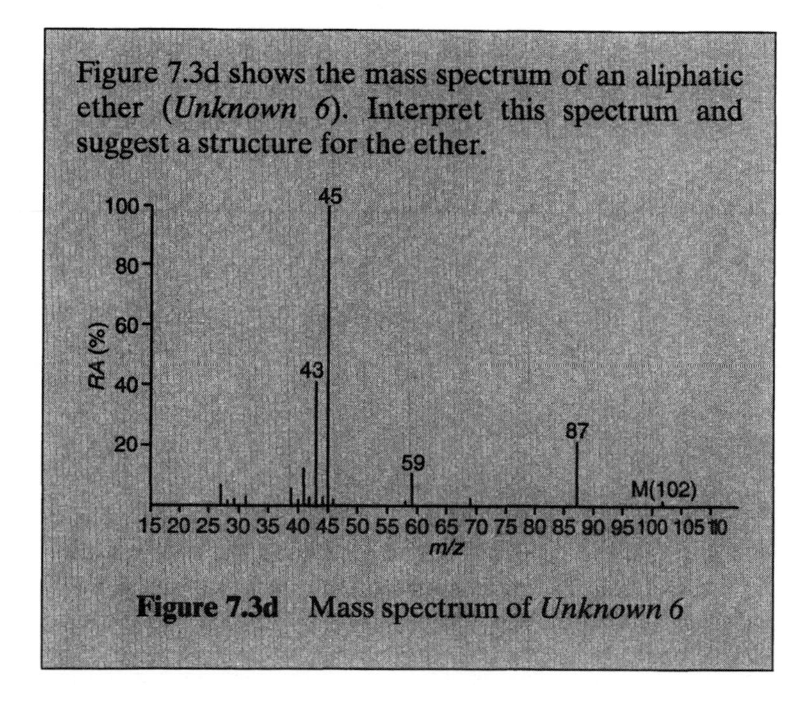

Figure 7.3d shows the mass spectrum of an aliphatic ether (*Unknown 6*). Interpret this spectrum and suggest a structure for the ether.



**Figure 7.3d** Mass spectrum of *Unknown 6*

**Response**

*Unknown 6* is diisopropyl ether, $(CH_3)_2CHOCH(CH_3)_2$. $M^{+\bullet}$ has *m/z* 102, and therefore the partial formula must be $C_6H_{14}O$, a saturated ether. Various combinations of alkyl groups are possible from $CH_3OC_5H_{11}$ to $(CH_3CH_2CH_2)_2O$ and *6* itself, but the simplicity of the spectrum points to a symmetrical structure which readily loses $CH_3^\bullet$ to give *m/z* 87, and one having at least one α-$CH_3$ group:

$$H_3C-\underset{H}{\overset{CH_3}{C}}-\overset{+\bullet}{O}-\underset{CH_3}{\overset{H}{C}}-CH_3 \longrightarrow \underset{H}{\overset{H_3C}{C}}=\overset{+}{O}-\underset{CH_3}{\overset{H}{C}}-CH_3 + CH_3^\bullet$$

$(M-CH_3)^+$, *m/z* 87

Next, *m/z* 87 loses $CH_3CH{=}CH_2$ to give the base peak, $CH_3CH{=}\overset{+}{O}$:

$$CH_3CH{=}\overset{+}{O}-\underset{H-CH_2}{CHCH_3} \longrightarrow CH_3CH{=}\overset{+}{O}H + CH_3CH{=}CH_2$$

*m/z* 45

The intensity of *m/z* 43 is quite high for the alkyl $R^+$ group derived from an ether ($R_2O$) so this would lead you to suspect it was a secondary isopropyl ion. The peak *m/z* 59, $(CH_3)_2CHO^+$, confirms the presence of $C_3H_7O$, complementary to *m/z* 43. If you had said *Unknown 6* was $(CH_3CH_2CH_2)_2O$, this would have been a fair answer based on the mass spectrum alone.

However, if you had suggested that *Unknown 6* could be $CH_3CH_2OC(CH_3)HCH_2CH_3$, note that this compound would tend to lose $CH_3CH_2^\bullet$ to give *m/z* 73 from $M^{+\bullet}$, rather than $CH_3^\bullet$ in the first α-cleavage, although it is true that *m/z* 73 would then give $CH_3CH{=}\overset{+}{O}H$ as the base peak by eliminating $H_2C{=}CH_2$ from the other ethyl group.

**SAQ 7.4**

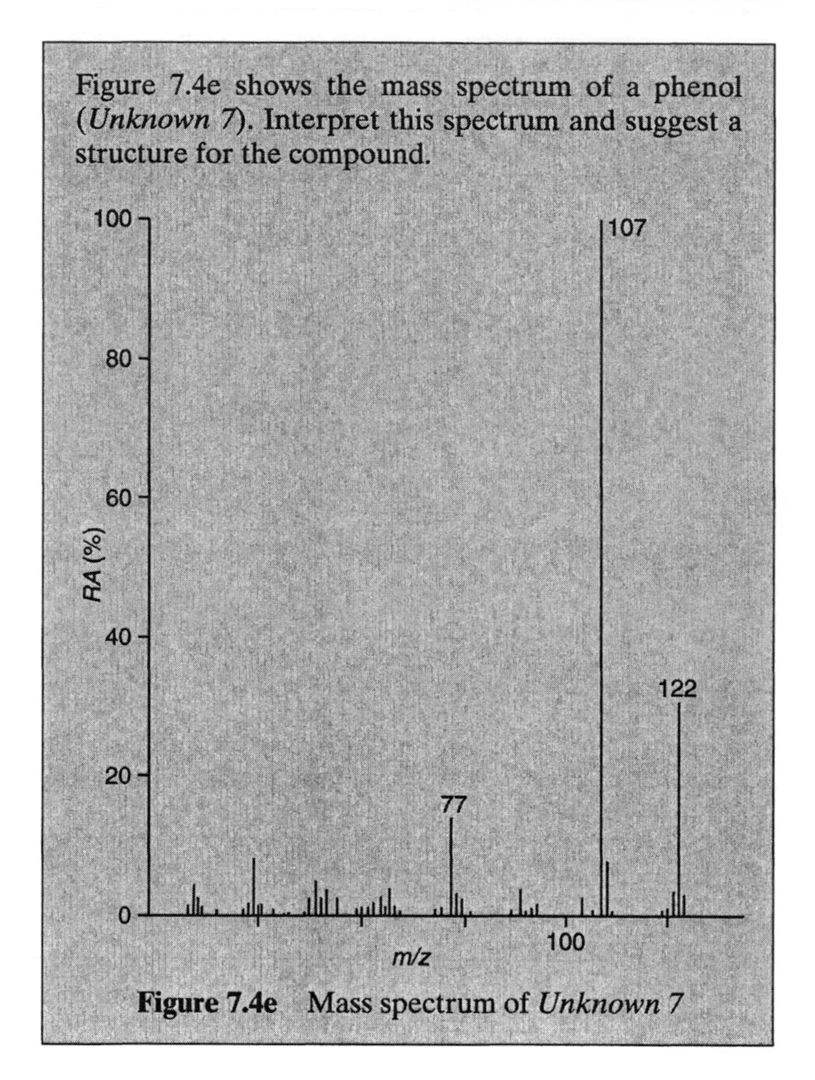

Figure 7.4e shows the mass spectrum of a phenol (*Unknown 7*). Interpret this spectrum and suggest a structure for the compound.



**Figure 7.4e** Mass spectrum of *Unknown 7*

## Response

*Unknown 7* is *p*-ethylphenol (4-ethylhydroxybenzene):

$$CH_3 \wr \underbrace{CH_2-C_6H_4-OH}_{107}$$

The spectrum shows a very intense $(M-CH_3)^+$ peak, typical of either benzylic cleavage as shown, or loss of a $CH_3$ from a dimethylphenol. However, a dimethyl isomer would be expected to show a significant $(M-H)^+$ at *m/z* 121. In this spectrum, *m/z* 121 is very weak, leading to the conclusion that it is an ethyl substituent. If you had said *Unknown* 7 was *m*-ethylphenol (3-ethylhydroxybenzene), this would have been an acceptable answer, because mass spectra do not permit the distinction of 3- and 4-isomers. However, you should have realised that it could not be either a 2-ethyl or a 2-methyl isomer because these would have shown loss of $H_2O$ via an *ortho*-effect. There is no *m/z* 104 $(M-H_2O)^{+\bullet}$ in this spectrum!

**SAQ 7.5a**

In which of the following compounds would you expect to see a 'Mcl' peak? Give its *m/z* value and structure.

(a) $CH_3COCH_2CH_3$;

(b) $CH_3COCH_2CH_2CH_3$;

(c) $CH_3COCH(CH_3)_2$;

(d) $(CH_3)_2CHCH_2CHO$;

(e) $CH_3CH_2COOCH_2CH_3$;

(f) $CH_3CH_2CH_2CONHCH_3$;

(g) $CH_3CH(CH_3)CH_2CH(CH_3)COOH$;

(h) $CH_3CH_2CH(CH_3)COC_6H_5$;

(i) $CH_3CH{=}CHCH_2COCH_2CH_3$;

(j) $(C_6H_5)_2P(=O)SCH_2CH_3$

**Response**

(a) $\overset{\alpha}{C}H_3CO\overset{\alpha}{C}H_2\overset{\beta}{C}H_3$ has no γ-hydrogens, so it has no 'McL' peak.

(b) $CH_3CO\overset{\alpha}{C}H_2\overset{\beta}{C}H_2\overset{\gamma}{C}H_3$ has a γ-hydrogen, so we can predict:

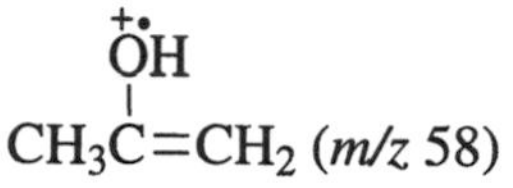

$$CH_3\overset{\overset{+\bullet}{O}H}{\overset{|}{C}}{=}CH_2 \quad (m/z\ 58)$$

(c) $\underset{\alpha}{C}H_3CO\underset{\alpha}{\overset{\overset{CH_3}{|}}{C}}H{-}\underset{\beta}{C}H_3$ has no γ-hydrogens, so it has no 'McL' peak.

(d) $\overset{\gamma}{C}H_3{-}\overset{\beta}{\underset{\underset{CH_3}{|}}{C}}H{-}\overset{\alpha}{C}H_2CHO$ has a γ-hydrogen, so we can predict:

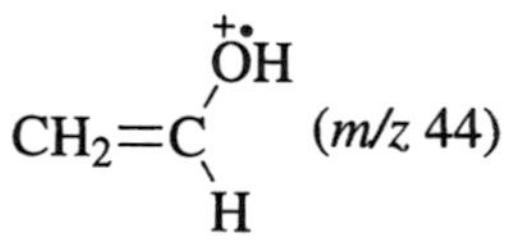

$$CH_2{=}C\begin{matrix} \diagup \overset{+\bullet}{O}H \\ \diagdown H \end{matrix} \quad (m/z\ 44)$$

(e) $CH_3CH_2\overset{\overset{O}{||}}{C}O\overset{\alpha}{\phantom{C}}\overset{\beta}{C}H_2\overset{\gamma}{C}H_3$ has a γ-hydrogen in the $-OCH_2CH_3$ group,

so we can predict:

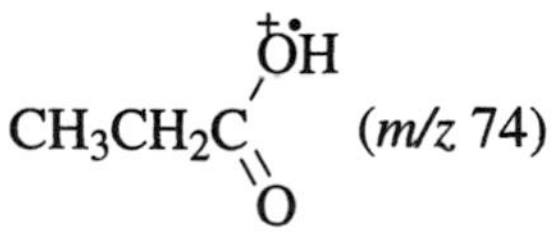

$$CH_3CH_2C\begin{matrix} \diagup \overset{+\bullet}{O}H \\ \diagdown\!\!\diagdown O \end{matrix} \quad (m/z\ 74)$$

Note that this is isomeric with the 'McL' peak of a methyl ester.

(f) $\overset{\gamma}{C}H_3\overset{\beta}{C}H_2\overset{\alpha}{C}H_2CONHCH_3$ has a γ-hydrogen in the $CH_3CH_2CH_2$ group, so we can predict:

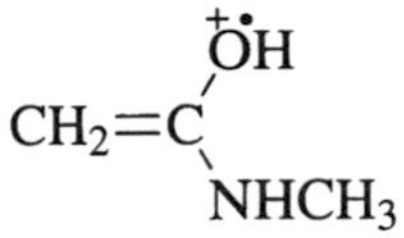

$$CH_2{=}C(\overset{+\bullet}{O}H)(NHCH_3) \quad (m/z\ 73)$$

(g) $\overset{\gamma}{C}H_3CH(CH_3)\overset{\beta}{C}H_2\overset{\alpha}{C}H(CH_3)COOH$ has a γ-hydrogen, so we can predict:

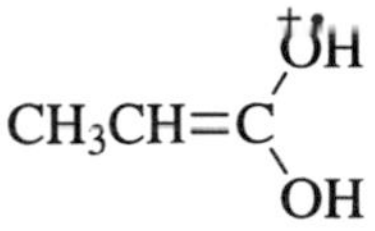

$$CH_3CH{=}C(\overset{+\bullet}{O}H)(OH) \quad (m/z\ 74)$$

(h) $\overset{\gamma}{C}H_3\overset{\beta}{C}H_2\overset{\alpha}{C}H(CH_3)COC_6H_5$ has a γ-hydrogen, so we can predict:

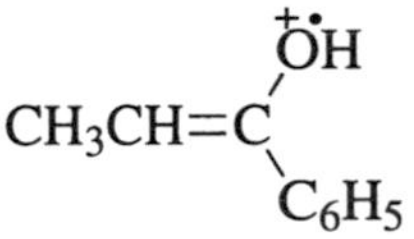

$$CH_3CH{=}C(\overset{+\bullet}{O}H)(C_6H_5) \quad (m/z\ 150)$$

(i) $\overset{\gamma}{C}H_3CH{=}\overset{\beta}{C}H\overset{\alpha}{C}H_2COCH_2CH_3$ has one γ-hydrogen, but this is on a double bond, so it has no 'McL' peak. If you obtained the following:

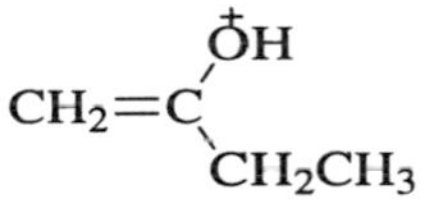

$$CH_2{=}C(\overset{+}{O}H)(CH_2CH_3) \quad (m/z\ 72)$$

then look at Section 7.5.1 again.

(j) $(C_6H_5)_2\overset{O}{\overset{\|}{P}}-\overset{\alpha}{S}-\overset{\beta}{C}H_2-\overset{\gamma}{C}H_3$ has a γ-hydrogen, so we can predict:

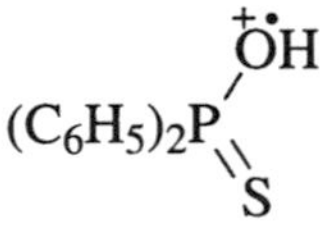

$$(C_6H_5)_2P(\overset{+\bullet}{O}H){=}S \quad (m/z\ 234)$$

Perhaps we may be 'pushing the boat out' a bit here, but P=O and S=O compounds *do* behave like C=O in the 'McL' manner.

**SAQ 7.5b**

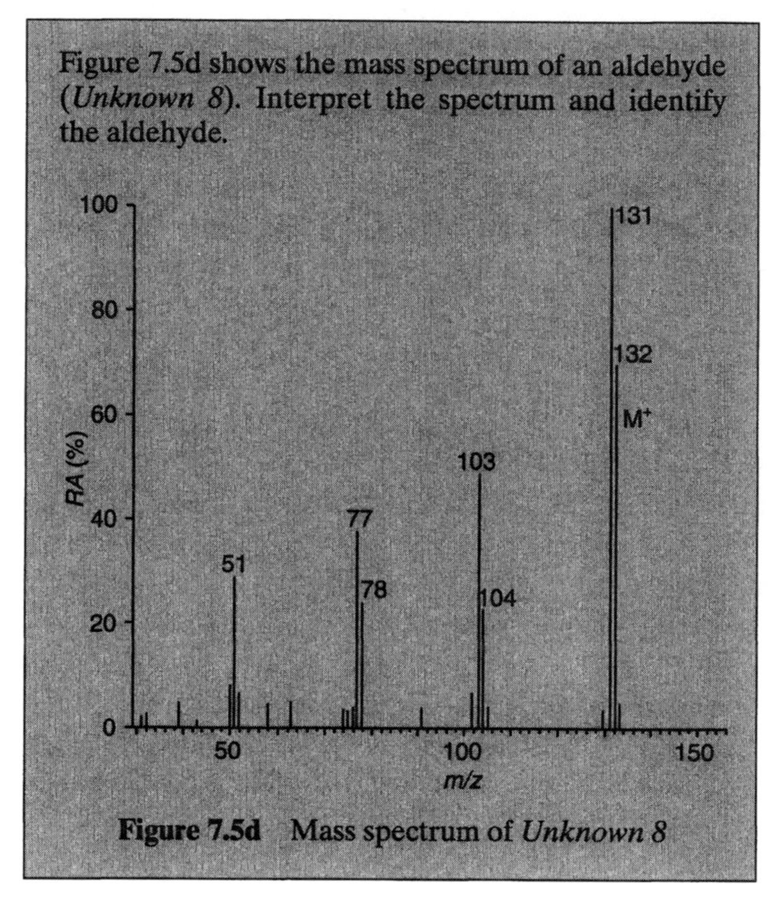

Figure 7.5d shows the mass spectrum of an aldehyde (*Unknown 8*). Interpret the spectrum and identify the aldehyde.



**Figure 7.5d** Mass spectrum of *Unknown 8*

## Response

*Unknown 8* is 3-phenylprop-2-en-1-al (cinnamaldehyde), $C_6H_5CH=CHCHO$. The small number of relatively intense high-mass ions show at once that this is an aromatic aldehyde, which is confirmed by the intense $(M-1)^+$, which is the base peak. This means that this ion is especially stabilised and must be conjugated with the benzene ring, a fact which is clearly indicated by the presence of *m/z* 77. Therefore, units of $C_6H_5$ and CHO must exist, leaving 26 amu to be accounted for. This cannot be CN, as the relative molecular mass would then be odd (the nitrogen rule), so it must be $C_2H_2$. This makes sense because it would contain a double bond and conjugate the —CHO to the benzene ring. The main fragmentations are shown below:

$C_6H_5CH{=}CH{-}C(H){=}\overset{+\bullet}{O}$ $\xrightarrow{-H^\bullet}$ $C_6H_5CH{=}CH{-}\overset{+}{C}{=}O$ ($m/z$ 131) $\longrightarrow$ $C_6H_5CH{=}CH^+ + CO$ ($m/z$ 103)

$m/z$ 103 $\xrightarrow{-HC\equiv CH}$ $C_6H_5^+$ ($m/z$ 77) $\xrightarrow{-HC\equiv CH}$ $C_4H_3^+$ ($m/z$ 51)

$\updownarrow$

(benzene ring)$^{+\bullet}$–CH=CH–C(H)=O $\xrightarrow{-CO}$ ring (+)–CH=$\dot{C}$H, with ring CH₂ ($m/z$ 104) $\xrightarrow{-HC\equiv CH}$ ring$^{+\bullet}$ with CH₂ ($m/z$ 78)

Did you notice that there were two parallel processes involving loss of CO, and then HC≡CH? One of these starts from $C_6H_5CH{=}CH\overset{+}{C}O$ and ends with $C_6H_5^+$ and is fairly obvious, while the other involves loss of CO from $M^{+\bullet}$ and leads to $m/z$ 78, an isomer of (benzene)$^{+\bullet}$. You should have been alerted to this by the intensity of $m/z$ 104 relative to 103. It is too intense to be a $^{13}C$ isotope peak of the latter. A six-centred transition state can be written for this, and it would be termed a McLafferty process, as mentioned in Section 7.5.1, even though the C=O group is not directly involved. It is remarkable how often even simple molecules like this can surprise us with the variety of their EI-induced fragmentations, but you must always bear in mind the large excess energy available to some $M^{+\bullet}$ ions. *Expect the unexpected!*

**SAQ 7.5c**

Figure 7.5f shows the mass spectrum of a ketone (*Unknown 9*). Interpret this spectrum and identify the ketone.

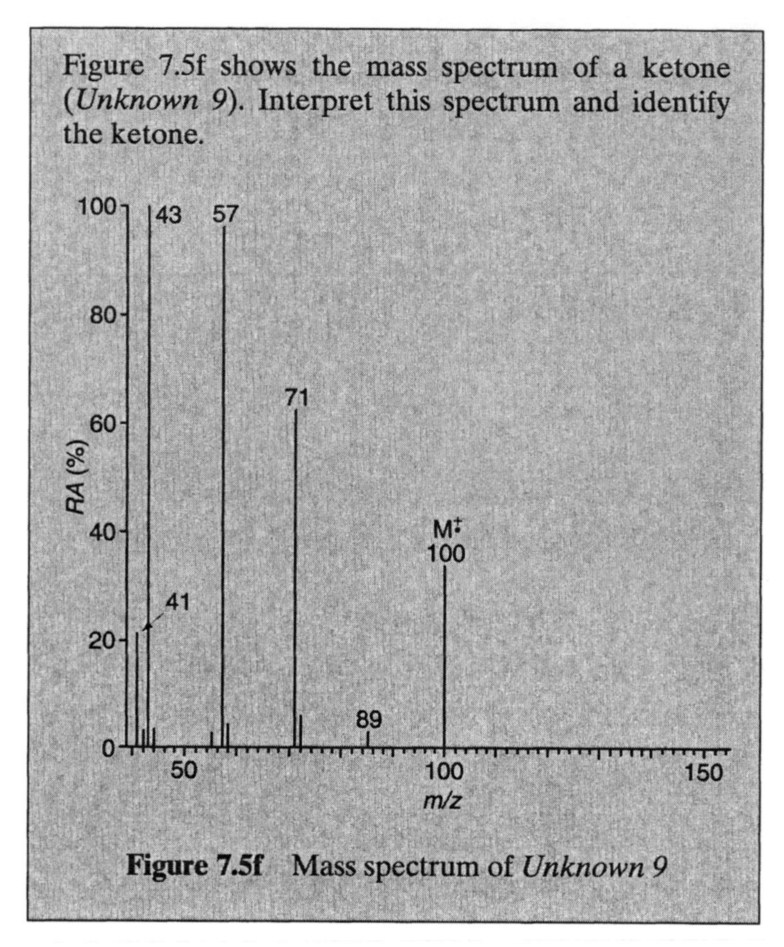


**Figure 7.5f** Mass spectrum of *Unknown 9*

## Response

*Unknown 9* is 3-hexanone, $CH_3CH_2COCH_2CH_2CH_3$, which is isomeric with 2-hexanone. The $M^{+\bullet}$ at *m/z* 100 shows it to be a saturated ketone. There are *no* even-mass ions of any consequence in the spectrum, therefore either there are *no* γ-hydrogens or they are in a $CH_3$ group which does not release them in the 'McL' process. We have seen this sometimes happens in propyl groups so would suspect this could be the case here. Therefore, the α-cleavage ions are $CH_3CH_2\overset{+}{C}O^+$, (*m/z* 57), and $CH_3CH_2CH_2CO^+$ (*m/z* 71), while the base peak is $CH_3CH_2\overset{+}{C}H_2$, which is derived from *m/z* 71 by the loss of :C=O. (M–$CH_3$), *m/z* 85, is present but is of lower relative abundance, which is typical of methyl compounds without branches.

**SAQ 7.5d**

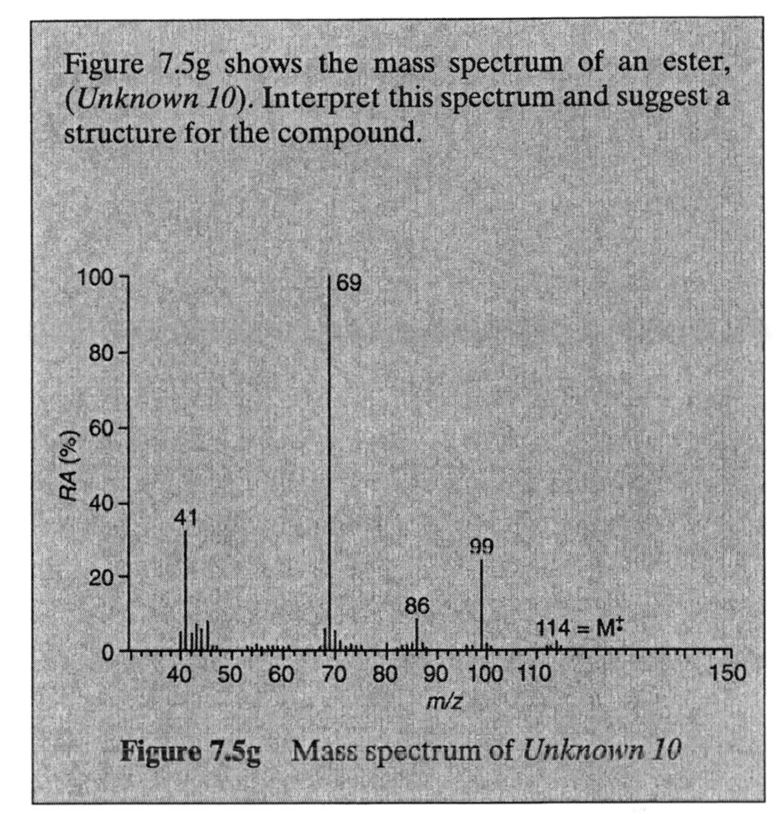

Figure 7.5g shows the mass spectrum of an ester, (*Unknown 10*). Interpret this spectrum and suggest a structure for the compound.



**Figure 7.5g** Mass spectrum of *Unknown 10*

**Response**

*Unknown 10* is ethyl 2-butenoate, $CH_3CH{=}CH{-}COOCH_2CH_3$. You might also have suggested its double bond isomer, $H_2C{=}CHCH_2COOCH_2CH_3$, but structures which reverse the acyl and alkyl groups, such as:

$$CH_3CH_2COOCH_2CH_2CH{=}CH_2$$

or

$$CH_3CH_2COOCH{=}CHCH_3$$

would either have other acyl-ion base peaks or ‘McL’/‘DHT’ peaks.

*Unknown 15* has a similar structure to ethyl 2-methylpropanoate ($CH_3COOCH_2CH_2CH_3$), but with a double bond since the $M_r$ is two less in value. The base peak *m/z* 69 must be an $RCO^+$ ion and this also is 2 amu less than the *m/z* 71 base peak in the propanoate ester, so the double bond must be in this R group. Possible structures are $CH_3CH{=}CHCO^+$ or $H_2C{=}CHCH_2CO^+$, but the relatively high intensity of the acylium ion would lead you to suspect that it might be conjugated and thus select the former structure. Loss of CO gives the *m/z* 41 peak:

$$CH_3CH{=}CH{-}\overset{+}{C}O \longrightarrow CH_3CH{=}CH^+ + CO$$

The peaks at *m/z* 86 and *m/z* 87 represent the 'McL' and 'DHT' ions derived from the loss of the ethyl group, with the 'McL' ion being the more intense, as is the usual case for an ethyl ester.

**SAQ 7.5e**

Figure 7.5i shows the mass spectrum of an amide (*Unknown 11*), which is isomeric with butanamide. Suggest a structure for this compound.

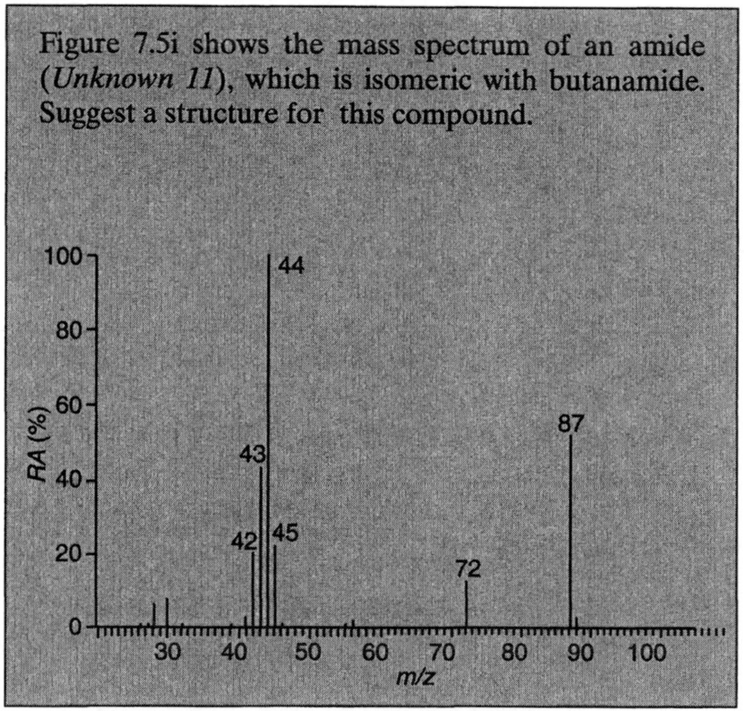


**Figure 7.5i** Mass spectrum of *Unknown 11*

**Response**

*Unknown 11* is *N,N*-dimethylethanamide, $CH_3CON(CH_3)_2$.

The base peak is $(CH_3)_2N^+$, *m/z* 44, and the next most intense ion, *m/z* 43, is $CH_3CO^+$, which are both formed by simple α-cleavages. Although *m/z* 44 is also characteristic of $OC{=}\overset{+}{N}H_2$, which can be formed from a primary amide, the absence of odd-mass 'McL' peaks rules out a long acid chain in this compound. The *m/z* 72 peak shows that $CH_3^{\bullet}$ loss occurs fairly readily from $M^{+\bullet}$, thus confirming the $CH_3CO$ group:

$$CH_3{-}\overset{O}{\overset{\|}{C}}{-}\overset{+\bullet}{N}(CH_3)_2 \longrightarrow CH_3^{\bullet} + O{=}C{=}\overset{+}{N}(CH_3)_2$$

*m/z* 72

If you concluded the *Unknown 11* was $(CH_3)_2CHCONH_2$, this was not a bad guess, because this would give a strong *m/z* 43 due to the $(CH_3)_2\overset{+}{C}H$, and no 'McL' peak. However, the $M^{+\bullet}$ in *Unknown 11* is quite intense, which is characteristic of *N*-substituted amides, while *m/z* 43 is not as intense as would be expected for the stable secondary $(CH_3)_2\overset{+}{C}H$ ion.

**SAQ 7.6a**

Figure 7.6b shows the mass spectrum of a tertiary amine (*Unknown 12*). Suggest a structure for this compound.

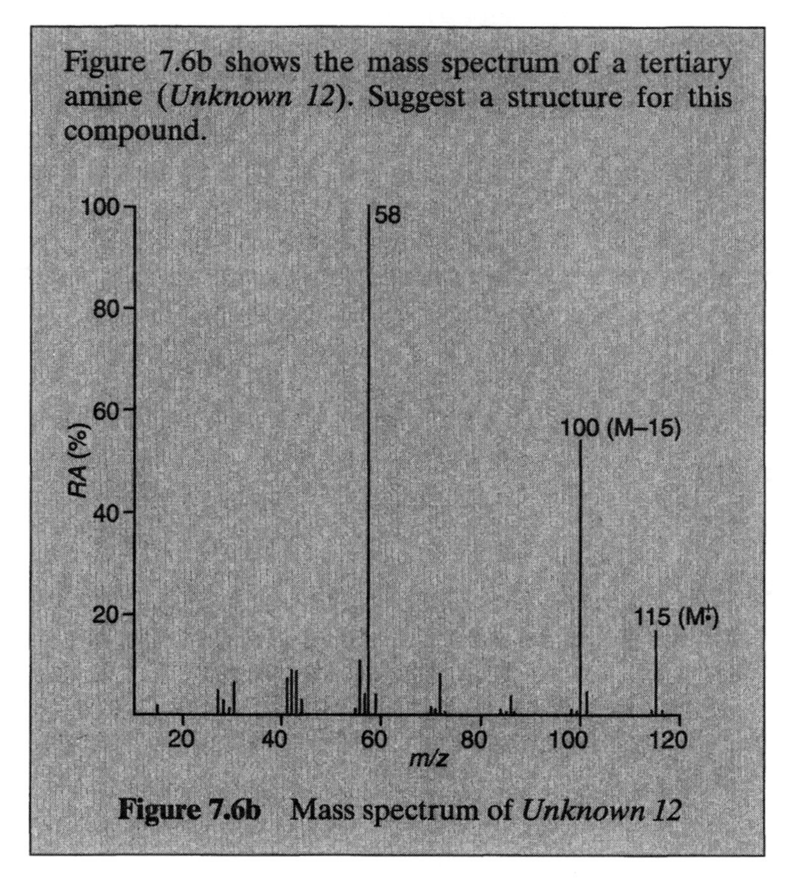


**Figure 7.6b** Mass spectrum of *Unknown 12*

## Response

*Unknown 12* is *N*-methyldiisopropylamine, $((CH_3)_2CH)_2NCH_3$, i.e. a saturated heptylamine.

The large $(M–CH_3)$ peak indicates the ready loss of a β-methyl group, which is typical of the $N—CH(CH_3)R$ structure. The *m/z* 58 ion must be formed from *m/z* 100 by alkene elimination, along with hydrogen transfer, so the alkene that is lost must be propene ($C_3H_6$ = 42 amu). The partial structure indicated for *m/z* 100 is therefore:

$$\begin{matrix} R_1 & & CH_3 \\ & \backslash & | \\ & C=\overset{+}{N}-CH-CH_3 \\ & / & | \\ R_2 & & CH_2 \\ & & H \end{matrix} \longrightarrow \begin{matrix} R_1 \\ \backslash \\ C=\overset{+}{N}HCH_3 \\ / \\ R_2 \end{matrix}$$

*m/z* 100      *m/z* 58

It follows that $R_1$ = H, and $R_2$ = $CH_3$ in the *m/z* 58 ion. If either or both of the isopropyl groups were propyl, then the original α-cleavage reaction would take place by loss of $CH_3CH_2^{\bullet}$ to give *m/z* 86 (by Stevenson's rule):

$$(CH_3)_2CH-\overset{CH_3}{\underset{+\bullet}{N}}-CH_2-CH_2-CH_3 \longrightarrow (CH_3)_2CH-\overset{CH_3}{\underset{+}{N}}=CH_2 + CH_3CH_2^{\bullet}$$

*m/z* 86

$\downarrow$ $-CH_3NH=CH_2$

$CH_3\overset{+}{N}H=CH_3$

*m/z* 44 (base peak)

The base peak would then be *m/z* 44, and *not m/z* 56.

**SAQ 7.6b**

Figure 7.6e shows the mass spectrum of an amine (*Unknown 13*). Suggest a structure for this compound.

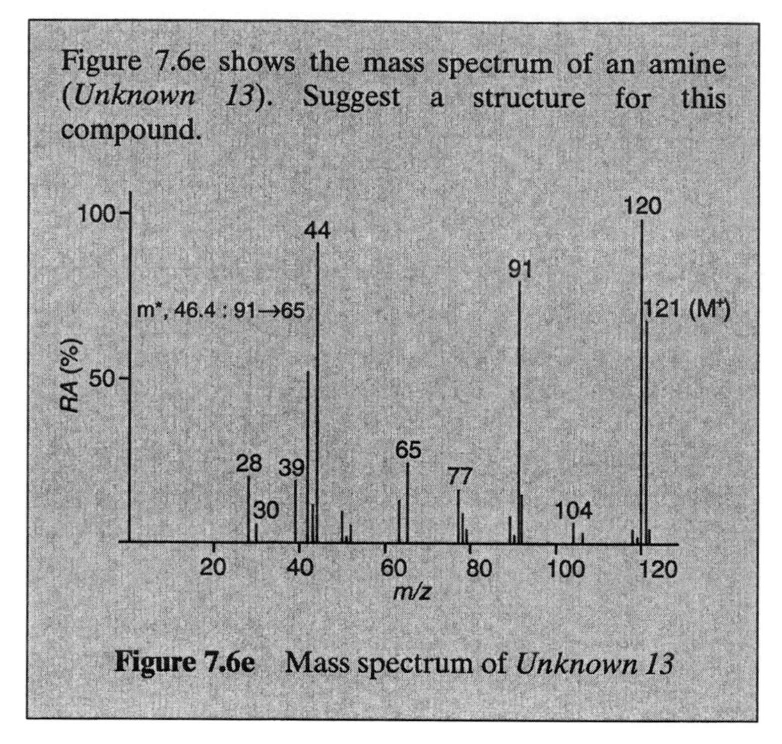


**Figure 7.6e** Mass spectrum of *Unknown 13*

## Response

*Unknown 13* is *N*-methylbenzylamine, $C_6H_5CH_2NHCH_3$.

The spectrum shows intense ions at (M–1), *m/z* 120, 91 and 44. There are also significant ions at *m/z* 77 and 65, and an *m** for 91 ⟶ 65. This strongly suggests a benzyl group, rather than an aniline derivative, such as $C_6H_5NHCH_2CH_3$ or $C_6H_5N(CH_3)_2$. The *m/z* 44 is $CH_2{=}\overset{+}{N}HCH_3$, arising from the loss of $C_6H_5^{\bullet}$. This is as expected from Stevenson's rule. Loss of a benzylic hydrogen from $M^{+\bullet}$ would give $C_6H_5CH{=}\overset{+}{N}HCH_3$, which is of course stabilised by resonance with the benzene ring and may indeed be a methylamino-substituted tropylium ion:

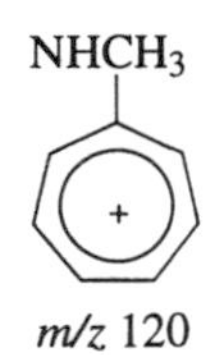


**SAQ 7.7a**

Figure 7.7b shows the mass spectrum of a hydrocarbon (*Unknown 14*). Try to identify this compound.

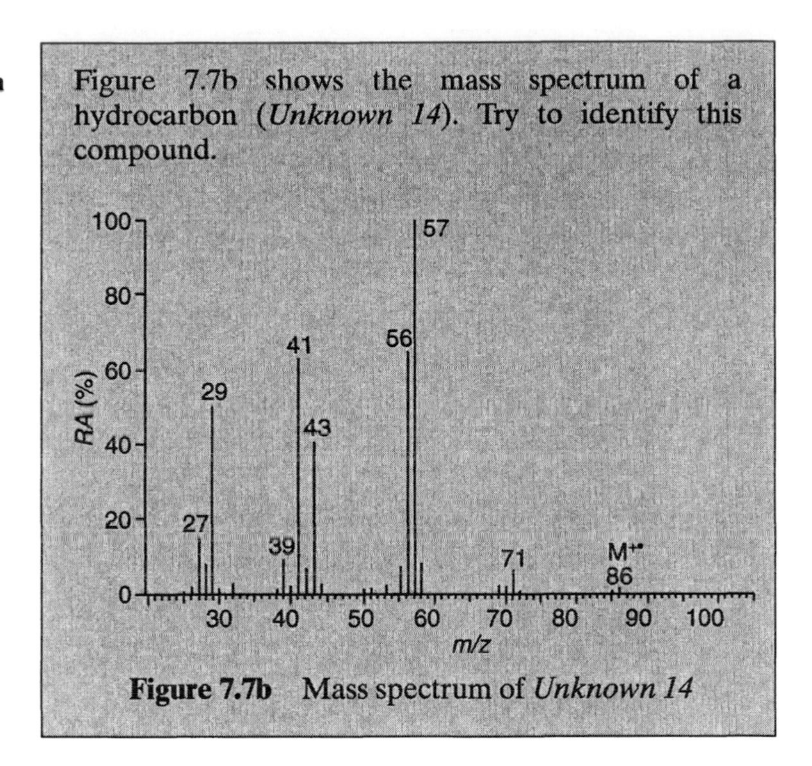


**Figure 7.7b** Mass spectrum of *Unknown 14*

**Response**

*Unknown 14* is $CH_3CH_2CH(CH_3)CH_2CH_3$, i.e. 3-methylpentane.

The molecular ion at *m/z* 86 shows it to be a hexane isomer, while the *m/z* 57 base peak is very intense compared to either 71 or 43, showing that the chain is branched so as to give preferential cleavage at $C_3—C_4$. The *m/z* 57 ion must be either:

$$CH_3\overset{+}{C}HCH_2CH_3 \text{ or } (CH_3)_3C^+$$

thus indicating:

$$CH_3CH_2CH(CH_3)CH_2CH_3 \text{ or } (CH_3)_3CCH_2CH_3$$

both of which would lose $CH_3CH_2^\bullet$ preferentially (Stevenson's rule).

If you got either of these, then well done. We can tell the difference between them by considering the $(M–CH_3)^+$ peak, *m/z* 71. This is quite small in this hydrocarbon, showing that the loss of $CH_3^\bullet$ is not favoured. This would be the case for 3-methylpentane, as the resulting *m/z* 71 is $(CH_3CH_2)_2CH^+$, i.e. a *secondary* ion. In the alternative structure, namely 2,2-dimethylbutane, the loss of $CH_3^\bullet$ would give $(CH_3)_2\overset{+}{C}CH_2CH_3$, namely a *tertiary* ion. This would give a more intense *m/z* 71.

**SAQ 7.7b**

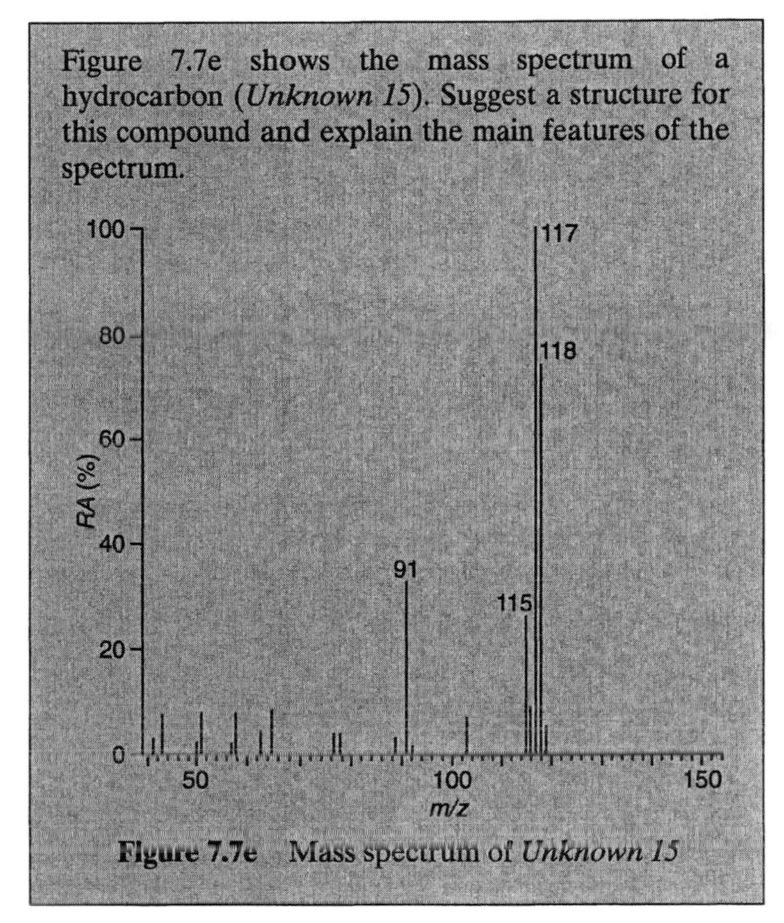

Figure 7.7e shows the mass spectrum of a hydrocarbon (*Unknown 15*). Suggest a structure for this compound and explain the main features of the spectrum.



**Figure 7.7e** Mass spectrum of *Unknown 15*

**Response**

*Unknown 15* is 3-phenyl propene (allylbenzene), $C_6H_5CH_2CH{=}CH_2$.

The presence of *m/z* 91 shows it to be a benzyl compound, leaving 27 amu to be assigned. This must be $C_2H_3$ and the compound has a double bond in the side-chain. Therefore, it must be $C_6H_5CH_2CH{=}CH_2$. The isomer, $C_6H_5CH{=}CHCH_3$, clearly could not fragment to give *m/z* 91 without hydrogen transfer, which would not be expected from an unsaturated side-chain. The (M–1) base peak would be due to the loss of a benzyl hydrogen and subsequent formation of a ethenyltropylium ion:

$C_6H_5-\overset{+}{C}H-CH=CH_2$ ⟶ (ring with + and $CH=CH_2$)

*m/z* 117

The peak at *m/z* 115 is difficult to explain on the basis of the information given in Section 7.7, so don't worry if you could not put a structure to this ion. It is, in fact, a bicyclic species formed by the loss of two further hydrogen atoms from *m/z* 117:

⟷ ⟷ etc. (*eight other forms!*)

**SAQ 7.8**

Figure 7.8b shows the mass spectrum of a haloalkane (*Unknown 16*). Suggest a structure for this compound, and assign structures (as far as you can) to the ions *m/z* 77/79, 63/65, 57, 56, and 41.

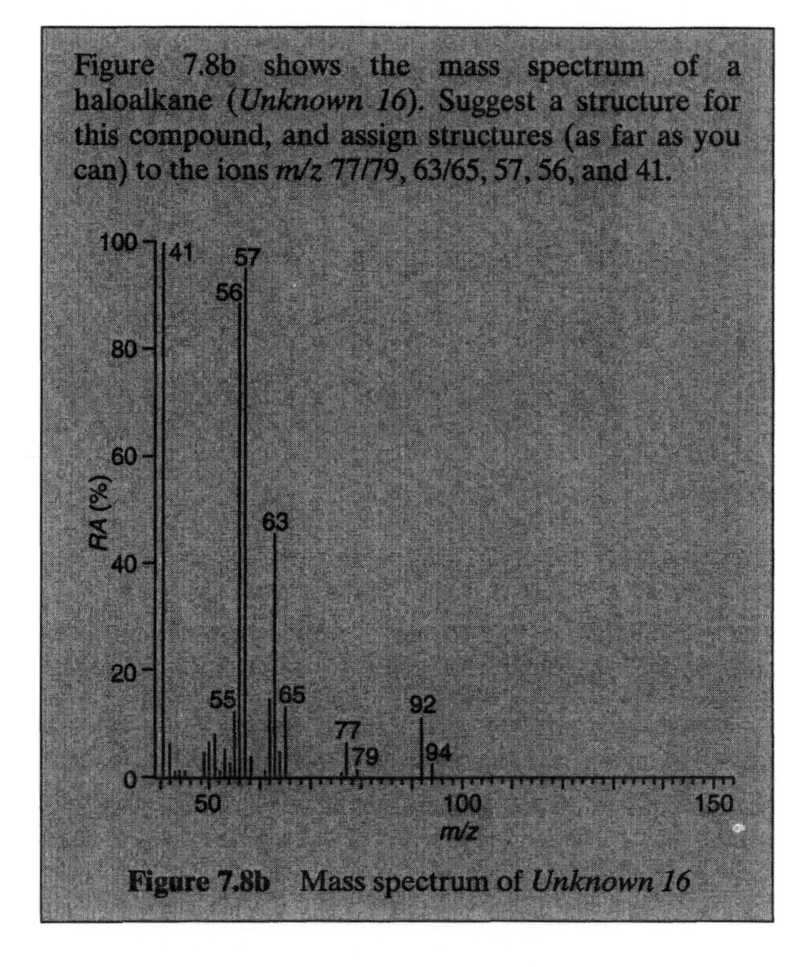


**Figure 7.8b** Mass spectrum of *Unknown 16*

**Response**

*Unknown 16* is 2-chlorobutane, $CH_3CHClCH_2CH_3$ ($M_r$ 92/94).

The 3/1 ratio shown by *m/z* 92/94 clearly shows this to be a monochloroalkane. Two other chlorine-containing ions appear at *m/z* 77/79 and 63/65, showing the losses of $CH_3^{\bullet}$ and $CH_3CH_2^{\bullet}$, respectively. This, coupled with the relatively high relative abundance of the molecular ion, shows that *Unknown 16* is a branched compound with $CH_3$ and $CH_3CH_2$ groups attached to the C—Cl carbon. $CH_3CH_2^{\bullet}$ is lost preferentially to give *m/z* 63/65 (50%), rather than *m/z* 77/79 (10%).

Loss of Cl• gives $CH_3\overset{+}{C}HCH_2CH_3$, *m/z* 57, while loss of HCl gives $CH_3\overset{+}{C}H\overset{\bullet}{C}HCH_3$, *m/z* 56.

The base peak, *m/z* 41, is $CH_3CH{=}\overset{+}{C}H$, which is formed by loss of $CH_3^{\bullet}$ from the latter. Note that *Unknown 16* cannot be the isomer, 2-chloro-2-methylpropane, $(CH_3)_3CCl$, because this could not give $(M{-}CH_3CH_2)^+$ ions.

**SAQ 7.10a** The mass spectrum of an unknown compound shows the following features: $M^{+\bullet}$ *m/z* 126 (82%), 111 (100%), 83 (36%), 57 (16%), 45 (18%), and 39 (25%), plus *m** 62.1 and 39.1. All of these ions have (M + 2) peaks of *ca.* 5%. Which of the following compounds do you think it is?

A $CH_3CO$–(2-thiazolyl)

B $CH_3CO_2$–(2-furyl)

C $CH_3CH_2CH_3$–(2-thiazolyl)

D $(CH_3)_2CH$–(2-thiazolyl)

**Response**

The compound is 2-acetylthiazole (A).

It cannot be B as this has no sulphur atom, which is clearly shown to be present by the (M + 2) peaks. In any case, B would give:

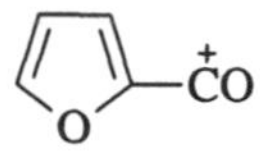

with *m/z* 95 as the base peak, by α-cleavage of the OC—$OCH_3$ bond.

Compound C would show easy β-cleavage to give the thiazole tropylium ion, *m/z* 97, as the base peak.

D would lose $CH_3$ to form the methyltropylium ion of thiazole, *m/z* 111, i.e. $C_4H_4S\overset{+}{C}HCH_3$, which would further fragment by loss of HC=CH to *m/z* 85, and not by loss of CO, as shown by the *m** at *m/z* 62.1.

The fragmentation of 2-acetylthiazole is as follows:

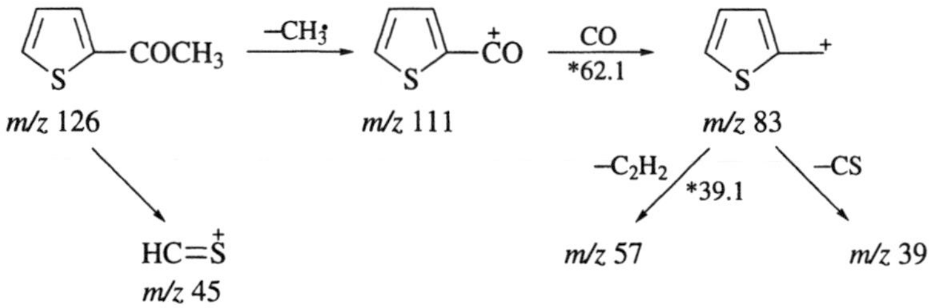

$-CH_3^{\bullet}$

*m/z* 126   *m/z* 111   *m/z* 83

CO, *62.1

$-C_2H_2$, *39.1   −CS

$HC{=}\overset{+}{S}$ *m/z* 45   *m/z* 57   *m/z* 39

**SAQ 7.10b**

An unknown compound contains one sulphur atom and its mass spectrum shows the following major ions (*m/z*): 138 (100%) ($M^{+\bullet}$); 123 (65%); 110 (66%); 109 (24%); 65 (21%); 51 (23%); 45 (32%). The sulphur atom is represented by peaks at *m/z* 138, 123, 110, 109 and 45.

The compound could be one of the following:

(a) $C_6H_5CH_2SCH_3$;

(b) 4-$CH_3C_6H_4SCH_3$;

(c) 4-$CH_3CH_2C_6H_4SH$;

(d) $C_6H_5SCH_2CH_3$.

Which do you think it is?

**Response**

The compound is ethyl phenyl sulphide, $C_6H_5SCH_2CH_3$ (d)

Isomer (a) is a benzylic thioether. This would be expected to give an intense *m/z* 91, and the usual fragments (*m/z* 65, 51, and 39) from that compound. It would not give a particularly stable $(M–CH_3)^+$ ion which our 'unknown' shows.

Isomer (b) is more plausible because the loss of $CH_3$ here could give a stable thiotropylium ion with a methyl substituent. This should then lose C=S to give *m/z* 79, but this is not found in our spectrum. Loss of $CH_3S$ would also be expected, but *m/z* 91 is not found. Hence (b) is unlikely.

Isomer (c) is thiophenol. It would also be expected to lose $CH_3^\bullet$ to give a sulphhydryl tropylium ion, and then lose CS from this, but the lack of *m/z* 79 throws doubt on this structure. It should also lose $HS^{\bullet}$ direct from $M^{+\bullet}$, and perhaps HC=CH, but *m/z* 105 and 112 are not found.

The fragmentation pattern of (d) is as follows:

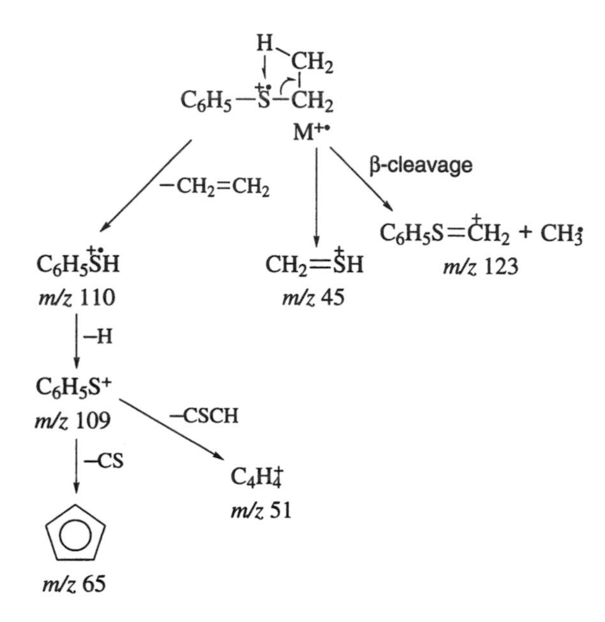


Apart from *m/z* 123 and 45, the ions in this spectrum result from the formation of the molecular ion of thiophenol (*m/z* 110) and its daughter ions. The (110–$C_2H_2$) and (110–SH) ions are present in the spectrum but are less than 10% relative abundance.

**SAQ 7.11a**

Describe the principles of operation of a MALDI ion source? Refer back to Chapter 2, if you are unsure.

**Response**

A MALDI instrument uses a nitrogen-UV laser to produce a pulse of a few ns, at a certain wavelength, through a neutral-density filter in conjunction with a variable iris (to vary the energy). The ion source is maintained at *ca.* 20 kV.

The solid sample is dissolved in a suitable matrix solvent and the solution is then evaporated. The laser pulse electronically excites the matrix molecules, thus resulting in desorption of the ions that are formed by proton transfer between the excited matrix–analyte mixture.

Two of the most commonly used MALDI matrices in peptide analysis are 2,5-dihydroxybenzoic acid (DHB) and α-cyano-4-hydroxycinnamic acid (4-HCCA).

**SAQ 7.11b**

Can you think of any other classes of biomolecules that could be analysed by mass spectrometry?

These could include oligosaccharides, lipids, and oligonucleotides, but there are many other examples. You could investigate this further, if your are interested, by carrying out a literature search, for example by using Chemical Abstracts.

**SAQ 8.1a**

What would be the effect on the apparent intensity of the molecular ion and fragment-ion peaks if the sample concentration of the component was rapidly increasing as the scan was made, i.e. from low-mass to high-mass?

**Response**

This might be the case if a GC peak was just entering the source. Here, the molecular ion and closely related ions would be recorded when the concentration was small and their intensities would be reduced relative to the higher-mass ions in the spectrum.

**SAQ 8.1b**

Suggest derivatives which would be suitable for GC/MS analysis of each of the following classes of compound:

(a) fatty acids;

(b) amino acids and peptides;

(c) carbohydrates;

(d) phenols;

(e) 2-hydroxyacids;

(f) steroid hormones (containing C=O and OH groups);

(g) fruit essences (containing alcohols, carbonyl compounds and esters).

In some cases there will be a choice — select which derivative you think would be best from the mass spectrometry point of view.

**Response**

(a) Methyl esters are most suitable for acids. TMS derivatives could be used, but they increase $M_r$ by 72 amu per acid group, which may be disadvantageous.

(b) Amino acids and peptides are best methylated to give the methyl esters from the free $-CO_2H$ groups, and either acetylated or trifluoroacetylated on the free $-NH_2$ groups. Neither alone is sufficient for good GC/MS analysis. TMS treatment would derivatise both $-CO_2H$ and $-NH_2$, but gives a large mass increment (144).

(c) Permethylation (using $OCH_3$ derivatives) is most suitable because of the number of $-OH$ groups in the carbohydrate. The use of acetates, trifluoroacetates and TMS derivatives would push $M_r$ up to levels beyond the usual range of GC/MS. These latter groups are sometimes used with monosaccharides which have a smaller number of $-OH$ groups.

(d) There is a real choice here unless the phenol has a high $M_r$ — any of the $-OH$ derivatives in Table 8.1a could be used as phenols are easily derivatised. Therefore, $-OCH_3$, $-OSi(CH_3)_3$, $-COCH_3$, and $-OCOCF_3$ could be used. At a pinch, the lower $M_r$ species, $-OCH_3$, would be best.

(e) Any of the three boronate derivatives could be your choice here, as a 2-hydroxyacid would form a five-membered cyclic boronate derivative.

(f) Steroid hormones contain $C{=}O$ groups, so your choice should fall on the *O*-methyloxime derivatives, plus perhaps using TMS for good measure for any $-OH$ groups that are present.

(g) Fruit essences usually contain mainly esters (which do not require derivatising), plus alcohols and limited amounts of aldehydes and ketones. The best strategy would be a TMS derivatisation for the alcohols, in order to help reduce tailing on the column and improve GC separation, while still providing observable $M^{+\bullet}$ ions.

**SAQ 8.1c**

Discuss the ways in which you think GC and MS are (i) compatible, and (ii) incompatible techniques.

**Response**

The compatibility of the two techniques can be considered under the following general aspects:

(a) *Volatility*—both can cater for gaseous, liquid and solid samples which have high vapour pressures.

(b) *Sensitivity*—mass spectrometers have high sensitivities, being able to detect down to pg levels of sample; this is adequate for capillary-column GC, which is where it counts.

(c) *Temperature*—the operating temperature range of GC columns is very similar to that of MS inlet systems and sources.

(d) *Speed*—mass spectrometers are available which can scan fast enough to produce several spectra, even across a narrow GC peak emerging from a capillary column.

(e) *Chromatogram Output*—a simple total-ion-current (TIC) device enables mass spectrometers to produce a mass chromatogram which is very similar to a flame-ionisation detector (FID) or other GC-detector chromatogram. Alternatively, the data system can reconstruct a TIC trace by the summation of all of the ion-peak intensities recorded in each scan. Splitting of the sample stream is unnecessary.

On the minus side, pressure difference is a major problem. Other disadvantages of combining the two techniques are column bleed, the need to derivatise some samples, e.g. alcohols, and the basic cost of the instrumentation.

**SAQ 8.1d**

> Think back to how a magnetic-sector mass spectrometer is scanned. What do you think is the best way of carrying out SIM? Are quadrupole analysers suitable for SIM work?

**Response**

Accelerating voltage switching is used. The magnetic-sector mass spectrometer equation is given by:

$$m/z = B^2r^2/2V$$

Normally, $B$ is scanned while $r$ and $V$ are constant. However, over a small mass difference, it is easier to switch from one accelerating voltage to another in order to focus selected ions (some machines have provision for eight ions to be monitored sequentially), while the magnetic field is held constant. This avoids the delay in recycling the magnet and enables more rapid switching between the selected ions, and is therefore the method which is usually adopted, provided that the $m/z$ values lie within a range of 30% or so. Quadrupoles focus ions by a suitable combination of RF and DC fields applied to their four rods in pairs. These can be switched from one selected pair of values to another very rapidly, and thus a quadrupole can also be an excellent SIM analyser.

**SAQ 8.2a**

What are the fundamental requirements for a LC/MS system?

**Response**

(a) An interface capable of handling at least 1 $cm^3 min^{-1}$ of an aqueous buffer solution, i.e. 200 $cm^3 min^{-1}$ of vapour.

(b) A mass spectrometer ionisation system capable of providing molecular ions or $MH^+$, and structural information on large, non-volatile, and thermally labile molecules.

**SAQ 8.2b**

What is the major problem in coupling a liquid chromatograph directly to a mass spectrometer and how has this problem been overcome?

**Response**

The major problem is the solvent volume and its rate of travel through the chromatograph. In typical LC separations this is 0.5–5 $cm^3$ $min^{-1}$. This translates into gas-volume flows of 100–3000 $cm^3$ $min^{-1}$, which is one or two order of magnitude greater than GC flows. The efficient removal and concentration of the sample is beyond the mass spectrometer pumping capacity if such flows are passed directly into the ion source. The presence of buffers only adds to this problem.

Various methods have been used to overcome this: These include the following:

(a) Direct liquid introduction splits the LC effluent into two streams, with *ca.* 1% being admitted to the mass spectrometer as a jet into the ion source. The vaporised solvent then acts as a reagent gas for CI. This has the disadvantages that 99% of the sample is not examined, only a limited range of solvents can be used, and only relative-molecular-mass information can be obtained when operating in the CI mode. With the advent of microbore columns this method might be used more widely but it lacks sensitivity and many compounds do not give $MH^+$.

(b) Thermospray (TS) can take flows up to a few $cm^3$ $min^{-1}$ directly into an ion source which is rather than like a CI source (see Figure 8.2b). Provided that at least 10% water is present in the solvent, plus ammonium salts as buffers, the electric field generated as the charged droplets shrink in the vacuum causes $MH^+$ ions to be formed. These are sampled by means of a cone inserted into the beam and admitted to the mass spectrometer analyser, which can be either a magnetic or quadrupole type. This is a 'soft' ionising technique rather like FD or CI and will readily produce ions from very polar biological molecules and charged species of high relative molecular mass. Therefore, it extends both the mass range and the types of compounds which can be examined by mass spectrometry very significantly. Some very polar molecules give multiply protonated molecular ions which analyse at *m/z* values of $M_r/n$, where $n$ is the number of protons acquired. In general, the mass spectra are very comparable to CI spectra, but more sample is required on column to give similar $MH^+$ intensities. Structurally related ions are sometimes, but not always, obtained. On balance, thermo spray is a very promising interface design, but it is not, at the moment, a universal solution to the problems posed by the LC solvent.

(c) Moving-belt interfaces can also take flow rates up to a few $cm^3$ $min^{-1}$, but here the idea is to evaporate the solvent before the belt carries the sample into the ion-source chamber. Provided that this can be achieved efficiently for a given solvent mixture, conventional ionisation methods, such as EI, CI and, most importantly, FAB, can be used to give $M^+$ and $MH^+$ species directly off of the belt. The matrix required for FAB can be introduced into a wash-bath situated where the belt leaves the vacuum system, so is constantly renewed while remaining buffer salts and sample residues are removed, thus reducing the 'memory' of previous samples. By spraying the LC effluent on to the belt at an angle, much more efficient evaporation can be achieved, and transfer efficiencies are quite good (40–80%). The method gives good results for drugs and other biomolecules of high relative molecular mass, down to the ng level, which is better than current thermosprays.

**SAQ 9.1a**

> Can you think of any more processes that might be involved?

**Response**

(a) Desorption and ionisation of the solid in the region immediately adjacent to the laser pulse;

(b) gas-phase, ion–molecule reactions;

(c) emission of neutral particles versus ion formation.

**SAQ 9.1b**

> Why can't PIXE be used to measure ions which are located a few mm below the surface of a sample?

**Response**

This is because of absorption of X-rays by the sample itself.

**SAQ 9.2a**

> What is a possible weakness of this method of analysis?

**Response**

The abundant-isotope current must remain relatively stable during pulsing between the stable- and rare-isotope measurements. Modern AMS machines use microsecond pulsing to overcome this potential problem.

**SAQ 9.2b**

How could you increase the conductivity of a solid?

**Response**

You could try mixing the sample with graphite powder, for example.

**SAQ 9.3**

Why is there a problem analysing for Ca in ICP–MS?

**Response**

Argon gas is used to produce the plasma, and this will give rise to a very intense ion of mass 40 which would mask any calcium ions.

# Units of Measurement

For historical reasons a number of different units of measurement have evolved to express quantities of the same thing. In the 1960s, many international scientific bodies recommended the standardisation of names and symbols and the adoption universally of a coherent set of units — the SI units (Système Internationale d'Unités) — based on the definition of seven basic units, namely metre (m), kilogram (kg), second (s), ampere (A), kelvin (K), mole (mol), and candela (cd).

The earlier literature references and some of the older text books naturally use the older units. Even today, many practising scientists have still not adopted SI units as their working units. It is, therefore, necessary to know of the older units and to be able to interconvert these with the SI units.

In this series of texts, SI units are used as standard practice. However, in areas of activity where their use has not become general practice, e.g. biologically based laboratories, the earlier defined units are used. This is explained in the study guide to each unit.

Table 1 shows some symbols and abbreviations commonly used in analytical chemistry, while Table 2 shows some of the alternative methods for expressing the values of physical quantities and their relationship to the values in SI units. In addition, Table 3 lists prefixes for SI units and Table 4 shows the recommended values of a selection of physical constants.

Further details and definitions of other units may be found in I. Mills, T. Cvitaš, K. Homann, N. Kallay and K. Kuchitsu, *Quantities, Units and Symbols in Physical Chemistry*, 2nd Edn, Blackwell Science, Oxford, 1993.

**Table 1** Symbols and abbreviations commonly used in analytical chemistry

| Symbol | Meaning |
|---|---|
| Å | angstrom |
| $A_r(X)$ | relative atomic mass of X |
| A | ampere |
| $E$ or $U$ | energy |
| $G$ | Gibbs free energy (function) |
| $H$ | enthalpy |
| $I$ (or $i$) | electric current |
| J | joule |
| K | kelvin (= 273.15 + $t$ (°C)) |
| $K$ | equilibrium constant (with subscripts p, c, etc.) |
| $K_a$, $K_b$ | acid and base ionisation constants |
| $M_r(X)$ | relative molecular mass of X |
| N | newton (SI unit of force) |
| $P$ | total pressure |
| $s$ | standard deviation |
| $T$ | temperature (K) |
| $V$ | volume |
| V | volt (J $A^{-1}$ $s^{-1}$) |
| $a$, $a$(A) | activity, activity of A |
| $c$ | concentration (mol $dm^{-3}$) |
| $e$ | electron |
| g | gram |
| s | second |
| $t$ | temperature (°C) |
| | |
| b.p. | boiling point |
| f.p. | freezing point |
| m.p. | melting point |
| | |
| ~ | approximately equal to |
| $<$ | less than |
| $>$ | greater than |
| e, exp($x$) | exponential of $x$ |
| ln $x$ | natural logarithm of $x$; ln $x$ = 2.303 log $x$ |
| log $x$ | common logarithm of $x$ to base 10 |

**Table 2** Summary of alternative methods of expressing physical quantities

(1) Mass (SI unit, kg)

| | | |
|---|---|---|
| g | = | $10^{-3}$ kg |
| mg | = | $10^{-3}$ g = $10^{-6}$ kg |
| μg | = | $10^{-6}$ g = $10^{-9}$ kg |

(2) Length (SI unit, m)

| | | |
|---|---|---|
| cm | = | $10^{-2}$ m |
| Å | = | $10^{-10}$ m |
| nm | = | $10^{-9}$ m = 10 Å |
| pm | = | $10^{-12}$ m = $10^{-2}$ Å |

(3) Volume (SI unit, $m^3$)

| | | |
|---|---|---|
| l | = | $dm^3$ = $10^{-3}$ $m^3$ |
| ml | = | $cm^3$ = $10^{-6}$ $m^3$ |
| μl | = | $10^{-3}$ $cm^3$ |

(4) Concentration (SI unit, mol $m^{-3}$)

| | | |
|---|---|---|
| M | = | mol $l^{-1}$ = mol $dm^{-3}$ = $10^3$ mol $m^{-3}$ |
| mg $l^{-1}$ | = | μg $cm^{-3}$ = ppm = $10^{-3}$ g $dm^{-3}$ |
| μg $g^{-1}$ | = | ppm = $10^{-6}$ g $g^{-1}$ |
| ng $cm^{-3}$ | = | ppb = $10^{-6}$ g $dm^{-3}$ |
| pg $g^{-1}$ | = | ppt = $10^{-12}$ g $g^{-1}$ |
| mg% | = | $10^{-2}$ g $dm^{-3}$ |
| μg% | = | $10^{-5}$ g $dm^{-3}$ |

(5) Pressure (SI unit, N $m^{-2}$ = kg $m^{-1}$ $s^{-2}$)

| | | |
|---|---|---|
| Pa | = | N $m^{-2}$ |
| atm | = | 101 325 N $m^{-2}$ |
| bar | = | $10^5$ N $m^{-2}$ |
| torr | = | mmHg = 133.322 N $m^{-2}$ |

(6) Energy (SI unit, J = kg $m^2$ $s^{-2}$)

| | | |
|---|---|---|
| cal | = | 4.184 J |
| erg | = | $10^{-7}$ J |
| eV | = | $1.602 \times 10^{-19}$ J |

**Table 3** Prefixes for SI units

| Fraction | Prefix | Symbol |
|---|---|---|
| $10^{-1}$ | deci | d |
| $10^{-2}$ | centi | c |
| $10^{-3}$ | milli | m |
| $10^{-6}$ | micro | μ |
| $10^{-9}$ | nano | n |
| $10^{-12}$ | pico | p |
| $10^{-15}$ | femto | f |
| $10^{-18}$ | atto | a |
| **Multiple** | **Prefix** | **Symbol** |
| 10 | deca | da |
| $10^{2}$ | hecto | h |
| $10^{3}$ | kilo | k |
| $10^{6}$ | mega | M |
| $10^{9}$ | giga | G |
| $10^{12}$ | tera | T |
| $10^{15}$ | peta | P |
| $10^{18}$ | exa | E |

**Table 4** Recommended values of physical constants

| Constant | Symbol | Value |
|---|---|---|
| Acceleration due to gravity | $g$ | $9.81\ m\ s^{-2}$ |
| Avogadro constant | $N_A$ | $6.022\ 14 \times 10^{23}\ mol^{-1}$ |
| Boltzmann constant | $k$ | $1.380\ 66 \times 10^{-23}\ J\ K^{-1}$ |
| Charge-to-mass ratio | $e/m$ | $1.758\ 796 \times 10^{11}\ C\ kg^{-1}$ |
| Electronic charge | $e$ | $1.602\ 18 \times 10^{-19}\ C$ |
| Faraday constant | $F$ | $9.648\ 46 \times 10^{4}\ C\ mol^{-1}$ |
| Gas constant | $R$ | $8.314\ J\ K^{-1}\ mol^{-1}$ |
| Ice-point temperature | $T_{ice}$ | $273.150\ K$ [a] |
| Molar volume of ideal gas (stp) | $V_m$ | $2.241\ 38 \times 10^{-2}\ m^3\ mol^{-1}$ |
| Permittivity of a vacuum | $\epsilon_0$ | $8.854\ 188 \times 10^{-12}$ $kg^{-1}\ m^{-3}\ s^4\ A^2\ (F\ m^{-1})$ |
| Planck constant | $h$ | $6.626\ 08 \times 10^{-34}\ J\ s$ |
| Standard atmosphere (pressure) | $p$ | $101\ 325\ N\ m^{-2}$ [a] |
| Atomic mass constant | $m_u$ | $1.660\ 54 \times 10^{-27}\ kg$ |
| Speed of light in a vacuum | $c$ | $2.997\ 925 \times 10^{8}\ m\ s^{-1}$ |

[a] Exact value

# Index

Entries marked with an asterix * are mass spectra. Those in **bold** refer to the analysis of that compound or group of compounds.

# The Modern Periodic Table of the Elements

Key: 1 — Atomic number; H; 1.0079 — Atomic mass

| PERIODS | IA | IIA | IIIB | IVB | VB | VIB | VIIB | VIII | VIII | VIII | IB | IIB | IIIA | IVA | VA | VIA | VIIA | NOBLE GASES 0 |
|---|---|---|---|---|---|---|---|---|---|---|---|---|---|---|---|---|---|---|
| 1 | 1<br>H<br>1.00797 | | | | | | | | | | | | | | | | | 2<br>He<br>4.00260 |
| 2 | 3<br>Li<br>6.941 | 4<br>Be<br>9.01218 | | | | | | | | | | | 5<br>B<br>10.81 | 6<br>C<br>12.01115 | 7<br>N<br>14.0067 | 8<br>O<br>15.9994 | 9<br>F<br>18.99840 | 10<br>Ne<br>20.179 |
| 3 | 11<br>Na<br>22.98977 | 12<br>Mg<br>24.305 | | | | | | | | | | | 13<br>Al<br>26.98154 | 14<br>Si<br>28.086 | 15<br>P<br>30.97376 | 16<br>S<br>32.06 | 17<br>Cl<br>35.453 | 18<br>Ar<br>39.948 |
| 4 | 19<br>K<br>39.098 | 20<br>Ca<br>40.08 | 21<br>Sc<br>44.9559 | 22<br>Ti<br>47.90 | 23<br>V<br>50.9414 | 24<br>Cr<br>51.996 | 25<br>Mn<br>54.9380 | 26<br>Fe<br>55.847 | 27<br>Co<br>58.9332 | 28<br>Ni<br>58.71 | 29<br>Cu<br>63.546 | 30<br>Zn<br>65.38 | 31<br>Ga<br>69.72 | 32<br>Ge<br>72.59 | 33<br>As<br>74.9216 | 34<br>Se<br>78.96 | 35<br>Br<br>79.904 | 36<br>Kr<br>83.80 |
| 5 | 37<br>Rb<br>85.4678 | 38<br>Sr<br>87.62 | 39<br>Y<br>88.9059 | 40<br>Zr<br>91.22 | 41<br>Nb<br>92.9064 | 42<br>Mo<br>95.94 | 43<br>Tc<br>98.9062 | 44<br>Ru<br>101.07 | 45<br>Rh<br>102.9055 | 46<br>Pd<br>106.4 | 47<br>Ag<br>107.868 | 48<br>Cd<br>112.40 | 49<br>In<br>114.82 | 50<br>Sn<br>118.69 | 51<br>Sb<br>121.75 | 52<br>Te<br>127.60 | 53<br>I<br>126.9045 | 54<br>Xe<br>131.30 |
| 6 | 55<br>Cs<br>132.9054 | 56<br>Ba<br>137.34 | 57<br>*La<br>138.9055 | 72<br>Hf<br>178.49 | 73<br>Ta<br>180.9479 | 74<br>W<br>183.85 | 75<br>Re<br>186.2 | 76<br>Os<br>190.2 | 77<br>Ir<br>192.22 | 78<br>Pt<br>195.09 | 79<br>Au<br>196.9665 | 80<br>Hg<br>200.59 | 81<br>Tl<br>204.37 | 82<br>Pb<br>207.19 | 83<br>Bi<br>208.9804 | 84<br>Po<br>(210) | 85<br>At<br>(210) | 86<br>Rn<br>(222) |
| 7 | 87<br>Fr<br>(223) | 88<br>Ra<br>226.0254 | 89<br>†Ac<br>(227) | 104<br>Ku<br>(261) | 105<br>Ha<br>(260) | 106 | 107 | | 109 | | | | | | | | | |

| | | | | | | | | | | | | | | |
|---|---|---|---|---|---|---|---|---|---|---|---|---|---|---|
| * | 58<br>Ce<br>140.32 | 59<br>Pr<br>140.9077 | 60<br>Nd<br>144.24 | 61<br>Pm<br>(147) | 62<br>Sm<br>150.4 | 63<br>Eu<br>151.96 | 64<br>Gd<br>157.25 | 65<br>Tb<br>158.9254 | 66<br>Dv<br>162.50 | 67<br>Ho<br>164.9304 | 68<br>Er<br>166.264 | 69<br>Tm<br>168.9342 | 70<br>Yb<br>174.06 | 71<br>Lu<br>174.97 |
| † | 90<br>Th<br>232.0381 | 91<br>Pa<br>231.0359 | 92<br>U<br>238.029 | 93<br>Np<br>237.0482 | 94<br>Pu<br>(244) | 95<br>Am<br>(243) | 96<br>Cm<br>(247) | 97<br>Bk<br>(247) | 98<br>Cf<br>(251) | 99<br>Es<br>(254) | 100<br>Fm<br>(257) | 101<br>Md<br>(258) | 102<br>No<br>(255) | 103<br>Lr<br>(256) |